CHAPMAN & HALL TEXTS IN STATISTICAL SCIENCE SERIES

OTHER TITLES IN THE SERIES INCLUDE

Practical Statistics for Medical Research
D.G. Altman

Interpreting Data
A.J.B. Anderson

Statistical Methods for SPC and TQM
D. Bissell

Statistics in Research and Development
Second edition
R. Caulcutt

The Analysis of Time Series
Fourth edition
C. Chatfield

Statistics in Engineering – A Practical Approach
A.V. Metcalfe

Statistics for Technology
Third edition
C. Chatfield

Introduction to Multivariate Analysis
C. Chatfield and A.J. Collins

Modelling Binary Data
D. Collett

Modelling Survival Data in Medical Research
D. Collett

Applied Statistics
D.R. Cox and E.J. Snell

Statistical Analysis of Reliability Data
M.J. Crowder, A.C. Kimber,
T.J. Sweeting and R.L. Smith

An Introduction to Generalized Linear Models
A.J. Dobson

Introduction to Optimization Methods and their Applications in Statistics
B.S. Everitt

Multivariate Studies – A Practical Approach
B. Flury and H. Riedwyl

Readings in Decision Analysis
S. French

Multivariate Analysis of Variance and Repeated Measures
D.J. Hand and C.C. Taylor

The Theory of Linear Models
B. Jorgensen

Statistical Theory
Fourth edition
B. Lindgren

Randomization and Monte Carlo Methods in Biology
B.F.J. Manly

Statistical Methods in Agriculture and Experimental Biology
Second edition
R. Mead, R.N. Curnow
and A.M. Hasted

Elements of Simulation
B. J. T. Morgan

Probability: Methods and Measurement
A.O'Hagan

Essential Statistics
Third edition
D.G. Rees

Large Sample Methods in Statistics
P.K. Sen and J.M. Singer

Decision Analysis: A Bayesian Approach
J.Q. Smith

Applied Nonparametric Statistical Methods
Second edition
P. Sprent

Elementary Applications of Probability Theory
Second edition
H.C. Tuckwell

Statistical Process Control: Theory and Practice
Third edition
G.B. Wetherill and D.W. Brown

Statistics for Accounts
S. Letchford

Full information on the complete range of Chapman & Hall statistics books is available from the publishers.

Practical Longitudinal Data Analysis

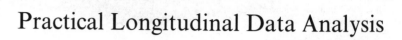

Practical Longitudinal Data Analysis

David Hand

Department of Statistics
The Open University
Milton Keynes
UK

and

Martin Crowder

Department of Mathematical and Computing Sciences
Surrey University
Guildford
UK

CHAPMAN & HALL

London · Glasgow · Weinheim · New York · Tokyo · Melbourne · Madras

Published by Chapman & Hall, 2–6 Boundary Row, London SE1 8HN, UK

Chapman & Hall, 2–6 Boundary Row, London SE1 8HN, UK

Blackie Academic & Professional, Wester Cleddens Road, Bishopbriggs, Glasgow G64 2NZ, UK

Chapman & Hall GmbH, Pappelallee 3, 69469 Weinheim, Germany

Chapman & Hall USA, 115 Fifth Avenue, New York, NY 10003, USA

Chapman & Hall Japan, ITP-Japan, Kyowa Building, 3F, 2-2-1 Hirakawacho, Chiyoda-ku, Tokyo 102, Japan

Chapman & Hall Australia, 102 Dodds Street, South Melbourne, Victoria 3205, Australia

Chapman & Hall India, R. Seshadri, 32 Second Main Road, CIT East, Madras 600 035, India

First edition 1996

Typeset in 10/12pt Times by Thompson Press (India) Ltd, Madras

Printed in England by St Edmundsbury Press Ltd, Bury St Edmunds, Suffolk

ISBN 0 412 59940 6

A catalogue record for this book is available from the British Library

∞ Printed on permanent acid-free text paper, manufactured in accordance with ANSI/NISO Z39.48-1992 and ANSI/NISO Z39.48-1984 (Permanence of Paper).

Contents

Preface

Our previous book on repeated measures data (Crowder and Hand, 1990) ranged fairly widely, bringing together a large and scattered literature on the subject. The present book originally began as a second edition to that. However, recognizing that several other books on the subject had appeared since 1990 (section 1.5), we decided instead to produce something more restricted: we decided to concentrate on regression-based models. Thus, although this book has much overlap with our previous one, it is intended to complement it: there are fewer topics here, but they are covered in more depth and with more discussion. Particular features of this book are:

- we emphasize statistical, rather than probabilistic, models. That is, models are regarded as adequate vehicles for inference about the general behaviour of the observations rather than as true descriptions of the underlying stochastic mechanisms responsible for generating the data.
- The text is illustrated with simple numerical examples called **illustrations** and most chapters contain larger real **examples**. The latter have been analyzed using commercial software when this is available (it does not yet exist for all of the methods described in the book), and illustrative computer commands and results are given. This is one reason why we have used the word 'practical' in the title.
- The first half of the book deals with normal regression models and the second half with increasingly non-normal data. We handle non-normal distributions by fitting normal type models–another reason for adopting the word 'practical'.
- This book presents a detailed discussion of the relationships between methods, making clear their assumptions and relative merits.

For convenience in locating data sets which are discussed in more than one example, we have collated all the data together into Appendix A.

A note about exercises is appropriate. This book is aimed at *practical* application of techniques for analysing longitudinal data. For this reason, rather than including *mathematical* exercises at the end of each chapter, we have included another data appendix (B) containing extra data sets not analysed in the text. We hope that readers will be encouraged to attempt to analyse these data sets using the tools described elsewhere in this book, so producing *practical* exercises in longitudinal data analysis.

The program segments are reproduced with permission from BMDP Statistical Software Inc., SAS Institute Inc. and SPSS Inc. We are grateful to the many researchers who gave us permission to use their real data sets, and especially to Irene Stratton and Russell Wolfinger for assistance in running the computer examples.

D.J. Hand, The Open University
M.J. Crowder, The University of Surrey
1995

CHAPTER 1

Introduction

1.1 Introduction

One of the attractive features of repeated measures data is that (for numerical data, at least) they can be displayed in a graphical plot which is readily interpretable, without requiring a great effort. Table A.1 (in Appendix A) shows measurements of the body weights of rats on three different diets, measured on 11 occasions. Figure 1.1 shows a plot of change in weight over time for the rats on diet 1. The horizontal axis shows time (in days) and the vertical axis weight (in grams). The values for individual rats are shown connected by straight lines.

We can clearly see that, in general, the rats gain weight with passing time (the overall positive slope of the plots) and that, although there is some irregularity, all of the rats follow the same basic pattern. There seems to be no suggestion that the variance of the weights increases with time. We can also see that one of the rats is an outlier, beginning with a low weight and remaining low – although also following a path more or less parallel to the others.

For none of this does one need any training in interpreting the plots.

But, of course, this does not go far enough. We want to be able to make more than general statements summarizing the apparent behaviour of the units being studied. We want to quantify this behaviour, we want to describe it accurately, and we want to compare the behaviour of different groups of units. This book describes methods for doing these things.

Before going into detail, let us define a few of the basic terms and some of the notation we shall use. The units being studied, each of which is measured on several occasions (or, conceivably, under several conditions) will be called (experimental) **units, individuals** or, sometimes, **subjects**. They will be measured at several **occasions** or **times**. Together the results of these measurements will form a **response profile** (or **curve** or, sometimes, **trend**) for each unit. And our aim is to model the **mean response profiles** in the groups. (The word 'clustered' is sometimes used for repeated measures data. However, with its usual English meaning it would seem to be more appropriate for groups of observations in classical split plots than for strings of observations taken over a time period.)

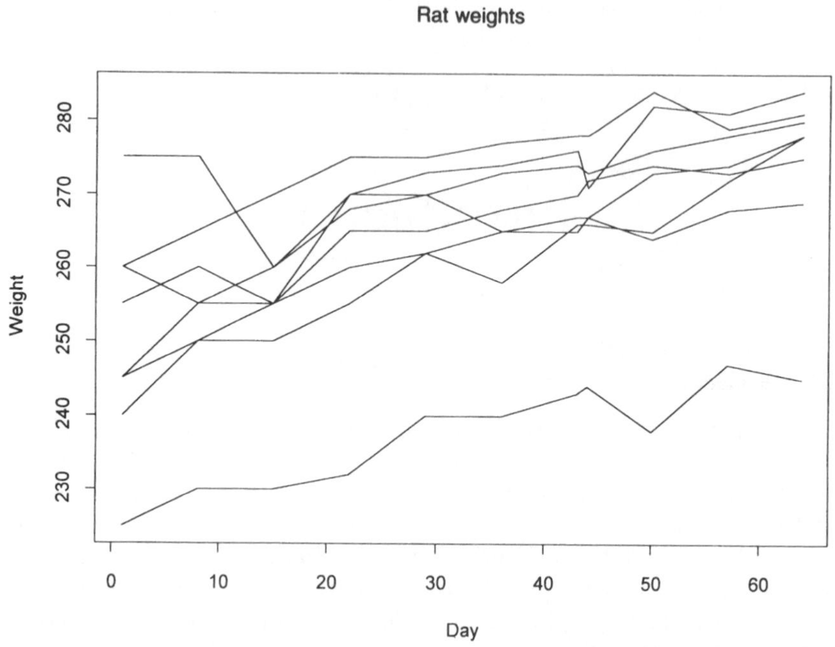

Figure 1.1. *Plots showing the change in rat weights over time. The data are given in Crowder and Hand (1990, Table 2.4), and are reproduced in Table A.1.*

The fact that we are discussing means will alert the reader to the main distinction between time series and the subject matter of this text. Instead of the single (long) series of observations which characterizes a time series, we shall assume that we have several (perhaps many) relatively short series of observations, one for each unit. The existence of multiple series gives us the advantage that we can more readily test (rather than assume) particular structures for the covariance matrix relating observations at subsequent times.

A word or two about notation is also appropriate. As in our previous book, a consistent notation will be used throughout. Thus, y_{ij} will denote the jth measurement made on the ith individual. Its mean is $\mu_{ij} = E(y_{ij})$, its variance is $\sigma_{ij}^2 = \text{var}\ (y_{ij})$, and it will often be accompanied by a vector $\mathbf{x}_{ij}$ of explanatory variables. The number of individuals in the sample will usually be n, the number of measures made on individual i will be p, or maybe p_i when these differ, and the dimension of $\mathbf{x}_{ij}$ will usually be q. The set of measures $y_{ij} (j = 1, \ldots, p)$ will often be collected into a $p \times 1$ vector $\mathbf{y}_i$ with mean vector $\boldsymbol{\mu}_i$ and covariance matrix $\boldsymbol{\Sigma}_i$.

Clearly, the problem we are addressing is intrinsically multivariate. However, it involves a restricted form of multivariate data. Three observations are of particular interest here. First, the same 'thing' is being measured at each time – so that the measurements are commensurate. This is not true of general multivariate data, where one variable might be height, another weight, and so on.

Second, the measures are typically taken at selected occasions on an underlying (time) continuum. Sometimes the individuals are measured on different occasions or on different numbers of occasions. This may complicate the analysis in practice, but it does not alter it in principle: the measurements are simply an attempt to get at the underlying continuous curve of change over time (or of probability of occupying a particular state in discrete cases). This is quite different from the general multivariate case, where there is no underlying continuum.

Third, the sequential nature of the observations means that particular kinds of covariance structures are likely to arise, unlike more general multivariate situations, where there may be few or no indications of the structure.

The inherent dependence, or the possibility of dependence, that is associated with longitudinal data introduces extra complications into the analysis. No longer can we rely on the simplifying properties arising from data which are independently and identically distributed. To yield conclusions in which we can have confidence, we must somehow take the possible dependence into account. Nevertheless, the simplicity of methods of analysis which assume independence is attractive. It is therefore hardly surprising that researchers have explored methods which modify the problem so that independence-based approaches can be used. In the next two sections of this chapter we describe two such methods. The first analyses each time separately, and we do not recommend this approach for several reasons. The second summarizes the profiles and works with these summaries. Section 1.4 summarizes the more sophisticated methods described in later chapters.

Before we leave this introductory section, however, to give the reader a flavour of the rich variety of problems which can arise in a longitudinal context, we show, in Figs 1.2 to 1.7, a series of plots of the response profiles from different problems. (Necessarily, these plots only show data which have arisen from numerical measurements. Discrete data yield another entire class of possible kinds of problem.) These figures illustrate the range of profiles which can arise, and the sorts of structures one might look for.

1.2 Comparisons at each time

As we have pointed out above, the distinctive feature of longitudinal data is that the observations on a particular individual at each time will not, in general, be independent. Somehow this has to be taken into account in any analysis – unless the particular circumstances of a problem mean that a justification for not making allowance for it can be found (Chapter 3). One approach which has been common in the past, and which has the merit of simplicity, is to analyse each time separately. Of course, this is restricted to those situations where all of the individuals are measured at the same time (or where the times can be grouped so that they are regarded as simultaneous). Groups can be compared at each time using t-tests, standard univariate analysis of variance, or nonparametric groups comparisons as appropriate.

While this approach may be attractive because of its simplicity, it does have some serious deficiencies. The only way it can shed light on the change of treatment effects with time is by a comparison of the separate analyses. A similar comparison could be made using independent groups at each time. At best, what can be obtained are statements about the change of averages – and not about the average of changes. The two might be quite different and, as we have remarked, it is typically the latter which is of interest. The advantages of undertaking a longitudinal study have been lost. Moreover, the tests are not independent, since the data have arisen from the same experimental units. The fact that one particular group scores significantly more highly than another on several occasions is not as strong evidence for the superiority of that group as would be the case if the tests were independent. (This is further complicated by the fact that several tests have been conducted.) Sometimes, also, the tests are compared with a view to deciding 'when' an effect occurs – the time being identified as that at which a

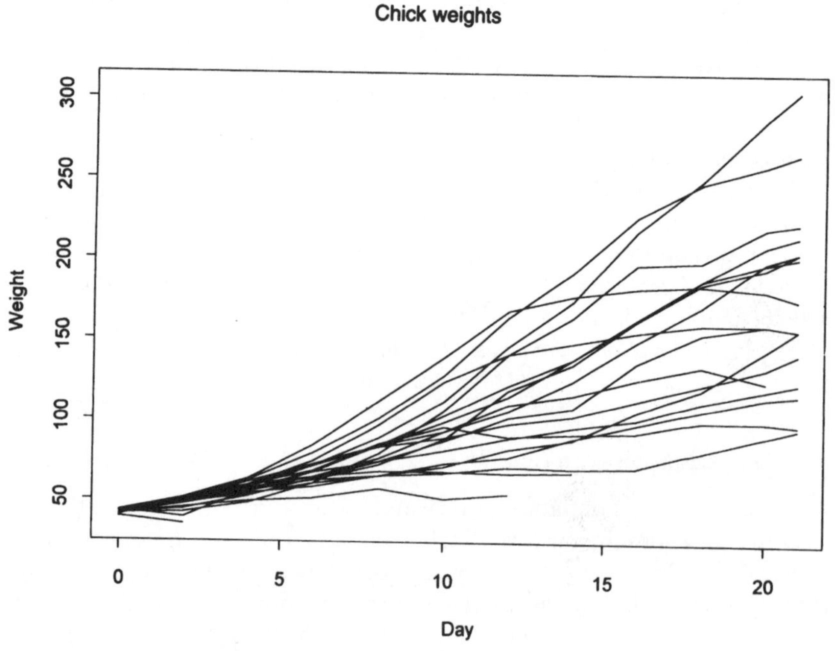

Figure 1.2. *Body weights of 20 chicks on a normal diet, measured on alternate days. The data are the first 20 rows of Table A.2. The fan shape, showing variance increasing with time, is typical of growth curves.*

significant difference first arises. This is, of course, usually meaningless because of the continuous nature of changes.

In summary, the approach based on testing the results at each time separately is typically invalid. More sophisticated methods, which take account of the relationships between observations at different times, need to be used. This book describes such methods.

1.3 Response feature analysis

Section 1.2 described a method of analysis which has been common in the past (though we would like to think it is less so nowadays since powerful computer software has become available which permits more valid analyses to be readily performed). This section describes another simple method which is quite common, and which is statistically valid. Its main weakness is that it may be not very powerful.

This text describes models which can be fitted to sequences of responses over time. Important aspects of the process generating the data are identified and these are used to produce a well-fitting model. Often, however, researchers' interests lie in a particular aspect of the change over time. For

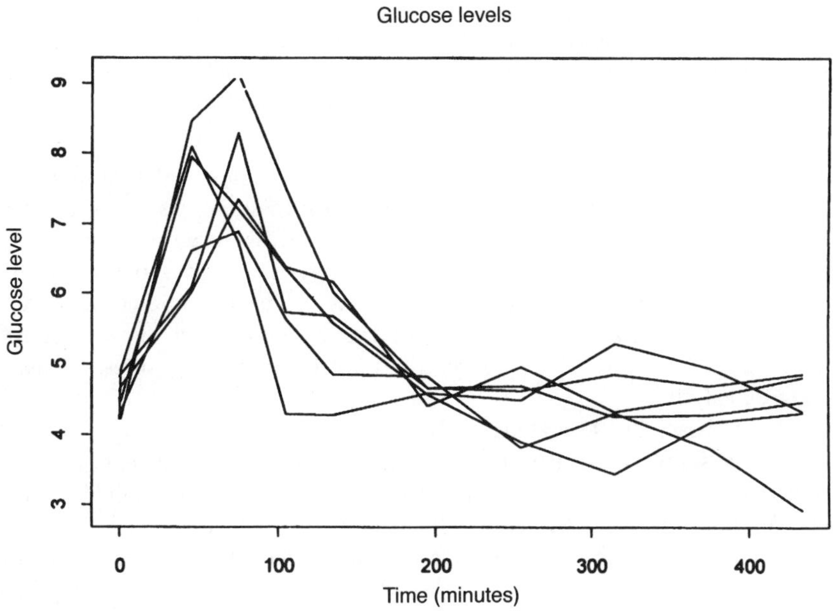

Figure 1.3. *Blood glucose levels following a meal taken at time 15 minutes. The data are given in Table A.3. The response builds up to a peak soon after the meal, and then gradually decays.*

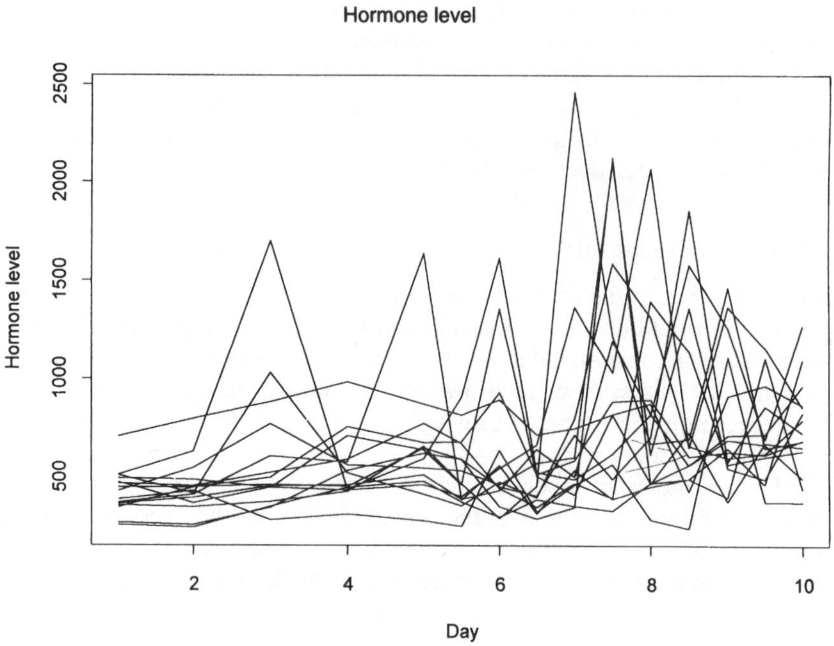

Figure 1.4. *Luteinizing hormone levels in non-suckling cows ($ng\,ml^{-1} \times 1000$), listed in Table A.4, with cow number 7 removed as it has a very large score at 9.5 days. Some of these profiles show marked variability: while we may draw conclusions about the mean profile, we must bear in mind that individuals may depart substantially from this.*

example, they might want to know the peak value achieved after administration of some treatment, the time taken to return to some baseline value, the difference between average post-treatment and average pre-treatment scores, the area under some response curve, or simply the slope showing rate of change over time. We call such an aspect of the profile a **response feature** beause it summarizes some aspect of the response over time, and the approach to analysis which concentrates solely on such features, **response feature analysis**.

The approach has several merits. It is easy to understand and explain, and it leads to a simple univariate analysis – the vector of measurements on each subject is reduced to a single summarizing score, focusing on the information of particular interest. Moreover, the summarizing features can often be calculated even if subjects have different numbers of measurements and are measured at different times. This is a powerful advantage since missing data are common in studies which extend over time. The method focuses attention on the shapes of curves for the individuals – and not on the curve through the means at each time, which might be quite different. (This is where the method described in section 1.2 failed.)

The chief disadvantage of the method is a loss in error degrees of freedom. If there are n individuals, then n summary features will be produced, one for each individual. Tests to compare groups must be based on these n derived observations. By computing a summary measure for each subject an accurate estimate is obtained for each subject but the variation *between* subjects is not reduced by this. There is a general point implicit in this: however large is p, the number of measurements on each individual, it cannot make up for small sample size, n. Between-individual variation is typically the dominant form, and this can only be estimated (and its average effect reduced) when n is large enough.

Moreover, although the method can handle subjects with different numbers of, or missing, measurements, things are not totally straightforward. If subjects' summary values are based on different numbers of scores then they will typically have different variances. Moreover, if the measurements occur at different times for different subjects then the summary values will normally have different variances. (Take, for example, a simple slope as the summary measure. Individuals with measurements at more extreme times will have more accurate slope estimates than individuals with measurements bunched together at an intermediate time, all other things being equal.) Modifications to the basic form of analysis have been suggested to take into account different numbers of and different distributions of raw measurements, but this means that the essential simplicity of the analysis is being sacrificed. If one is going to do this, then one might as well adopt one of the more powerful and correct methods of analysis described later in this book.

Response feature analysis can be extended to more than one feature. For example, one might summarize change over time into linear and quadratic components. These might be of interest separately, and so analysed individually, or they might be brought together to summarize the shapes of the response profiles, and so analysed simultaneously in a multivariate analysis. At an extreme the p observed measurements are transformed into p derived variables which are equivalent to the raw data but which describe the profiles in ways which are of more relevance. This leads to the multivariate analysis of variance approach described in Chapter 2.

An example of response feature analysis is given Crowder and Hand (1990, section 2.2).

1.4 Outline of the rest of the book

Part One describes approaches based on an assumption of normal error distributions. The normal case is important for a number of reasons. One is that it provides an adequate approximation to many real situations. Another is that a great deal of theory and many different methods and models have been developed for this case. All of the major standard computer packages contain routines for normal errors longitudinal data analysis, and in many

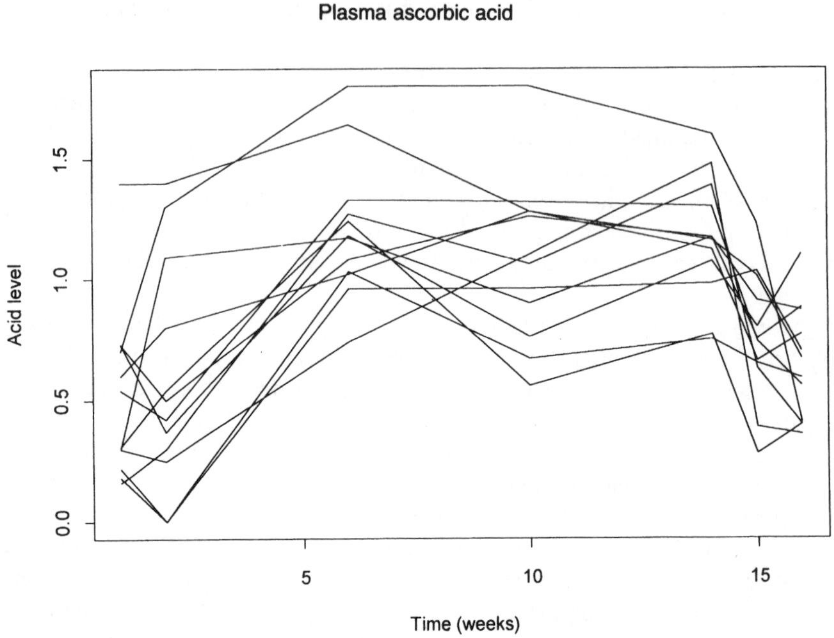

Figure 1.5. *Plasma ascorbic acid of 12 patients measured on seven occasions. The data are reproduced in Table A.5. Weeks 1 and 2 are before treatment, weeks 6, 10 and 14 are during treatment, and weeks 15 and 16 are after treatment. The profiles show a plateau effect, although there is also substantial inter-subject variability.*

cases this is the only distribution they do cater for. Historically, a vast amount of research has taken as the basic form.

In presenting the material based on normal theory, we commenced with analysis of variance approaches because of their importance and widespread use. We also adopted the unconventional order of presenting multivariate analysis of variance methods before univariate analysis of variance methods. This is because the former is conceptually simpler in that it imposes no restrictive assumptions on the form of the error covariance matrix–it assumes this matrix to be unstructured. As is explained in Chapter 3, univariate analysis of variance methods do make certain restrictive assumptions.

Similarly, we describe general regression models before random effects models. The former allow arbitrary covariance matrices to be adopted, whereas, in the latter, a natural form for the matrices emerges from the model. The chapters discussing these models are followed by a chapter discussing particular forms for covariance matrices.

Part Two of the book turns to non-normal error distributions. Here we discuss Gaussian estimation as a general method for fitting such models,

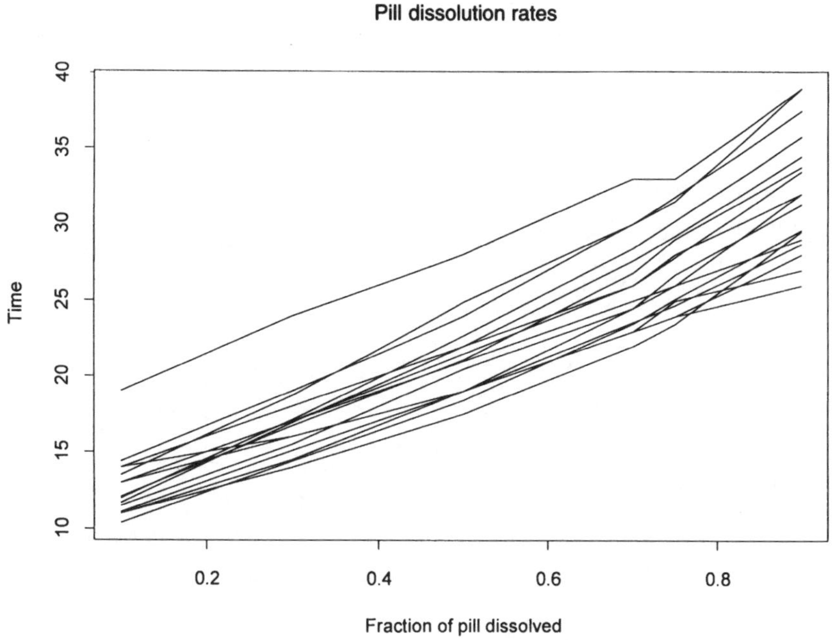

Figure 1.6. *These data listed in Table A.6, show the times (on the vertical axis) by which a specified proportion of a pill has dissolved (on the horizontal axis), for each of 17 pills. Here the 'underlying continuum' from which occasions are chosen at which to take measurement is 'proportion dissolved'. The grouping is ignored in this plot.*

whatever the error distribution involved. There follows an account of generalized linear models and maximum quasi-likelihood estimation, particularly for binary and categorical data. These latter methods have been the subject of vigorous research activity in recent years and are being applied widely, particularly in the health sciences.

Where possible, we have illustrated this book using commercially available software. But our aim here is not to write a software manual, so we have not gone into great detail. In any case, software evolves with time, and placing too great an emphasis on software details would detract from our main objective–which is to present the theoretical underpinning and show how the methods are applied in practice. Statisticians with a good grasp of the theory of the methods can, of course, write special-purpose routines in any of a large variety of statistical languages. Researchers in other disciplines, however, who simply wish to apply the methods to their data, may not have the requisite expertise to write such routines and may want to use commercial software specially written for analysing repeated measures data.

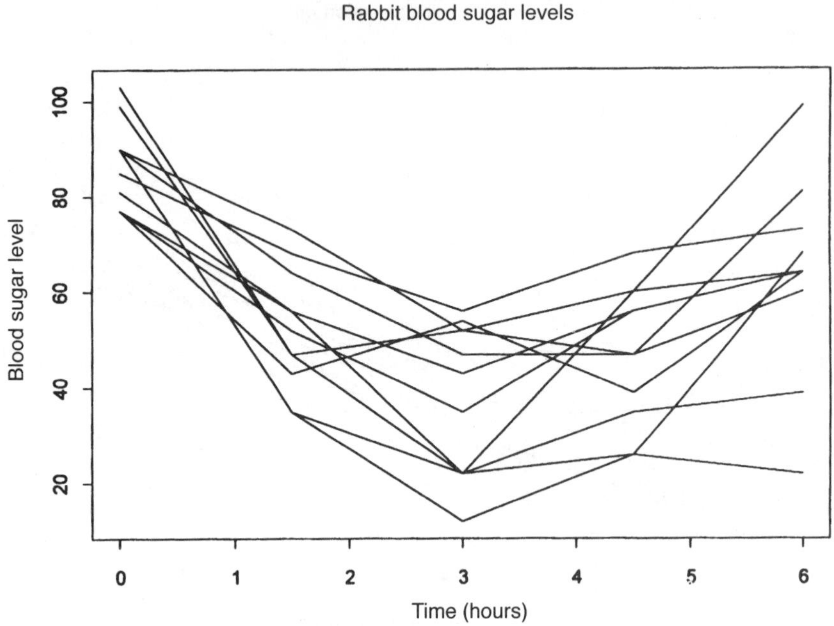

Figure 1.7. *Blood sugar levels of rabbits following an injection of insulin. The data are reproduced in Table A.7.*

Until relatively recently the only specially written software for repeated measures analysis adopted the univariate or multivariate analysis of variance approaches. Then BMDP5V and SAS PROC MIXED appeared, which could fit models having structured covariance matrices. No specially written software is yet generally available for the more advanced techniques outlined in the second part of the book, but this is just a question of time. The second author has used his own Fortran routines for the examples there, and a copy can be obtained from him free of charge (and of guarantee). Further details of the commercial packages we used here can be obtained from: BMDP Statistical Software, Inc., 1440 Sepulveda Boulevard, Suite 316, Los Angeles, CA 90025, USA; SPSS Inc., 444 N. Michigan Avenue, Chicago, IL 60611, USA; SAS Institute, Inc., SAS Circle, Box 8000, Cary, NC 27512-8000, USA.

1.5 Further reading

Response feature analysis has been described, though not by that name, in Wishart (1938), Rowell and Walters (1976), Walters and Rowell (1982), Yates (1982), and, more recently, by Matthews *et al.* (1990).

In recent years several books have been written on the topic of longitudinal data analysis and repeated measures. We summarize their style and content below.

Hand and Taylor (1987) is aimed at researchers from the behavioural sciences who have repeated measures problems, rather than at statisticians. It tries to explain analysis of variance, multivariate analysis of variance, and corresponding approches to repeated measures analysis without going into mathematical detail. It does this by focusing on *contrasts* between groups and between measurements, rather than by building global models. A series of eight real case studies which arose in the authors' daily work is described in the second half of the book, along with detailed illustrations of how to use appropriate computer software. From our perspective it is limited in that it only describes univariate and multivariate approaches to repeated measures. These are both popular in psychology and related disciplines, but there are now entire classes of alternatives which have advantages in some circumstances.

The first book to provide a comprehensive overview of methods of repeated measures analysis was Crowder and Hand (1990), though even this focuses mainly on continuous normally distributed data. It covers the methods described in sections 1.2 and 1.3 of this chapter, univariate and multivariate analysis of variance, regression models, and two-stage models. It also has chapters on categorical data, crossover designs, and miscellaneous topics (including some discussion of nonlinear growth curves). Chapter 10 consists of a series of annotated computer listings showing examples of SAS, BMDP and SPSSX analyses of repeated measures problems. This book has a higher mathematical level than Hand and Taylor (1987), and is aimed more at statisticians than at their clients. It includes an extensive bibliography of work on repeated measures.

Jones (1993) describes situations where normal error distributions may be assumed and builds Laird–Ware models making allowance for intra-subject correlations and random between-subject effects. In particular, the book adopts a recursive method of calculating likelihoods based on state-space representations and the Kalman filter. Because of the unusual approach, the book does not rely on commercially available software, but instead gives Fortran subroutine listings in an appendix. Chapter 7 extends the work to nonlinear situations.

Lindsey (1993) is a wide-ranging book. The first 88 pages present an impressive general description of statistical modelling, before narrowing down to the particular context of repeated measurements. In keeping with the title, the emphasis seems to be on modelling rather than data analysis. Equal space is devoted to normal distribution models and categorical data. The final two chapters of the book discuss frailty models and event history models. An extensive bibliography is included. Various programs have been used for the analysis, including GLIM and Matlab.

Diggle, Liang and Zeger (1994), like much of the work in the area, places a strong emphasis on examples from the biological and health sciences. It includes discussion of marginal models, random effects models, and transition models. Again, the book goes further than current commercially available software, and so does not adopt an existing package. Instead, the authors used S and wrote their own routines as necessary. Particularly unusual and noteworthy features of this book are its discussion of design considerations, of exploratory methods for longitudinal data, and of missing values. The book adopts a careful style and would make a suitable text.

Longford (1993) is restricted to random coefficient models, so that the covariance matrix has an induced structure arising from the random effects, but independence beyond that. (In our terms, in Chapter 5, he takes the error matrix to be diagonal.) Most of the discussion is about models based on normal theory and the substantive background is primarily educational.

Part One: Normal Error Distributions

CHAPTER 2

Multivariate analysis of variance

2.1 Basic ideas

Let $\mathbf{y}_i$ be a row vector representing the p scores for the ith individual. If there are n individuals, we can stack these vectors to form an $n \times p$ data matrix

$$\underset{n \times p}{\mathbf{Y}} = \begin{bmatrix} \mathbf{y}_1 \\ \cdots \\ \mathbf{y}_n \end{bmatrix}$$

In general, the $\mathbf{y}_i$'s will arise from different groups (there may be a treatment structure, with individuals being assigned to different treatment conditions) and we can model the observed data in terms of the group means and random deviations from those means:

$$\mathbf{Y} = \mathbf{X\Xi} + \mathbf{U}$$

Here the ith row of $\mathbf{X\Xi}$ gives the mean responses for the ith individual and the ith row of $\mathbf{U}$ gives the random deviations from those means. That is, the rows of $\mathbf{X\Xi}$ give the mean scores that would be observed if the study was repeated many times for each individual, while the rows of $\mathbf{U}$ give the departures from those means for the study actually conducted. Note also that *all* sources of random variation are included in $\mathbf{U}$ (so $\mathbf{U}$ stands for 'umbrella', if you like). Different rows of $\mathbf{U}$ correspond to different individuals and so, are independent. Since, in this chapter we are assuming the error distributions to be multivariate normal, we have that the ith row of $\mathbf{U}$ is distributed as a multivariate normal distribution, $N_p(\mathbf{0}, \mathbf{\Sigma}_p)$. Our model for the mean responses thus has the form

$$\underset{n \times p}{E(\mathbf{Y})} = \underset{n \times q}{\mathbf{X}} \; \underset{q \times p}{\mathbf{\Xi}} \tag{2.1}$$

If $p = 1$, this is simply the univariate general linear model, with a different set of parameters for each group.

■ *Illustration 2.1* Suppose $p = 1$ and $n = 5$, with two groups, three individuals in the first group and two in the second. Then equation (2.1) becomes

$$E\begin{bmatrix} y_1 \\ y_2 \\ y_3 \\ y_4 \\ y_5 \end{bmatrix} = \begin{bmatrix} 1 & 0 \\ 1 & 0 \\ 1 & 0 \\ 0 & 1 \\ 0 & 1 \end{bmatrix} \begin{bmatrix} \xi_1 \\ \xi_2 \end{bmatrix}$$

Here ξ_1 would be the mean response of group 1 and ξ_2 that of group 2. ■

■ *Illustration 2.2* An alternative parametrization for Illustration 2.1 would be

$$E\begin{bmatrix} y_1 \\ y_2 \\ y_3 \\ y_4 \\ y_5 \end{bmatrix} = \begin{bmatrix} 1 & 0 \\ 1 & 0 \\ 1 & 0 \\ 1 & 1 \\ 1 & 1 \end{bmatrix} \begin{bmatrix} \xi_1 \\ \xi_2 \end{bmatrix}$$

Now ξ_1 represents the mean response of group 1 and ξ_2 the difference in mean response between group 1 and group 2. ■

Model (2.1) above extends this to the p-variate case simply by expanding the Ξ matrix so that it has p columns.

■ *Illustration 2.3* Take Illustration 2.1 above and suppose that measurements are taken at $p = 3$ times. Then the expected scores are

$$E\begin{bmatrix} y_{11} & y_{12} & y_{13} \\ y_{21} & y_{22} & y_{23} \\ y_{31} & y_{32} & y_{33} \\ y_{41} & y_{42} & y_{43} \\ y_{51} & y_{52} & y_{53} \end{bmatrix} = \begin{bmatrix} 1 & 0 \\ 1 & 0 \\ 1 & 0 \\ 0 & 1 \\ 0 & 1 \end{bmatrix} \begin{bmatrix} \xi_{11} & \xi_{12} & \xi_{13} \\ \xi_{21} & \xi_{22} & \xi_{23} \end{bmatrix}$$

So the column

$$\begin{bmatrix} \xi_{1j} \\ \xi_{2j} \end{bmatrix}$$

corresponds to the jth occasion. Note that the **X** matrix has remained the same. ■

In model (2.1) **X** describes the group membership structure of the subjects. It can also include other covariates which do not change with time.

■ *Illustration 2.4* Suppose that age at the start of the study is thought to be relevant to the response in Illustration 2.3. Then, letting x_i be the age of the

ith individual at the start of the study, we have

$$E\begin{bmatrix} y_{11} & y_{12} & y_{13} \\ y_{21} & y_{22} & y_{23} \\ y_{31} & y_{32} & y_{33} \\ y_{41} & y_{42} & y_{43} \\ y_{51} & y_{52} & y_{53} \end{bmatrix} = \begin{bmatrix} 1 & 0 & x_1 \\ 1 & 0 & x_2 \\ 1 & 0 & x_3 \\ 0 & 1 & x_4 \\ 0 & 1 & x_5 \end{bmatrix} \begin{bmatrix} \xi_{11} & \xi_{12} & \xi_{13} \\ \xi_{21} & \xi_{22} & \xi_{23} \\ \xi_{31} & \xi_{32} & \xi_{33} \end{bmatrix}$$

Here ξ_{3j} is the coefficient for age at the jth occasion – that is, the effect of age on the ith subject on the jth occasion is given by $x_i\xi_{3j}$. ■

Since **X** is just concerned with the distinctions between individuals, and since its rows apply to all occasions, it is called the **between-individuals matrix** or **between-groups matrix**.

Model (2.1), which we have developed so far, is the basic multivariate analysis of variance (manova) model. It defines a between-individuals structure for the expected values of the observations, but does not define any simple model for the relationships between the variables. But it is the latter which is of primary interest with longitudinal data – we want to model the mean response profiles. Put another way, we want to impose some restrictions relating the observations at each occasion. The obvious way to do this is by assuming the Ξ matrix arises from some model

$$\Xi = \Gamma\mathbf{B}.$$

The complete model is then

$$\underset{n\times p}{E(\mathbf{Y})} = \underset{n\times q}{\mathbf{X}}\ \underset{q\times r}{\Gamma}\ \underset{r\times p}{\mathbf{B}} \tag{2.2}$$

Here **B** describes the model for the pattern of change over occasion – the profile of expected scores within an individual. It is therefore called the **within-individuals matrix** or the **within-groups matrix**.

■ *Illustration 2.5* Suppose we have just a single group of five subjects. This means that each row of the **X** matrix in (2.1) and (2.2) will have the same form. Suppose, now, that in Illustration 2.3 we believe that response is linearly related to time. We measured the response in that illustration on three occasions – times t_1, t_2, and t_3, say. Now, according to the new model, the expected response at time t *is* $\gamma_1 + \gamma_2 t$. We can write this as

$$E\begin{bmatrix} y_{11} & y_{12} & y_{13} \\ y_{21} & y_{22} & y_{23} \\ y_{31} & y_{32} & y_{33} \\ y_{41} & y_{42} & y_{43} \\ y_{51} & y_{52} & y_{53} \end{bmatrix} = \begin{bmatrix} 1 \\ 1 \\ 1 \\ 1 \\ 1 \end{bmatrix} [\gamma_1 \quad \gamma_2] \begin{bmatrix} 1 & 1 & 1 \\ t_1 & t_2 & t_3 \end{bmatrix}$$

■

In this model t_j acts as a covariate which takes different values at different times.

■ *Illustration 2.6* With two groups, of three and two individuals respectively, as in Illustration 2.3, we would have

$$E\begin{bmatrix} y_{11} & y_{12} & y_{13} \\ y_{21} & y_{22} & y_{23} \\ y_{31} & y_{32} & y_{33} \\ y_{41} & y_{42} & y_{43} \\ y_{51} & y_{52} & y_{53} \end{bmatrix} = \begin{bmatrix} 1 & 0 \\ 1 & 0 \\ 1 & 0 \\ 0 & 1 \\ 0 & 1 \end{bmatrix} \begin{bmatrix} \gamma_{11} & \gamma_{12} \\ \gamma_{21} & \gamma_{22} \end{bmatrix} \begin{bmatrix} 1 & 1 & 1 \\ t_1 & t_2 & t_3 \end{bmatrix}$$

Although here there is a linear relationship over time within each of the groups, the slopes and intercepts may differ between the groups. Thus, individuals from group 1 have regression $\gamma_{11} + \gamma_{12}t$ while individuals from group 2 have regression $\gamma_{21} + \gamma_{22}t$. ■

Other covariates which change with measurement occasion can also be included in just such a straightforward manner and alternative parametrizations, perhaps more convenient or more central to the hypotheses under investigation, can easily be adopted.

■ *Illustration 2.7* A different way to parametrize the example in Illustration 2.6 is as follows.

$$E\begin{bmatrix} y_{11} & y_{12} & y_{13} \\ y_{21} & y_{22} & y_{23} \\ y_{31} & y_{32} & y_{33} \\ y_{41} & y_{42} & y_{43} \\ y_{51} & y_{52} & y_{53} \end{bmatrix} = \begin{bmatrix} 1 & 0 \\ 1 & 0 \\ 1 & 0 \\ 0 & 1 \\ 0 & 1 \end{bmatrix} \begin{bmatrix} \mu_{11} & \mu_{12} & \mu_{13} \\ \mu_{21} & \mu_{22} & \mu_{23} \end{bmatrix} \begin{bmatrix} 1 & 1 & 1 \\ 1 & 0 & -1 \\ 0 & 1 & -1 \end{bmatrix}$$

Here, instead of describing the response as a linear trend over time, we are describing it in terms of deviations from an overall mean, μ_{k1}, for group k. Thus, the means for group k over the three occasions are $\mu_{k1} + \mu_{k2}$, $\mu_{k1} + \mu_{k3}$, and $\mu_{k1} - \mu_{k2} - \mu_{k3}$. The form for the third occasion is devised to make the average of the three means equal to μ_{k1}. ■

2.2 Parameter estimates

The maximum likelihood (ML) estimates of the parameters in model (2.2) are given by (Khatri, 1966)

$$\hat{\mathbf{\Gamma}} = (\mathbf{X}^{\mathrm{T}}\mathbf{X})^{-1}\mathbf{X}^{\mathrm{T}}\mathbf{Y}\hat{\mathbf{\Sigma}}_p^{-1}\mathbf{B}^{\mathrm{T}}(\mathbf{B}\hat{\mathbf{\Sigma}}_p^{-1}\mathbf{B}^{\mathrm{T}})^{-1} \tag{2.3}$$

where $\hat{\mathbf{\Sigma}}_p$ is the usual ML estimate of the covariance matrix of the rows of $\mathbf{U}$. (Of course, for $\mathbf{B}\hat{\mathbf{\Sigma}}_p^{-1}\mathbf{B}^{\mathrm{T}}$ to be nonsingular we require rank $\mathbf{B} = r$ and for $\mathbf{X}^{\mathrm{T}}\mathbf{X}$ to be nonsingular we require rank $\mathbf{X} = q$.)

The structure of (2.3) is perhaps worth remarking upon, because it can seem daunting. First, suppose that there was a single observation on each individual. Then

$$\hat{\mathbf{\Gamma}} = (\mathbf{X}^{\mathrm{T}}\mathbf{X})^{-1}\mathbf{X}^{\mathrm{T}}\mathbf{Y}$$

which is the usual least-squares solution for the model $E(\mathbf{Y}) = \mathbf{X\Gamma}$.

Similarly, if instead we suppose that there is but a single individual measured on p occasions, then the estimate is

$$\hat{\mathbf{\Gamma}} = \mathbf{Y}\hat{\mathbf{\Sigma}}_p^{-1}\mathbf{B}^{\mathrm{T}}(\mathbf{B}\hat{\mathbf{\Sigma}}_p^{-1}\mathbf{B}^{\mathrm{T}})^{-1}$$

which is the weighted least-squares solution taking account of the correlations between the scores at the different times. Putting these two solutions together yields (2.3).

2.3 Derived variables

The maximized likelihood values alone can be used for likelihood ratio tests to assess the adequacy of a model, but a more common approach is as follows. The rows of the matrix $\mathbf{\Gamma B}$ consist of linear combinations of the row vectors of $\mathbf{B}$. That is, for a given row of $\mathbf{\Gamma}$, the product yields a point in the row space of $\mathbf{B}$. A different, but equivalent, description of these points can be found by changing the basis of the row space of $\mathbf{B}$ by post-multiplying $\mathbf{\Gamma B}$ by a matrix $\mathbf{H}_1$, of order $p \times r$, with columns which span the row space of $\mathbf{B}$. The p-component rows of $E(\mathbf{Y})$ lie in an r-dimensional subspace of the p-space. By post-multiplying by $\mathbf{H}_1$ we transform to $E(\mathbf{Y})\mathbf{H}_1$, with r-component rows. If, now $\mathbf{H}_1$ is expanded to a $p \times p$ matrix $\mathbf{H} = (\mathbf{H}_1, \mathbf{H}_2)$, of full rank, where the columns of $\mathbf{H}_2$ are orthogonal to those of $\mathbf{H}_1$ (and, so, to the rows of $\mathbf{B}$) then, defining $\mathbf{Y}_1 = \mathbf{YH}_1$ and $\mathbf{Y}_2 = \mathbf{YH}_2$, we have

$$E(\mathbf{Y}_1) = \mathbf{X\Gamma BH}_1 = \mathbf{X\Theta} \qquad \text{(say)}$$

and

$$E(\mathbf{Y}_2) = \mathbf{X\Gamma BH}_2 = \mathbf{0}$$

$\mathbf{H}_1$ will be chosen so that particular columns summarize the pattern of change over time in useful ways. For example, the first column could consist of coefficients such that the first column of $\mathbf{YH}_1$ contains the means of the p observations for each individual, the second column could yield the linear trend, and so on. The columns of $\mathbf{H}_2$ will provide those linear combinations of the p observations for each individual which are thought to have zero expectation.

To test whether the model provides an adequate fit to the profile of means we can now use standard manova tests (see below – Hotelling's T^2-test in the two-group case) to test $E(\mathbf{Y}_2) = E(\mathbf{YH}_2) = \mathbf{0}$. That is, we transform the observation matrix $\mathbf{Y}$ by $\mathbf{H}_2$ and then see if the means of the derived variables are zero.

■ *Illustration 2.8* Suppose (cf. Illustration 2.5)

$$\mathbf{X} = \begin{bmatrix} 1 \\ 1 \\ 1 \end{bmatrix}, \mathbf{\Gamma} = [\gamma_1 \quad \gamma_2], \quad \text{and} \quad \mathbf{B} = \begin{bmatrix} 1 & 1 & 1 \\ 1 & 2 & 3 \end{bmatrix}$$

Then

$$E(\mathbf{Y}) = \mathbf{X\Gamma B} = \begin{bmatrix} 1 \\ 1 \\ 1 \end{bmatrix} [\gamma_1 \quad \gamma_2] \begin{bmatrix} 1 & 1 & 1 \\ 1 & 2 & 3 \end{bmatrix} = \begin{bmatrix} \gamma_1 + \gamma_2 & \gamma_1 + 2\gamma_2 & \gamma_1 + 3\gamma_2 \\ \gamma_1 + \gamma_2 & \gamma_1 + 2\gamma_2 & \gamma_1 + 3\gamma_2 \\ \gamma_1 + \gamma_2 & \gamma_1 + 2\gamma_2 & \gamma_1 + 3\gamma_2 \end{bmatrix}$$

so the model is that there is a single group of three subjects, with the responses linearly related to occasion, with intercept γ_1 and regression coefficient γ_2.

Now define

$$\mathbf{H}_1 = \begin{bmatrix} 1 & -1 \\ 1 & 0 \\ 1 & 1 \end{bmatrix} \quad \text{and} \quad \mathbf{H}_2 = \begin{bmatrix} 1 \\ -2 \\ 1 \end{bmatrix}$$

($\mathbf{H}_1$ could be chosen in other ways, of course. We shall discuss this below). First, let us verify that the columns of $\mathbf{H}_1$ span the row space of $\mathbf{B}$. We have the first row of $\mathbf{B}$

$$[1 \quad 1 \quad 1] = 1.[1 \quad 1 \quad 1] + 0.[-1 \quad 0 \quad 1]$$

where the row vectors on the right-hand side are the columns of $\mathbf{H}_1$. Similarly, for the second row of $\mathbf{B}$,

$$[1 \quad 2 \quad 3] = 2.[1 \quad 1 \quad 1] + 1.[-1 \quad 0 \quad 1]$$

And the columns of $\mathbf{H}_2$ are ('is', in this case, there being only one of them) orthogonal to those of $\mathbf{H}_1$:

$$[1 \quad -2 \quad 1] \begin{bmatrix} 1 \\ 1 \\ 1 \end{bmatrix} = 0, \quad [1 \quad -2 \quad 1] \begin{bmatrix} -1 \\ 0 \\ 1 \end{bmatrix} = 0$$

Now we can test the adequacy of the model – that the responses increase linearly with occasion – by comparing the departures from linearity with zero – that is, by comparing $\mathbf{X\Gamma BH}_2$ with zero. This is $E(\mathbf{Y}_2)$ or $E(\mathbf{YH}_2)$. So we transform the rows of $\mathbf{Y}$ by post-multiplying by $\mathbf{H}_2$. In this case, since $\mathbf{H}_2$ is a single column, the transformation results in a single score for each individual. To see if the mean of these scores is zero we can use a one-sample *t*-test. ■

We will see more complicated and interesting examples of this when we have explored multivariate test statistics.

We have described the above starting with a model for $E(\mathbf{X\Gamma B})$. However, having arrived at the stage of fitting such models via a transformation of the observation vectors, we can shift the emphasis to these transformations. This is what the descriptions accompanying most of the major computer packages do and is the approach described in many multivariate statistics texts. So, we can take the transformation as our starting point, yielding a set of 'derived' variables, and work from there.

■ *Illustration 2.9* Suppose we have two groups of subjects, each measured at three times. Then there are three fundamental questions we might consider:

1. Are the profiles of the means of the groups at the same level? (Or, on the other hand, is there a 'group effect'?)
2. Are the profiles flat? (Or, on the other hand, is there a 'time effect'?)
3. Are the profiles parallel? (Or, on the other hand, is there a 'group by time interaction'?)

In practice, of course, one should address interaction questions first because, if there does appear to be an interaction present, it probably means that the lower-order questions (the main-effect questions in this case) do not make sense. However, for simplicity in this example, since it is the first of its kind that we have encountered, we shall address the questions in the above order.

To produce derived variables which will enable us to answer the above questions, we will post-multiply the data matrix by

$$\mathbf{H} = \begin{bmatrix} 1 & 1 & 0 \\ 1 & -1 & 1 \\ 1 & 0 & -1 \end{bmatrix}$$

Consider, as an example, a particular subject's vector of scores $\mathbf{y} = [y_1 \ y_2 \ y_3]$. Then

$$\begin{aligned} \mathbf{yH} &= (y_1 \ y_2 \ y_3) \begin{bmatrix} 1 & 1 & 0 \\ 1 & -1 & 1 \\ 1 & 0 & -1 \end{bmatrix} \\ &= [y_1 + y_2 + y_3 \ \ y_1 - y_2 \ \ y_2 - y_3] \\ &= [z_1 \ z_2 \ z_3] \end{aligned}$$

say. The first of these derived variables, z_1, is proportional to the mean of the three scores. The second and third together summarize the possible differences between the three scores. Question 1 is concerned with mean scores – it asks whether the group means of these mean scores are the same.

That is, in terms of the derived variables, the null hypothesis for question 1 is $H_0\colon E(z_1)_{\text{Group 1}} = E(z_1)_{\text{Group 2}}$. Since, in this illustration, we are supposing there to be only two groups, this can be tested using a two-sample t-test.

Question 2 is concerned with the *shape* of the profiles – with the pattern of differences between the y_1, y_2, and y_3 scores. The derived variables z_1 and z_2 summarize these differences. If derived variables z_1 and z_2 both have zero expectation in each of the two groups then there is no time trend – the profiles are flat. Thus, to answer Question 2 we test the null hypothesis $H_0\colon E(z_2, z_3) = (0, 0)$. This, being a bivariate null hypothesis, involves Hotelling's T^2-test, a multivariate extension of the ordinary t-test. We discuss this below.

Finally, Question 3 asks if the patterns of change over occasion are the same in the two groups. As with Question 2, derived variables z_2 and z_3 summarize the patterns of change. So now the relevant null hypothesis is $H_0\colon E(z_2, z_3)_{\text{Group 1}} = E(z_2, z_3)_{\text{Group 2}}$. This is clearly a multivariate analogue of the univariate two-sample t-test. Again, this is discussed below. ■

In Illustration 2.9 we defined the derived variables z_2 and z_3 using $[1 \ \ -1 \ \ 0]$ and $[0 \ \ 1 \ \ -1]$. Together these derived variables span the space of possible differences. Other derived variables, also spanning this space, could equally be used. For example, we could use $[1 \ \ 0 \ \ -1]$ and $[1 \ \ -2 \ \ 1]$. The two sets are equivalent since

$$[1 \ \ 0 \ \ -1] = [1 \ \ -1 \ \ 0] + [0 \ \ 1 \ \ -1]$$

and

$$[1 \ \ -2 \ \ 1] = [1 \ \ -1 \ \ 0] - [0 \ \ 1 \ \ -1]$$

In general, if $\mathbf{H}_i$ is one such set, then $\mathbf{H}_i\mathbf{G}$ is another such set, where $\mathbf{G}$ is a $p \times p$ invertible matrix. Which set one chooses will depend on the problem. In the above, the set $[1 \ \ 0 \ \ -1]$ and $[1 \ \ -2 \ \ 1]$ are coefficients of orthogonal polynomials – they correspond to linear and quadratic terms, respectively. These would be useful if one wanted to describe time change in such terms. By sequentially examining such polynomial components, one can determine the degree of a polynomial fit to the responses over time. (If one was merely interested in whether such profiles were flat or not, as in Illustration 2.9, then it would not matter which was chosen.)

The ML solution in (2.3) is the generalized least-squares solution in which the correlations between the variables are taken into account. Using Lemma 1 of Khatri (1966), this can be alternatively expressed as

$$\hat{\Gamma} = (\mathbf{X}^{\mathrm{T}}\mathbf{X})^{-1}\mathbf{X}^{\mathrm{T}}\mathbf{YRB}^{\mathrm{T}}(\mathbf{BB}^{\mathrm{T}})^{-1}$$

where $\mathbf{YR}$ are the residuals after covarying out $\mathbf{YH}_2$, where $(\mathbf{H}_2)_{p \times p-r}$ satisfies $\mathbf{BH}_2 = 0$ and $\mathbf{H}_2^{\mathrm{T}}\mathbf{H}_2 = \mathbf{I}_{p-r}$ i.e.

$$\mathbf{YR} = \mathbf{Y}(\mathbf{I} - \mathbf{P}_2) = \mathbf{Y}\{\mathbf{I} - \mathbf{H}_2\,(\mathbf{H}_2^{\mathrm{T}}\boldsymbol{\Sigma}_p\mathbf{H}_2)^{-1}\,\mathbf{H}_2^{\mathrm{T}}\boldsymbol{\Sigma}_p\}$$

Hence the ML estimator is equivalent to a simple least-squares estimator based on the derived variables formed by covarying out those which have zero expectation under the model. This equivalence suggests a generalization in which some, but not all, of these derived variables might be covaried out. We shall not pursue this here, but interested readers may refer to Kenward (1986).

2.4 More complex repeated measures structures

Often there is structure to the dependent variables beyond the simple repeated structure. For example, **multiple repeated measurements** often occur, in which several variables are measured on each occasion and where there are relationships between these variables. Thus, one might measure response at each of four times after treatment under five different drugs or five different drug doses, yielding 20 observations on each subject in all. In a sense this is a doubly repeated measures problem. Designs like this are common in some areas of psychology, where repeated measurements are often made at short intervals (minutes, say) and the whole exercise repeated at long intervals (days, say), yielding a minute by day structure for the within-subjects measurements. Example 2.2 (page 33) illustrates such a multiple repeated measurements problem.

In terms of the above presentation, the **H** matrix serves to re-express the raw variables in terms of derived variables which represent main effects of and interactions between the within-subjects factors.

■ *Illustration 2.10* Suppose that six measurements are taken on each individual, the first three being measurements at 1 hour, 2 hours and 3 hours after treatment with dose 1, and the last three being measurements at 1 hour, 2 hours and 3 hours after treatment with dose 2. Then a suitable **H** matrix by which to multiply the raw data matrix to yield appropriate derived variables would be

$$\mathbf{H} = \begin{bmatrix} 1 & 1 & 1 & 1 & 1 & 1 \\ 1 & 1 & 0 & -2 & 0 & -2 \\ 1 & 1 & -1 & 1 & -1 & 1 \\ 1 & -1 & 1 & 1 & -1 & -1 \\ 1 & -1 & 0 & -2 & 0 & 2 \\ 1 & -1 & -1 & 1 & 1 & -1 \end{bmatrix}$$

The first column of **H** gives the overall response level. Column 2 produces a derived variable summarizing the effect of dose. Columns 3 and 4 together measure the effect of time – column 3 being the linear effect and column 4 the quadratic effect. And columns 5 and 6 summarize the interaction of dose by time – column 5 (computed as column 2 × column 3) the linear component of the interaction and column 6 (computed as column 2 × column 4) the quadratic effect.

As we will see in the examples below, manova computer packages group these derived variables into their natural sets and then conduct separate tests for each (possibly multivariate) main effect and each (possibly multivariate) interaction. ■

2.5 Multivariate tests

In the preceding section we showed how we begin with the vector of raw observations and transform them to a set of 'derived variables' which focus on particular questions of interest. In Illustration 2.9 we produced two such derived sets. The first set consisted of a single derived variable (z_1) and was used to answer questions relating to mean response level. The second set consisted of two derived variables (z_2 and z_3) and was used to answer questions relating to pattern of change over time. Example 2.2 below shows a case where four sets of derived variables arise. We pointed out, after Illustration 2.9, that what was significant were the spaces spanned by each set of derived variables, and not the particular variables themselves.

If a particular question is answered by a single derived variable (such as Question 1 in Illustration 2.9) then the analysis reduces to a univariate analysis of variance (or a t-test in the case of questions involving just one or two groups). However, if more than one derived variable is involved then multivariate test statistics are needed.

The univariate case involves F-statistics. These are ratios of **hypothesis sums of squares** to **residual** or **error sums of squares**. The multivariate case involves straightforward extensions in which the sums of squares are replaced by matrices. The diagonal elements of these matrices consist of the univariate hypothesis sums of squares and error sums of squares for each of the p variables separately. The off-diagonal elements consist of the corresponding cross-product terms. And, of course, in place of *ratios* of matrices, which are not defined, we use the product of the hypothesis matrix and the inverse of the error matrix.

To see how these matrices arise, let us start with the simplest case: that of a single variable, y, distributed as $N(\mu, \sigma^2)$, and measured on individuals within a single group of size n. The null hypothesis to be tested is $H_0{:}\mu = \mu_0$. The test statistic is the familiar t-statistic:

$$t = \frac{\bar{y} - \mu_0}{s/\sqrt{n}}$$

Note (for what is about to follow) that this can be rewritten as

$$t^2 = (\bar{y} - \mu_0)s^{-2}(\bar{y} - \mu_0)n \tag{2.4}$$

Now let $\mathbf{y}$ be a p-variate vector, with corresponding mean vector $\boldsymbol{\mu}$, hypothesized to take the values given by the vector $\boldsymbol{\mu}_0$. This can be reduced to the univariate case by forming a linear combination of its components, $\mathbf{a}^T\mathbf{y}$, and this could be subjected to a t-test, as above.

Now, if $\mathbf{a}^T\boldsymbol{\mu} = \mathbf{a}^T\boldsymbol{\mu}_0$ for *all* (non-null) linear combinations $\mathbf{a}$, then $\boldsymbol{\mu} = \boldsymbol{\mu}_0$. So we can test the latter by testing the former.

The former of these hypotheses appears, superficially, to be quite tough to test. However, if we can show that $\mathbf{a}$ leading to the largest such t-statistic does not lead to rejection of $\mathbf{a}^T\boldsymbol{\mu} = \mathbf{a}^T\boldsymbol{\mu}_0$, then no other $\mathbf{a}$ will lead to rejection (this is called the **union-intersection principle**).

The sample mean of the $\mathbf{a}^T\mathbf{y}_i$ is $\mathbf{a}^T\bar{\mathbf{y}}$, with hypothesized mean $\mathbf{a}^T\boldsymbol{\mu}_0$ and estimated standard deviation $\sqrt{(\mathbf{a}^T\mathbf{S}_p\mathbf{a})}$, yielding t-statistic

$$t(\mathbf{a}) = \frac{\mathbf{a}^T\bar{\mathbf{y}} - \mathbf{a}^T\boldsymbol{\mu}_0}{\sqrt{\mathbf{a}^T\mathbf{S}_p\mathbf{a}}/\sqrt{n}}$$

We wish to find the $\mathbf{a}$ which maximizes this. Unfortunately, it can be easily seen that this does not uniquely determine $\mathbf{a}$ – multiplying by an arbitrary constant would also yield a solution. This indeterminancy is avoided by imposing the constraint that $\mathbf{a}^T\mathbf{S}_p\mathbf{a} = 1$ and maximizing $t(\mathbf{a})$ subject to this. (Another way of looking at this is that the metric through which distances in the $\mathbf{y}$-space are measured is distorted so that the data have a unit covariance matrix. Then, in this distorted space, the maximum t-statistic is found.)

Using a Lagrange multiplier to impose this constraint leads to $\max_{\mathbf{a}} t^2(\mathbf{a})$ being the single non-zero value of λ such that

$$|(\bar{\mathbf{y}} - \boldsymbol{\mu}_0)(\bar{\mathbf{y}} - \boldsymbol{\mu}_0)^T n - \lambda\mathbf{S}_p| = 0$$

which is

$$T^2 = (\bar{\mathbf{y}} - \boldsymbol{\mu}_0)^T\mathbf{S}_p^{-1}(\bar{\mathbf{y}} - \boldsymbol{\mu}_0)n$$

a multivariate version of (2.4). This is **Hotelling's T^2-statistic**. Under the null hypothesis,

$$\frac{n-p}{p(n-1)}T^2 \sim F(p, n-p)$$

The case of two groups can be developed in an analogous way. Here $H_0{:}\boldsymbol{\mu}_1 = \boldsymbol{\mu}_2$ and the solution is given by the root of

$$\left|(\bar{\mathbf{y}}_1 - \bar{\mathbf{y}}_2)(\bar{\mathbf{y}}_1 - \bar{\mathbf{y}}_2)^T\frac{n_1 n_2}{n_1 + n_2} - \lambda\mathbf{S}_p\right| = 0$$

yielding

$$T^2 = (\bar{\mathbf{y}}_1 - \bar{\mathbf{y}}_2)^T\mathbf{S}_p^{-1}(\bar{\mathbf{y}}_1 - \bar{\mathbf{y}}_2)\frac{n_1 n_2}{n_1 + n_2}$$

with

$$\frac{n_1 + n_2 - p - 1}{(n_1 + n_2 - 2)p} T^2 \sim F(p, n_1 + n_2 - p - 1)$$

Here the observations in group 1 are assumed to be distributed as $N(\boldsymbol{\mu}_1, \boldsymbol{\Sigma}_p)$ and those in group 2 as $N(\boldsymbol{\mu}_2, \boldsymbol{\Sigma}_p)$ – with the same covariance matrix. This latter is often a risky assumption, but it is known that when n_1 and n_2 are large and equal then unequal covariance matrices have little effect on type I error.

The above two cases have solutions given by the roots of equations of the form

$$|\mathbf{T} - \lambda\mathbf{E}| = 0$$

where $\mathbf{E}$ is the within-groups error sums of squares matrix and $\mathbf{T}$ is the between-groups hypothesis sums of squares matrix. And this allows us to generalize to more than two groups.

This generalization to multiple classes (so that the hypothesis matrix is of rank greater than 1) introduces the additional complication that there is more than one root. Each of these roots, denoted $\lambda_1, \ldots, \lambda_r$ below, represents a component of the between-groups variation. Different test statistics arise from combining these characteristic roots, or eigenvalues, in different ways.

- **Roy's largest eigenvalue statistic** is λ_1, the largest of the roots. Often this is described, not in terms of the largest eigenvalue of $\mathbf{TE}^{-1}$, as we have done above, but in the equivalent terms of the largest eigenvalue of $\mathbf{T}(\mathbf{T} + \mathbf{E})^{-1}$.
- The **Hotelling–Lawley trace** is $\sum\lambda_i$.
- The **likelihood ratio test statistic**, often called **Wilks's lambda**, is defined as $\Pi(1 + \lambda_i)^{-1}$.
- The **Pillai–Bartlett trace** is $\sum \lambda_i/(1 + \lambda_i)$. It is the trace of $\mathbf{T}(\mathbf{T} + \mathbf{E})^{-1}$.

The F-statistic which arises in the univariate case follows the F-distribution when the null hypothesis is true. Unfortunately, in general the above test statistics do not follow such simple distributions (except in certain special cases). However, it is possible to apply transformations (which we shall not go into here since it would lead us too far from our main objective) such that the result is approximately F-distributed. These transformations can involve interpolation – which explains why the F values given as output in software packages are often associated with fractional degrees of freedom.

Many software packages for manova give all four of these test statistics, leaving the user to choose among them. Ideally, one would like to pick the most powerful test, but unfortunately their relative power depends on the nature of the departures from the null hypothesis of equal mean vectors. If the difference lies along a univariate continuum then Roy's largest eigenvalue statistic seems to be the most powerful (perhaps not surprisingly), and power decreases as we move down the list above. Otherwise, the order of powers seems to be reversed. However, it seems that the concentration into a single dimension must be fairly extreme before it changes the order of powers – so some authors favour the Pillai–Bartlett trace. This test statistic also seems to have most robust type I error rates arising from occasional outliers.

When a comparison between only two groups is being made (or, more generally, when only a single degree of freedom space of contrasts between groups of subjects is being studied), then the $\mathbf{TE}^{-1}$ matrix is of rank 1. Consequently, all of the above test statistics will then yield the same result.

2.6 Tests of assumptions

The manova approach assumes that the observation vectors for each individual arise from multivariate normal distributions, and that the distributions for each class (defined by the between-subjects factors) have the same covariance matrix. This latter assumption is the natural extension of the equal variance assumption in univariate analysis of variance (anova). And, as in that case, the error rates of tests are less affected if the assumption is false when the sample sizes are approximately equal (provided they are not too small).

A multivariate normal distribution necessarily has normal marginals, and checks of these can be made using histograms, box and whisker plots, stem and leaf plots, or probability plots of the observations on each occasion separately. These can be extended to bivariate scattergrams, allowing pairs of variables to be examined simultaneously. However, normality of marginals does not imply multivariate normality. Tests of multivariate normality have been developed (for example, Mardia, 1975), but one must be wary of rejecting a robust comparison of means on the basis of an over-sensitive test of departures from normality.

Two approximations are in common use to test the hypothesis of equality of the covariance matrices. Both are based on the likelihood ratio criterion.

The null hypothesis is that $\mathbf{\Sigma}_{p1} = \mathbf{\Sigma}_{p2} = \cdots \mathbf{\Sigma}_{pg}$ (where $\mathbf{\Sigma}_{pi}$ denotes the covariance matrix of the *ith* group) and the alternative is that $\mathbf{\Sigma}_{pr} \neq \mathbf{\Sigma}_{ps}$ for some r and s. Let $\mathbf{S}_{pr}$ be the usual unbiased estimator of $\mathbf{\Sigma}_{pr}$ and let $\mathbf{S}_p = \sum (n_r - 1)\mathbf{S}_{pr}/\sum (n_r - 1)$ be the estimate of the assumed common covariance matrix. The likelihood ratio test statistic is

$$M = \sum (n_r - 1) \ln |\mathbf{S}_p| - \sum (n_r - 1) \ln |\mathbf{S}_{pr}|$$

The first approximation leads to Mk following approximately a χ^2 distribution with $(g-1)p(p+1)/2$ degrees of freedom, where k is defined as

$$k = 1 - \frac{2p^2 + 3p - 1}{6(p+1)(g-1)} \left(\sum_r \frac{1}{(n_r - 1)} - \frac{1}{\sum (n_r - 1)} \right)$$

The second approximation is rather more complicated and leads to an approximate F-distribution. Details are given in Box (1949).

Some computer programs (such as SPSSPC MANOVA) will carry out these tests if requested.

2.7 Examples

Example 2.1

The data analysed in this example are a classic data set first presented by Potthof and Roy (1964). The data (shown in Table A.8) were collected at the University of North Carolina Dental School and show the distance, in millimetres, from the centre of the pituitary to the pteryomaxillary fissure. This was measured at ages 8, 10, 12 and 14 years for 11 girls and 16 boys. (Potthof and Roy comment that 'the reason why there is an occasional instance where this distance decreases with age is that the distance represents the relative position of two points'.) We shall use this data set to illustrate the three profile questions in Illustration 2.9. That is, we want to know:

1. Are the mean profiles of the boys and girls at the same level? (Is there a 'group effect'?)
2. Are the profiles flat? (Is there a 'time effect'?)
3. Are the profiles parallel? (Is there a 'group by time interaction'?)

As noted in Illustration 2.9, we shall begin by considering the interaction question 3.

The first thing is to look at a plot of the data. This is shown in Fig. 2.1. There is nothing terribly anomalous about these data. One of the boys looks as if he might be a multivariate outlier in that he moves from having the smallest measurement at age 10 to the joint largest measurement at age 12 and then back down to almost the smallest. This curious behaviour could be explained in terms of something odd about the measurement at age 12 (a misreading of 21 for 31, for example). However, since we have no external evidence to support this we shall analyse the data as presented. The boys' data also hint at positive skewness. Certainly, there seems to be a cluster of profiles at the bottom of the diagram. However, the girls' data do not suggest asymmetry, so we have not attempted to transform the data.

For this analysis, we used the MANOVA routine in SPSSPC+, presented in Table 2.1. The MANOVA command itself says that there are four dependent variables, which we have called AGE8 to AGE14, and that the groups are classified by a categorical variable SEX, which has two levels.

The WSFACTORS command specifies that there is a single within-subjects factor, AGE, which has four levels. So that we can explore the nature of any change over time, we have transformed the four response variables into four polynomial components using a CONTRAST command, and these are renamed in the RENAME command so that the output can be readily interpreted. The WSDESIGN command specifies the within-subjects design – in this case a saturated design on the single within-subjects factor, AGE.

The PRINT command requests that the transformation matrix (shown in Table 2.2) should be printed (always a good idea, so that one can see that the transformation is actually doing what one intended) and that the estimated parameters should be printed. The first column of the matrix will generate a derived variable

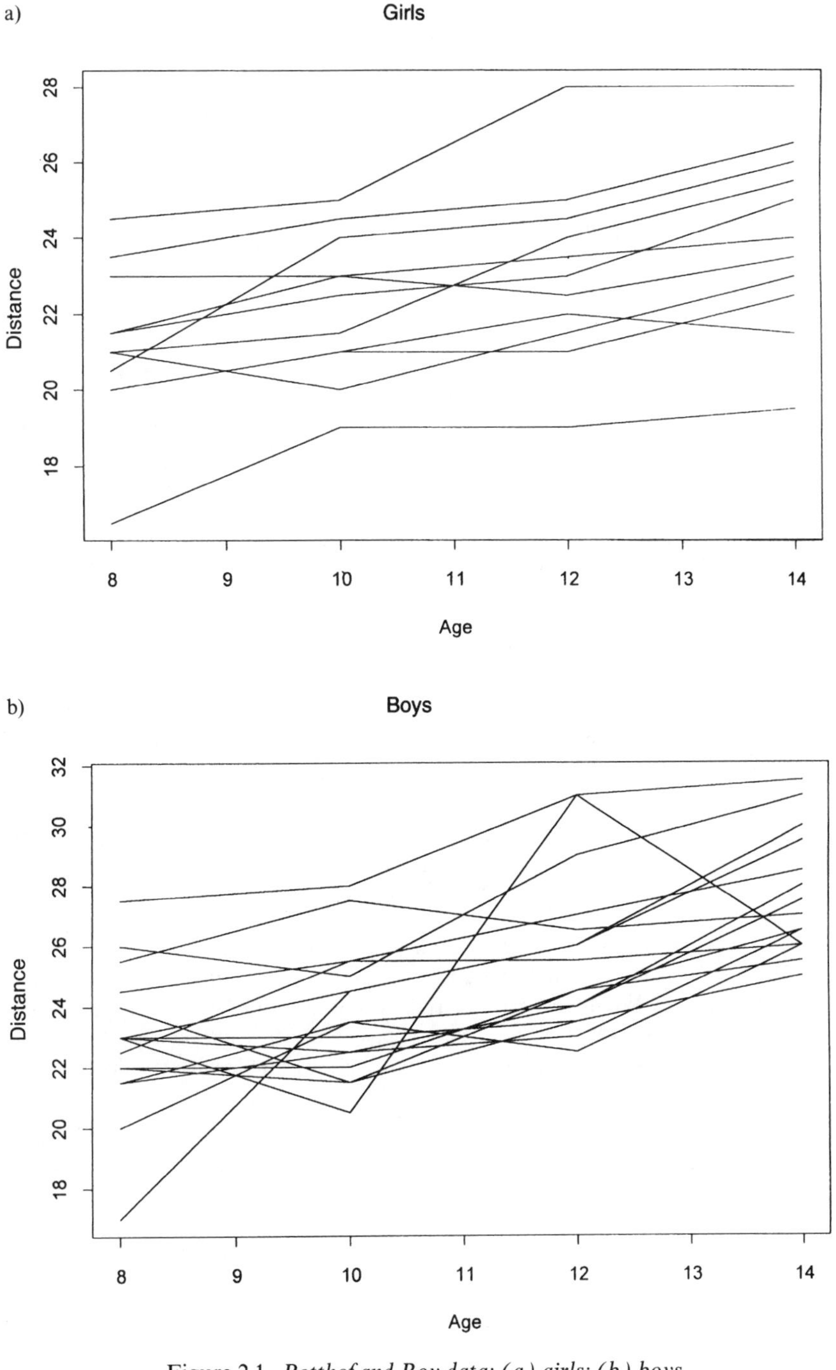

Figure 2.1. *Potthof and Roy data: (a) girls; (b) boys.*

```
MANOVA AGE8 AGE10 AGE12 AGE14 BY SEX (1,2)
   /WSFACTORS = AGE (4)
   /CONTRAST (AGE) = POLYNOMIAL
   /RENAME = CONS LIN QUAD CUB
   /WSDESIGN
   /PRINT TRANSFORM PARAMETERS (ESTIM)
   /DESIGN.
```

Table 2.1. *The SPSSPC+ MANOVA commands for Example 2.1*

proportional to the mean of the four scores for each subject, the second column will generate a derived variable representing the linear trend of the curves over time, the third column the quadratic trend, and the fourth the cubic trend. The columns have been normalized so that univariate test statistics can be readily extracted, as explained in section 2.3.

Since, in a typical manova repeated measures analysis there will be several hypotheses tested (like those in Questions 1 – 3 of Illustration 2.9), there is plenty of scope for presenting the results in different ways – and different computer packages do, indeed, present their results in different ways. SPSSPC + MANOVA adopts a particular order, which in this example becomes:

- Using the mean derived variable
  Averaged over groups and compared with 0 (this hypothesis is only rarely of interest)
  Comparison between groups (in this case, this results in a comparison of the average levels of the two sexes)
- Using the linear/quadratic/cubic derived variables
  Comparison between groups (this is thus an examination of the age by sex interaction)
  Averaged over the two groups (yielding a test of the age or time effect).

Tests involving the mean derived variable are, of course, univariate.

Question 3 of Illustration 2.9 addressed the interaction of time by group – in this case of age by sex. We should really consider this interaction before we look at main effects since a significant interaction means that care needs to be taken in interpreting main effects (for example, perhaps we should look at simple effects – effects of one

Orthonormalized Transformation Matrix (Transposed)

	CONS	LIN	QUAD	CUB
AGE8	.500	−.671	.500	−.224
AGE10	.500	−.224	−.500	.671
AGE12	.500	.224	−.500	−.671
AGE14	.500	.671	.500	.224

Table 2.2. *The orthonormalized transformation matrix*

factor conditional on a given level of the other – instead of overall main effects). Table 2.3 shows the SPSSPC + MANOVA results testing this interaction. As noted above, several test statistics are given. In this case, although the statistics differ, they all lead to the same approximate F-statistic and the same p value since there are only two groups.

The approximate F value of 2.7 is not significant at the 5% level, so we shall suppose there to be no sex by age interaction. This means that we can go on to look at the age and sex main effects. Let us look at sex first.

The anova table for tests involving the single derived variable measuring the constant (or mean) level, is shown in Table 2.4. Averaged over all subjects and groups this gives a highly significant CONSTANT effect. As noted above, this is seldom of interest when the responses can only take positive values, since it merely indicates that the mean of a set of positive values is significantly different from 0.

Of more interest is the SEX row of the table, which compares the mean levels of the two groups. The significant result here shows that the two groups have different levels – that there is a significant sex effect.

The parameter estimates for this derived variable are given in Table 2.5. The first is for the overall average scores of the two groups. It is thus obtained by multiplying the

EFFECT .. SEX BY AGE

Multivariate Tests of Significance (S = 1, M = 1/2, N = 10 1/2)

Test Name	Value	Approx. F	Hypoth. DF	Error DF	Sig. of F
Pillais	.26011	2.69527	3.00	23.00	.070
Hotellings	.35156	2.69527	3.00	23.00	.070
Wilks	.73989	2.69527	3.00	23.00	.070
Roys	.26011				

Table 2.3. *Multivariate tests of the sex by age interaction effects*

Source of Variation	SS	DF	MS	F	Sig. of F
WITHIN CELLS	377.91	25	15.12		
CONSTANT	59118.50	1	59118.50	3910.84	.000
SEX	140.46	1	140.46	9.29	.005

Table 2.4. *Tests involving the constant (mean) derived variable*

Estimates for CONS

CONSTANT

Parameter	Coeff.	Std. Err.	t-Value	Sig. t	Lower-95%	CL-Upper
1	47.6164773	.76142	62.53667	.000	46.04831	49.18464

SEX

Parameter	Coeff.	Std. Err.	t-Value	Sig. t	Lower-95%	CL-Upper
2	−2.3210227	.76142	−3.04829	.005	−3.88919	−.75286

Table 2.5. *Parameter estimates for the constant (mean) variable*

means of each sex group at each of the four ages by the corresponding coefficients from the first column of the transformation matrix given in Table 2.2 and then averaging over sex groups. In fact, since we are here discussing the 'constant' derived variable, we multiply the four means of each group by 0.5. We thus obtain:

Means for girls	21.182	22.227	23.091	24.091
Means for boys	22.875	23.813	25.719	27.469

This yields $0.5 \times 21.182 + 0.5 \times 22.227 + 0.5 \times 23.091 + 0.5 \times 24.091 = 45.296$ for the girls and, similarly, 49.938 for the boys. The average of these is $(45.296 + 49.938)/2 = 47.62$ – the result in Table 2.5.

The second, more interesting, parameter estimate in Table 2.5 tells us the size of the overall SEX effect. This parameter estimate is obtained by subtracting the mean for the boys (49.938) from the overall average given above (47.62), yielding −2.32 as in Table 2.5. It is thus a measure of the difference between the two sexes, but expressed as a deviation from their overall mean.

These results just relate to overall levels for the two sexes, and do not tell us about age effects. To see these we look at Tables 2.6 and 2.7.

EFFECT .. AGE
Multivariate Tests of Significance (S = 1, M = 1/2, N = 10 1/2)

Test Name	Value	Approx. F	Hypoth. DF	Error DF	Sig. of F
Pillais	.80521	31.69110	3.00	23.00	.000
Hotellings	4.13362	31.69110	3.00	23.00	.000
Wilks	.19479	31.69110	3.00	23.00	.000
Roys	.80521				

Table 2.6. *Tests for the (multivariate) effect of age*

Estimates for LIN
AGE

Parameter	Coeff.	Std. Err.	t-Value	Sig. t	Lower-95%	CL-Upper
1	2.82621205	.30128	9.38076	.000	2.20572	3.44670

Estimates for QUAD
AGE

Parameter	Coeff.	Std. Err.	t-Value	Sig. t	Lower-95%	CL-Upper
1	.191761364	.19987	.95941	.347	−.21989	.60341

Estimates for CUB
AGE

Parameter	Coeff.	Std. Err.	t-Value	Sig. t	Lower-95%	CL-Upper
1	−.09020502	.31068	−.29034	.774	−.73007	.54966

Table 2.7. *Parameter estimates for the derived variables showing change over age–the linear, quadratic, and cubic orthogonal polynomial components*

Table 2.6 shows the multivariate test of the hypothesis that there is no age effect. That is, it simultaneously tests the linear, quadratic and cubic derived variables to see if their means differ significantly from zero. If any such difference is found it means that the scores do not remain constant over age. And it is clear from Table 2.6 that there is a highly significant effect – perhaps as we would expect from the figures. Note that again the various multivariate test statistics lead to the same result as this is a comparison of a single group of scores with zero.

Table 2.7 shows the parameter estimates for the linear, quadratic, and cubic orthogonal polynomial components of change with age. In fact, the SPSSPC+ MANOVA output also gives corresponding parameter estimates for the interaction effects of each of these components by sex. Since, however, we have decided that there is no interaction effect, these have been omitted from the output below.

These estimates are obtained in the same way as for the constant (mean) effects above, but using the second, third and fourth columns of the transformation matrix in Table 2.2. For example, for the girls we obtain $-0.671 \times 21.182 - 0.224 \times 22.227 + 0.224 \times 23.091 + 0.671 \times 24.091 = 2.145$. (This way of obtaining the result – applying the transformation coefficients to the means – is equivalent to averaging the derived variables.) Using the same coefficients for the boys, we obtain 3.510. The average of these is $(2.145 + 3.510)/2 = 2.83$, as is shown as the parameter estimate for the linear effect of age in the first row of Table 2.7.

Table 2.7 shows that neither the cubic nor the quadratic component is significant when simple univariate tests are conducted on these derived variables. The highly significant pattern of change over age is obtained from the linear component. Figure 2.2 shows a plot of the means against age for the two groups. The fact that the curves do not depart a great deal from straight lines, and yet are not horizontal, is apparent.

This example involved a single 'within-subjects factor' – namely age – which gave the levels at which each of the repeated measures occurred. It also involved a single 'between-subjects' or grouping factor – sex – with only two levels. In Example 2.2 we consider a situation with just one group of subjects but where there are two within-subjects factors.

Example 2.2

The data for this example come from Crowder and Hand (1990, p. 30) and are reproduced in Hand *et al.* (1994, Table 397). They are presented in Table A.9. The question of interest was whether the response time of eyes to a visual stimulus (a light flash) varies with the power of a lens through which the stimulus is viewed. The response time was the number of milliseconds between the light flash and the electrical response at the back of the cortex. Each of seven subjects was measured with lens powers of 6/6, 6/18, 6/36 and 6/60, and this was done for both left and right eyes. (A power of 6/18 means that the magnification is such that the eye will perceive as being at 6 feet an object which is actually positioned at a distance of 18 feet.) The within-subjects factors are eye (with two levels: left and right) and lens strength (with four levels).

We performed the analysis of these data using BMDP4V. An SPSSX MANOVA analysis is presented in Crowder and Hand (1990).

The profiles of the seven subjects are shown in Fig. 2.3. There seems to be quite a lot of variability, both within and between subjects. From the limited information

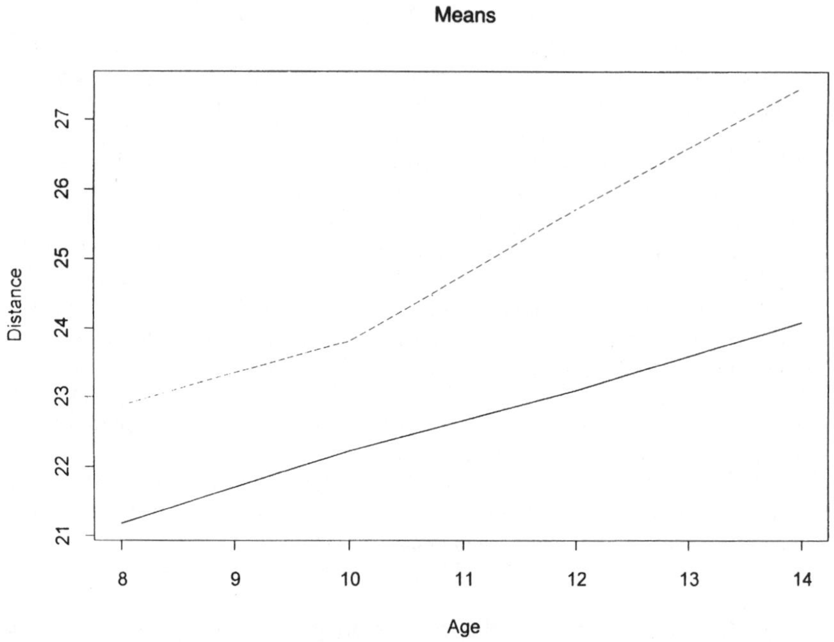

Figure 2.2. *The means of the two groups by age. The broken line is for the boys.*

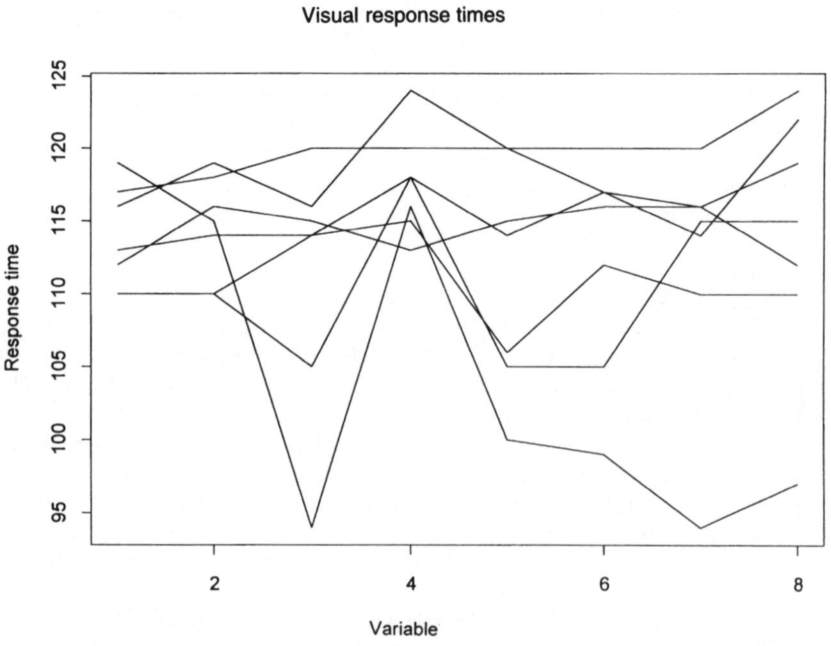

Figure 2.3. *Profiles of the responses for the seven subjects in the visual response time study. Variables 1–4 are left eye, 5–8 right eye.*

in the figure there seems to be little difference between the two eyes. Similarly, there seems to be little trend over lens strength. We shall see if these observations are reflected in the formal analysis.

The BMDP4V command lines are shown in Table 2.8. These specify where the data are to be found, how many variables there are, the format of the data, the names to be given to the variables (we used the mnemonic notation using L for left and R for right and a number to indicate the lens power). This data set has only a single group of subjects, so there is no /BETWEEN command line, which would give details of the grouping structure. The /WITHIN paragraph shows that there are two within-subjects factors, which we have called SIDE and POWER. SIDE has two levels, which we have called LEFT and RIGHT. POWER has four levels, named P6, P18, P36 and P60.

This program allows one to specify the weights to be given to each cell of the within-individual cross-classification to represent their importance. We have specified equal weights.

The default **H** matrix generated was

$$\mathbf{H} = \begin{bmatrix} 1 & 1 & 1 & 0 & 0 & 1 & 0 & 0 \\ 1 & 1 & 0 & 1 & 0 & 0 & 1 & 0 \\ 1 & 1 & 0 & 0 & 1 & 0 & 0 & 1 \\ 1 & 1 & -1 & -1 & -1 & -1 & -1 & -1 \\ 1 & -1 & 1 & 0 & 0 & -1 & 0 & 0 \\ 1 & -1 & 0 & 1 & 0 & 0 & -1 & 0 \\ 1 & -1 & 0 & 0 & 1 & 0 & 0 & -1 \\ 1 & -1 & -1 & -1 & -1 & 1 & 1 & 1 \end{bmatrix}$$

```
/INPUT FILE = 'Data file name goes here'.
    VARIABLES = 8.
    FORMAT = FREE.
/VARIABLE NAMES = L6, L18, L36, L60, R6, R18, R36, R60.
/WITHIN FACTOR = SIDE, POWER.
    CODES(SIDE) = 1 TO 2.
    NAMES(SIDE) = LEFT, RIGHT.
    CODES(POWER) = 1 TO 4.
    NAMES(POWER) = P6, P18, P36, P60.
/WEIGHTS WITHIN = EQUAL.
/PRINT LINESIZE = 80.
/END
ANALYSIS PROCEDURE = FACTORIAL.
    DISP.
    PRIN.
/END
```

Table 2.8. *BMDP4V command file for the visual response time data*

The first column of this produces a derived variable showing the overall level of the responses. Tests on this derived variable will be of little interest and are not reproduced below. The second column produces a derived variable which describes the SIDE effect. Columns 3, 4 and 5 yield derived variables which describe the POWER effect. Note that these are not expressed as orthogonal polynomials here. If we wanted to study the linear, quadratic and cubic trends over increasing lens power it would be necessary to use appropriate polynomial coefficients to produce suitable derived variables. (Note also that these coefficients would have to be such as to make allowance for the unequal gaps between the four lens powers used in the study.) However, if our aim is simply to see if there is a SIDE effect, a POWER effect, and a SIDE by POWER interaction, then it does not matter what sets of derived variables are used, provided they span the appropriate subspaces, as explained above.

We should first look to see if there is a SIDE by POWER interaction effect. The BMDP4V results for this test are shown in Table 2.9. It is a multivariate test since the interaction is described by three components (the three differences (left minus right) for each of the three derived POWER variables). In this analysis the multivariate test statistic is shown in the row beginning TSQ – standing for Hotelling's T^2-test on the three variables, as outlined above. From the above, this T^2 leads to a statistic of $(7-3)\times 2.485/3\times(7-1)=0.55$, as shown in Table 2.9. Under the model of a multivariate normal distribution, assuming the null hypothesis of no effect to be true, this will follow an F-distribution on 3 and $(7-3)=4$ degrees of freedom – again as given in the table. The corresponding p value is 0.67 – not significant. Thus, the effect of lens power does not appear to differ between left and right eyes.

The other rows of this table refer to the univariate tests and are discussed in Chapter 3.

WITHIN EFFECT: SP

EFFECT	VARIATE	STATISTIC	F	DF	P
SP					
	DEP_VAR				
	TSQ =	2.48540	0.55	3, 4	0.6733
	WCP SS =	40.625000			
	WCP MS =	13.541667	1.06	3, 18	0.3925
	GREENHOUSE-GEISSER ADJ.DF		1.06	1.65, 9.89	0.3700
	HUYNH-FELDT ADJUSTED DF		1.06	2.19, 13.14	0.3819
ERROR					
	DEP_VAR				
	WCP SS =	231.00000000			
	WCP MS =	12.83333333			
	GGI EPSILON	0.54927			
	H-F EPSILON	0.73025			

Table 2.9. *Multivariate test of the interaction of SIDE by POWER*

Having determined that there is no interaction, we can now go on to examine each of the SIDE and POWER main effects. The results of the test on the SIDE effect are shown in Table 2.10. Note that this is a univariate test since the derived variable is summarizing the single difference in overall levels between the left and right sides. This effect also appears not to be significant.

Finally, we turn to the multivariate test of POWER. This, like the interaction test, is trivariate and the test statistic will be compared with an F- distribution with 3 and 4 degrees of freedom. Table 2.11 shows the results. Hotelling's T^2–test statistic yields a p value of 0.0063 – significant at the 1% level.

WITHIN EFFECT: S:SIDE

EFFECT	VARIATE	STATISTIC		F	DF	P
S						
	DEP_VAR					
		SS =	46.446429			
		MS =	46.446429	0.78	1, 6	0.4112
ERROR						
	DEP_VAR					
		SS =	357.42857143			
		MS =	59.57142857			

Table 2.10. *Analysis of variance table for the effect of SIDE*

WITHIN EFFECT: P:POWER

EFFECT	VARIATE	STATISTIC		F	DF	P
P						
	DEP_VAR					
		TSQ =	96.7055	21.49	3, 4	0.0063
		WCP SS =	140.767857			
		WCP MS =	46.922619	2.25	3, 18	0.1177
	GREENHOUSE-GEISSER ADJ.DF			2.25	1.49, 8.94	0.1665
	HUYNH-FELDT ADJUSTED DF			2.25	1.87, 11.21	0.1528
ERROR						
	DEP_VAR					
		WCP SS =	375.85714286			
		WCP MS =	20.88095238			
		GGI EPSILON	0.49661			
		H-F EPSILON	0.62294			

Table 2.11. *The test of POWER*

2.8 Further reading

Multivariate analysis of variance is described in many texts on multivariate statistics. Most, however, present it from the general viewpoint of non-commensurate variables, rather than from the particular viewpoint of repeated measures. Exceptions to this include Bock (1975), the sections on 'profile analysis' in Morrison (1976), Hand and Taylor (1987) and Chapter 4 of Crowder and Hand (1990).

We point out in Chapter 11 that different sciences favour different approaches to the analysis of repeated measures data. Psychology has made extensive use of the manova approach. References to papers in the psychological literature discussing such usage are given in that chapter.

Rao (1965; 1966; 1967), Grizzle and Allen (1969) and Kenward (1986) discuss how the ML estimator covaries out those derived variables with zero expectation. Olson (1974; 1976; 1979) discusses the comparative power of multivariate test statistics. Box (1949) gives the χ^2 and F approximations to the likelihood ratio statistic for testing the hypothesis that the populations have identical covariance matrices.

CHAPTER 3

Univariate analysis of variance

3.1 Basic ideas, compound symmetry and sphericity

The univariate anova approach to analysing repeated measures data is popular in some disciplines (in psychology, for example, like the manova approach) where there is a strong history of use of analysis of variance methods. Superficially, the n individuals by p occasions structure of the data resembles that of a randomized block or split plot design, so one might be tempted to carry out a standard two-factor analysis of variance. In such an analysis the np observations are regarded as comprising the elements of a single $np \times 1$ observation vector: the first set of p values being the p scores from the first individual, the second set of p values being the p scores from the second individual, and so on.

■ *Illustration 3.1* This describes the same data as Illustration 2.7, but from the univariate anova viewpoint. Consider two groups, of three and two individuals respectively, with the group k expected response profile defined in terms of parameters μ_{kj} as $[\mu_{k1} + \mu_{k2} \;\; \mu_{k1} + \mu_{k3} \;\; \mu_{k1} - \mu_{k2} - \mu_{k3}]$. Then

$$E\begin{bmatrix} y_{11} \\ y_{12} \\ y_{13} \\ y_{21} \\ y_{22} \\ y_{23} \\ y_{31} \\ y_{32} \\ y_{33} \\ y_{41} \\ y_{42} \\ y_{43} \\ y_{51} \\ y_{52} \\ y_{53} \end{bmatrix} = \begin{bmatrix} 1 & 1 & 0 & 0 & 0 & 0 \\ 1 & 0 & 1 & 0 & 0 & 0 \\ 1 & -1 & -1 & 0 & 0 & 0 \\ 1 & 1 & 0 & 0 & 0 & 0 \\ 1 & 0 & 1 & 0 & 0 & 0 \\ 1 & -1 & -1 & 0 & 0 & 0 \\ 1 & 1 & 0 & 0 & 0 & 0 \\ 1 & 0 & 1 & 0 & 0 & 0 \\ 1 & -1 & -1 & 0 & 0 & 0 \\ 0 & 0 & 0 & 1 & 1 & 0 \\ 0 & 0 & 0 & 1 & 0 & 1 \\ 0 & 0 & 0 & 1 & -1 & -1 \\ 0 & 0 & 0 & 1 & 1 & 0 \\ 0 & 0 & 0 & 1 & 0 & 1 \\ 0 & 0 & 0 & 1 & -1 & -1 \end{bmatrix} \begin{bmatrix} \mu_{11} \\ \mu_{12} \\ \mu_{13} \\ \mu_{21} \\ \mu_{22} \\ \mu_{23} \end{bmatrix}$$

■

The sort of model we might adopt to describe such a data structure is

$$y_{ij} = \alpha_i + \beta_j + u_{ij} \tag{3.1}$$

where y_{ij} is the observation on the ith individual on occasion j, α_i is the individual effect, β_j is the occasion effect, and u_{ij} is random error. Unfortunately, however, there are two problems in applying the model directly.

The first is that, in a randomized block or split plot experiment, the α_i belong to pre-defined categories – they are **fixed** effects. As a consequence, the np observations will be independent and the associated covariance matrix of the np observations will be diagonal. In our case, however, the α_i are randomly selected from a distribution of individual effects and constitute a **random** effect – with the consequence that the covariance matrix of the np observations becomes block diagonal.

To see this more formally, let

$$y_{ij} = \alpha + a_i + \beta_j + e_{ij} \qquad (i = 1, \ldots, n; j = 1, \ldots, p) \tag{3.2}$$

where we have changed α_i to $\alpha + a_i$, where a_i is a random effect and α is a constant for all i, j and where we have changed u to e to indicate that it is not now the only source of random variation. Differences between individuals now arise because of random between-individual variation, modelled by a_i.

Let the distribution from which the a_i are drawn have zero mean and variance σ_a^2, so that $E(a) = 0$, $\text{var}(a) = \sigma_a^2$, and let the distribution from which the e_{ij} are drawn have $E(e) = 0$, $\text{var}(e) = \sigma^2$. Finally, assume that the a and e distributions are independent. Then

$$\text{cov}(y_{ij}, y_{ij'}) = E[(a_i + \beta_j + e_{ij})(a_i + \beta_{j'} + e_{ij'})]$$

$$- E(a_i + \beta_j + e_{ij})E(a_i + \beta_{j'} + e_{ij'}) = \sigma_a^2$$

$$\text{cov}(y_{ij}, y_{ij}) = \sigma_a^2 + \sigma^2$$

$$\text{cov}(y_{ij}, y_{i'j}) = 0$$

The covariance matrix $\boldsymbol{\Sigma}$ of the np observations thus consists of a diagonal series of n $p \times p$ blocks, each one of which has $\sigma_a^2 + \sigma^2$ as its diagonal elements and σ_a^2 as its off-diagonal elements. Each of these n submatrices is said to be **compound symmetric**. We can write them conveniently using vector notation as

$$\boldsymbol{\Sigma}_p = \sigma_a^2 \mathbf{1}\mathbf{1}^{\mathrm{T}} + \sigma^2 \mathbf{I}$$

where **1** is a vector of p 1's.

■ *Illustration 3.2* A compound symmetric covariance matrix for four variables has the structure

$$\begin{bmatrix} \sigma^2+\sigma_a^2 & \sigma_a^2 & \sigma_a^2 & \sigma_a^2 \\ \sigma_a^2 & \sigma^2+\sigma_a^2 & \sigma_a^2 & \sigma_a^2 \\ \sigma_a^2 & \sigma_a^2 & \sigma^2+\sigma_a^2 & \sigma_a^2 \\ \sigma_a^2 & \sigma_a^2 & \sigma_a^2 & \sigma^2+\sigma_a^2 \end{bmatrix}$$ ■

As it happens, although $\mathbf{\Sigma}$ is no longer diagonal when the individuals constitute a random effect, the usual anova F-tests are still valid. Indeed, they remain valid under the more general condition that $\text{var}(y_{ij}-y_{ij'})$ is constant for $j \neq j'$. Such a condition is termed **sphericity** or **circularity**. It may alternatively be expressed as the (equivalent) condition that the covariance matrix of any set of $p-1$ orthonormal contrasts (Crowder and Hand, 1990, section 3.5) over the occasions should be proportional to the identity matrix. That compound symmetry is a special case of this may be easily seen by rewriting $\text{var}(y_{ij}-y_{ij'})$ as $\text{var}(y_{ij})+\text{var}(y_{ij'})-2\text{cov}(y_{ij}, y_{ij'})$. When compound symmetry holds, this reduces to $(\sigma_a^2+\sigma^2)+(\sigma_a^2+\sigma^2)-2\sigma_a^2=2\sigma^2$ which is constant, as required.

Each term on the leading diagonal of the covariance matrix of orthonormal contrasts estimates the denominator mean square in an F-test of that contrast. Hence, when sphericity holds, they each estimate the same thing. In such a case one can get a better estimate of their common value by averaging them – which is exactly what the usual univariate F-test does. This explains why the univariate F-test is sometimes called an 'averaged F-test' in the output of repeated measures software. If sphericity does not hold, one will be averaging different things, so that the resulting tests on individual contrasts may have denominators which are too large or small, leading to biased tests. If only certain contrasts are of interest, then one can restrict one's sphericity requirements to those. This will mean fewer degrees of freedom in the denominator sum of squares, resulting in less power, but it also means that a less stringent set of conditions has to be met.

The observant reader will recall that, at the start of this section, we said that there were two problems in applying model (3.1) to repeated measures studies – but so far we have discussed only one, the random nature of the subjects effects.

The second problem concerns the ordered nature of the observations. Unlike a randomized block or split plot experiment, there is no scope for randomizing the occasions. The first measurement *is* the first measurement, it cannot be taken third! Similarly, the first and second measurements are inevitably closer in time (or whatever the ordering continuum is) than the first and third. The consequence is that, in many situations, one might expect measurements which are close together to show a higher correlation than measurements which are far apart. The compound symmetry structure described above may have to be replaced by some other structure – such as a banded diagonal matrix, in which the correlation decreases with increasing distance from the leading diagonal. In general, these extra complications mean that the simple univariate anova F-tests will no longer be valid.

Model (3.2) expresses the observations as a linear combination of random effects (the a_i and e_{ij}) and fixed effects (the β_j). The random effects each had particularly simple covariance structures. We could, alternatively, express (3.2) as $y_{ij} = \beta_j + u_{ij}$, where β_j is still a fixed effect and where the single random effect $[u_{i1} \ldots u_{ip}]$ (u for 'umbrella' again, since it covers all sources of random variation) now has a compound symmetric covariance matrix. This leads to the natural generalization, conveniently written in matrix terms, as

$$\mathbf{y} = \mathbf{X}\boldsymbol{\beta} + \mathbf{u} \tag{3.3}$$

where $\mathbf{X}$ is the $np \times q$ design matrix, $\boldsymbol{\beta}$ is a $q \times 1$ fixed parameter vector, and $\mathbf{u}$ is an $np \times 1$ vector of random terms with zero mean and covariance matrix $\boldsymbol{\Sigma}$, block diagonal with compound symmetric blocks. This in turn yields parameter estimates

$$\hat{\boldsymbol{\beta}} = (\mathbf{X}^{\mathrm{T}}\hat{\boldsymbol{\Sigma}}^{-1}\mathbf{X})^{-1}\mathbf{X}^{\mathrm{T}}\hat{\boldsymbol{\Sigma}}^{-1}\mathbf{y}$$

In Chapter 4 we assume all of the random variation arises from $\mathbf{u}$ in this way, but generalize so that we no longer require Σ_p to be compound symmetric.

Given that the univariate anova approach is legitimate when sphericity (or compound symmetry) holds, we are naturally led to two questions. First, how do we test to see if the matrix satisfies sphericity, and second, what should we do if it does not satisfy it (and we still want to use a univariate anova approach)?

3.2 Tests for and measures of sphericity

Tests do exist for compound symmetry (for example, Rouanet and Lepine, 1970), but the major statistical software packages seem to place most emphasis on the more general condition of sphericity. Indeed, most of the theoretical work seems to address this more general condition.

Many computer programs report the results of the **Mauchly test** of sphericity. The test statistic is

$$W = \frac{|\mathbf{CS}_p\mathbf{C}^{\mathrm{T}}|}{|\mathrm{tr}(\mathbf{CS}_p\mathbf{C}^{\mathrm{T}})/(p-1)|^{p-1}}$$

where $\mathbf{S}_p$ is the pooled within-subject sample covariance matrix and $\mathbf{C}$ is a matrix of $p-1$ orthonormal contrasts.

The Mauchly test is based on $\{\sum_g (n_g - 1) - (p-2)(p+1)/2\} \log(W)$, (with n_g the number of cases in group g) which is asymptotically distributed as χ^2_v with $v = (p-2)(p+1)/2$ when the null hypothesis is true. Of course, many computer packages give the results of such tests automatically or on demand.

If sphericity does not hold then the ordinary univariate anova F-tests reject too many true null hypotheses. To overcome this, a measure of deviation from sphericity, denoted ε, has been developed and adjustments to the ordinary F-tests have been devised using this measure.

Let $\mathbf{\Sigma}_c = \mathbf{C}\mathbf{\Sigma}_p\mathbf{C}^{\mathrm{T}}$ be the covariance matrix for a set of $p-1$ orthonormal contrasts. Then

$$\varepsilon = \frac{\{\operatorname{tr}(\mathbf{\Sigma}_c)\}^2}{(p-1)\operatorname{tr}(\mathbf{\Sigma}_c^2)} = \frac{(\sum \theta_j)^2}{(p-1)\sum \theta_j^2} \tag{3.4}$$

where the θ_j are the $p-1$ eigenvalues of $\mathbf{\Sigma}_c$. (Note an analogy with variance here. Variance is the *difference* between the average of the squared scores $\sum x_i^2/n$ and the square of the average scores $(\sum x/n)^2$. ε is the *ratio* of the square of the average eigenvalue $(\sum\theta_j/(p-1))^2$ to the average of the squared eigenvalues $\sum\theta_j^2/(p-1)$.)

When sphericity holds, $\varepsilon = 1.0$, otherwise $\varepsilon < 1.0$ because the θ_j are all equal if and only if sphericity holds (just as variance is zero if, and only if, the components x_j are all equal).

3.3 Adjusting for non-sphericity

In the case of G groups, with a total of n individuals each measured on p occasions, the usual overall anova test for an occasions effect relates the ratio of mean squares test statistic to the $F(p-1, (p-1)(n-G))$ distribution. This is appropriate when sphericity holds. When the matrices are non-spherical, however, the appropriate distribution is approximated by an F-distribution as above but with the degrees of freedom adjusted by multiplying by ε, yielding $F(\varepsilon(p-1), \varepsilon(p-1)(n-G))$. This can, of course, result in fractional degrees of freedom, so that if one is using tables to look up significance levels by hand, interpolation may be necessary. This adjustment only applies to tests involving occasions effects – tests on group effects remain unchanged.

Of course, this is all very well, but it does not completely resolve things: ε is unknown and will need to be estimated. This estimation will introduce inaccuracies which will affect the distributions.

An obvious estimate is obtained by substituting the sample estimate of the covariance matrix in equation (3.4) above, and one could use this directly in one's tests. This is the Greenhouse–Geisser estimate reported in many programs. In fact, Greenhouse and Geisser (1959) suggested the following strategy to overcome the problem of the variation intrinsic in this estimate:

1. First conduct the F-test with unadjusted degrees of freedom. If this yields a non-significant result then stop (since adjustment then cannot yield significance). Otherwise go to step 2.

2. If the unadjusted F-test yields significance, carry out a conservative adjusted F-test, adjusting the degrees of freedom by the factor $1/(p-1)$, which is the smallest value that ε can take. If this is significant then stop (the ε-adjusted F-test must yield a result between the tests in step 1 and step 2. If both are significant then so must it be.). Otherwise go to step 3.
3. Estimate ε from the sample covariance matrix and conduct the ε adjusted test.

Later work suggested that, for $\varepsilon > 0.75$ and $n < 2p$, the simple estimate of ε may be seriously biased so that it tends to overcorrect the degrees of freedom and produce a conservative test. Because of this, Huynh and Feldt (1976) devised a less biased estimator, defined as min $(1, a/b)$, where $a = n\ (p-1)\hat{\varepsilon} - 2$ and $b = (p-1)\{n - G - (p-1)\hat{\varepsilon}\}$, where $\hat{\varepsilon}$ is the simple estimator using the sample covariance matrix. Both the Greenhouse–Geisser and Huynh–Feldt estimates are often reported by standard computer programs.

A natural inclination would be to use the Mauchly test to see if a sphericity assumption was justified and, if not, to carry out some adjustment as outlined above. It seems, however, that Mauchly's test is weak at detecting small departures from sphericity–and yet that small departures can produce substantial bias in the standard F-tests. This seems to suggest that adjusted tests should be performed unless one has theoretical reasons to suppose that sphericity holds.

3.4 Examples

Example 3.1

The data for this example come from Hand and Taylor (1987, Table G.1). They arose during an investigation into Alzheimer's disease. Two groups of patients were compared, one of which received a placebo and the other treatment with lecithin. In the data reported here, each of the subjects, 26 in the placebo group and 22 in the lecithin group, were measured on five occasions. The measurements were the number of words that the patients could recall from lists of words. (This does mean that the normality assumption made by analysis of variance methods might be questioned. If one is uneasy about the effect of this on the analysis then one can consider the methods described later in this book, though they, too, have their limitations.) The data are reproduced in Table A.10 and plots of the raw responses are given in Fig. 3.1. There seems to be some suspicion of clustering in the profiles of the placebo group towards the lower scores, but this is not evident in the lecithin group. Closer examination, in Fig. 3.2, led us to ignore this possibility, and we analysed the data as presented.

The BMDP2V commands to analyse these data using the univariate anova approach are shown in Table 3.1 Once the data have been located and read in (the first three lines of commands), the names of the variables are given. In this case the first variable indicates the group membership and the remainder the scores at months 0, 1, 2, 4 and 6. The GROUP command tells the system which variable is to

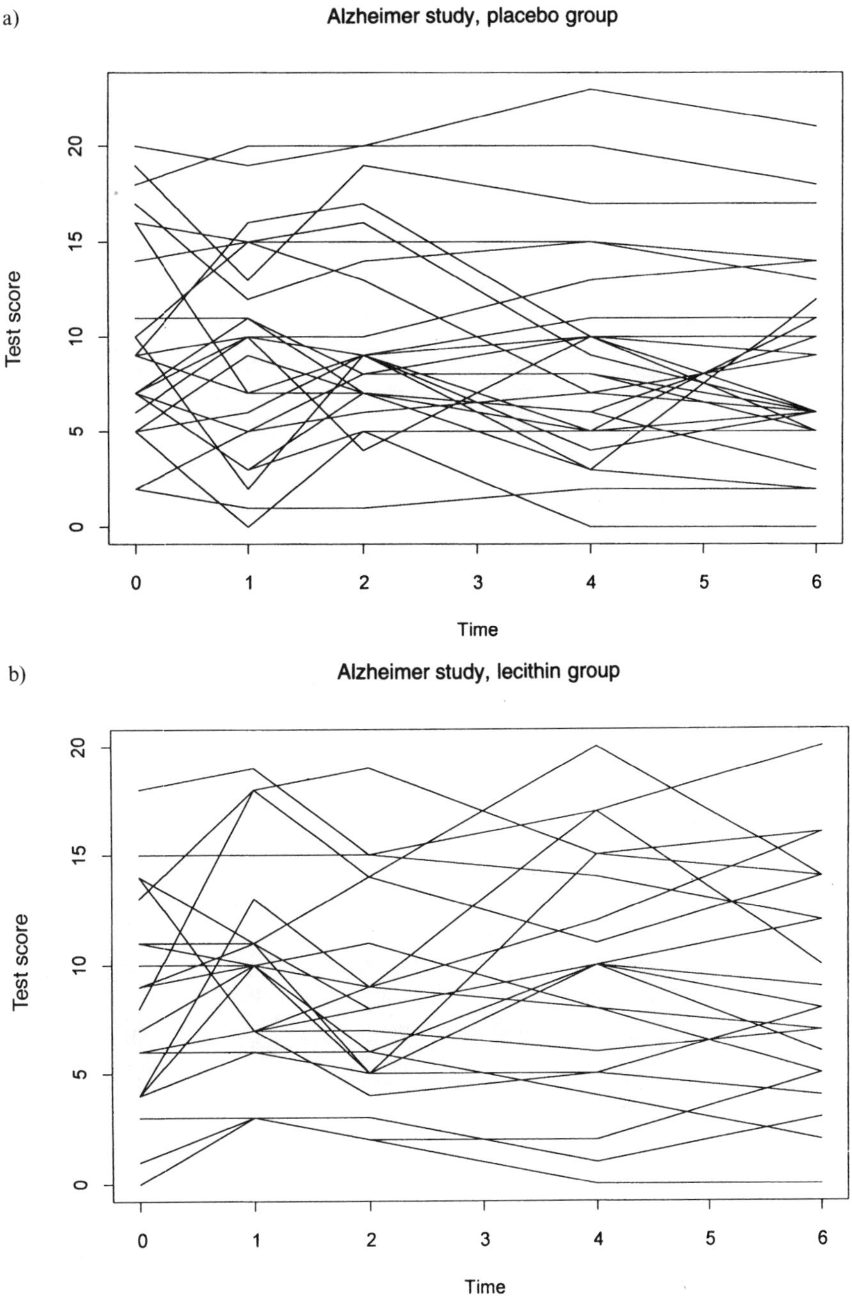

Figure 3.1. *Alzheimer study response profiles: (a) placebo group; (b) lecithin group.*

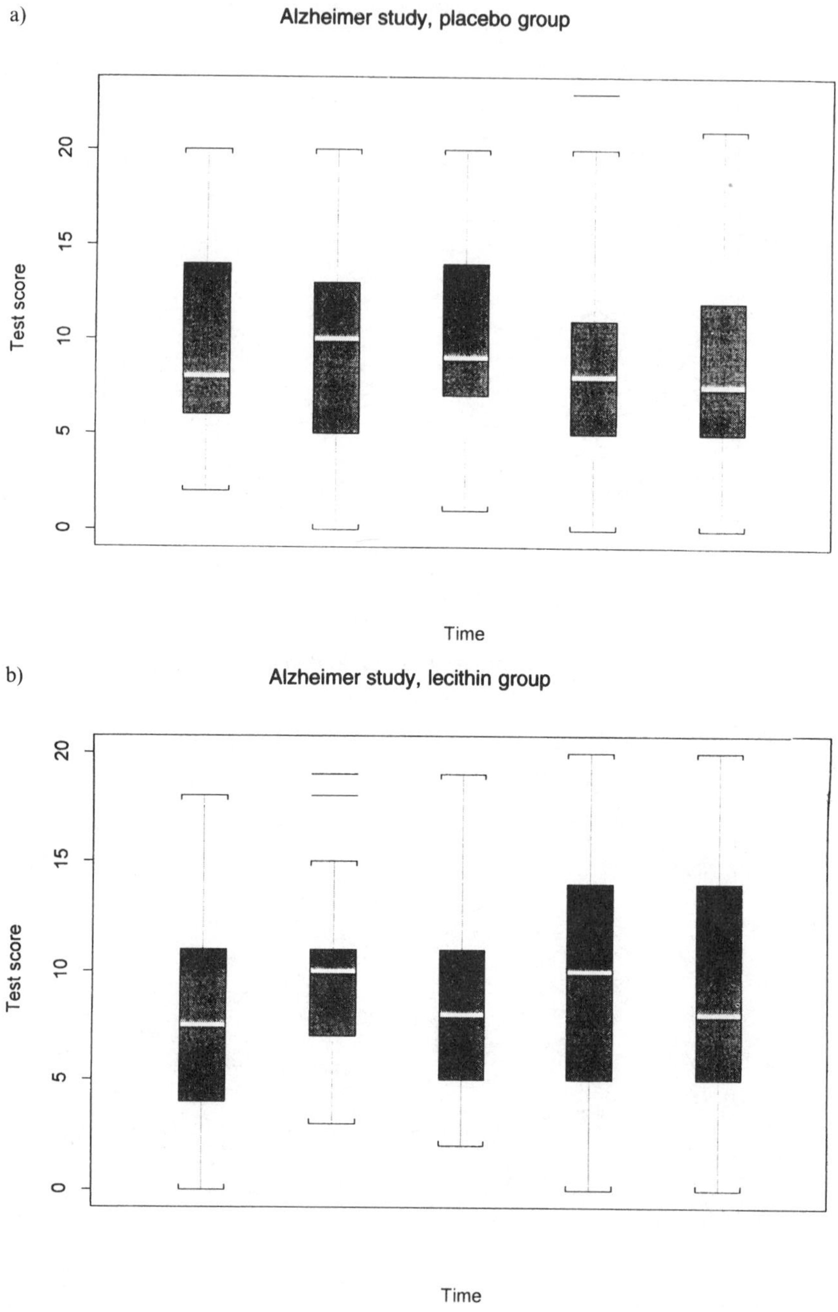

Figure 3.2. *Box plots of the profiles at each time for: (a) the placebo group; (b) the lecithin group.*

```
/INPUT FILE = '\bmdp\eg5.dat'.
    VARIABLES = 6.
    FORMAT = FREE.
/VARIABLE NAMES = GROUP, TIME0, TIME1, TIME2, TIME4, TIME6.
/GROUP CODES(GROUP) = 1, 2.
    NAMES(GROUP) = PLACEB, LECITH.
/DESIGN GROUPING = GROUP.
    DEPENDENT = TIME0 TO TIME6.
    LEVEL = 5.
    NAME = TIME.
    ORTHOGONAL.
    POINT(1) = 0, 1, 2, 4, 6.
/PRINT LINESIZE = 80.
/END
```

Table 3.1. *The BMDP2V commands for Example 3.1*

be used to define the groups and how many levels it has. To make the output easier to interpret, we have explicitly named the levels of the group factor as PLACEB and LECITH and named the single within-subjects factor as TIME. The ORTHOGONAL command in the DESIGN paragraph requests that an orthogonal polynomial transformation of the five time variables is given so that we can explore the nature of any time trends. However, since the five measurements are not taken at equally spaced times we specify, in the POINT command, the appropriate metric to use.

For an unmodified univariate anova to be legitimate, the data must have arisen from a distribution such that orthonormal polynomial components satisfy sphericity. Table 3.2 shows the sums of squares and the correlations for these components. The sums of squares show substantial differences, suggesting that the covariance matrix may be far from proportional to the identity matrix. The correlations also show substantial differences. The last two lines in this table show the results of

SUMS OF SQUARES AND CORRELATION MATRIC OF THE ORTHOGONAL COMPONENTS POOLED FOR ERROR 2 IN ANOVA TABLE BELOW

448.61621	1.000			
327.66295	−0.052	1.000		
262.65953	−0.047	−0.308	1.000	
206.26550	−0.371	0.131	−0.062	1.000

SPHERICITY TEST APPLIED TO ORTHOGONAL COMPONENTS – TAIL PROBABILITY 0.0220

Table 3.2. *The sums of squares and the correlation matrix of the orthogonal polynominal components*

a sphericity test. It seems that our suspicions were justified – one would be uneasy about accepting these data as satisfying sphericity. It seems wise to conduct adjusted F-tests.

Table 3.3 shows the analysis of variance results. In this piece of output, the first block (MEAN, GROUP) gives the results of the means of the five scores – to produce an overall test that the average of the two groups is zero (of little interest) and a comparison of the group means. There appears to be no overall difference between the levels of the two groups. T(1) to T(4) give tests for the linear to quartic effects, first averaged over groups and second (in the T(i) G lines) compared between groups. There is a suggestion that the quartic effect may differ between the groups, but this result has to be taken in the context of the large number of tests which have been carried out.

Of perhaps more interest is the final block of results. This shows the overall time effect (not splitting it into separate components) and the overall time by group interaction effect. These results might, in fact, be the ones first looked at, since if there was no overall time effect there would be little point in examining the separate polynomial components. There seems to be little evidence of an effect here.

All of this is all very well, but we have already noted that sphericity is not supported by the data. This means that we cannot be very confident about

SOURCE	SUM OF SQUARES	D.F.	MEAN SQUARE	F	TAIL PROB.
MEAN	19924.89802	1	19924.89802	188.81	0.0000
GROUP	7.69802	1	7.69802	0.07	0.7883
1 ERROR	4854.28531	46	105.52794		
T(1)	1.28864	1	1.28864	0.13	0.7179
T(1)G	17.83778	1	17.83778	1.83	0.1829
ERROR	448.61621	46	9.75253		
T(2)	6.18914	1	6.18914	0.87	0.3561
T(2)G	0.45612	1	0.45612	0.06	0.8014
ERROR	327.66295	46	7.12311		
T(3)	2.07348	1	2.07348	0.36	0.5497
T(3)G	0.02566	1	0.02566	0.00	0.9468
ERROR	262.65953	46	5.70999		
T(4)	3.18621	1	3.18621	0.71	0.4036
T(4)G	42.61792	1	42.61792	9.50	0.0035
ERROR	206.26550	46	4.48403		
TIME	12.73747	4	3.18437	0.47	0.7573
TG	60.93747	4	15.23437	2.25	0.0653
2 ERROR	1245.20420	184	6.76741		

Table 3.3. *The analysis of variance results for each polynomial component separately and for the overall pattern of change over time*

these test results. However, if we are following the strategy outlined above, we can comfortably stop now: a non-significant unadjusted test result implies a non-significant adjusted test result.

For purposes of exposition, however, we shall look at the adjusted test results. The program gives these, automatically. For the TIME by GROUP interaction the Greenhouse–Geisser and Huynh–Feldt adjusted test p values are, respectively, 0.0780 and 0.0707 (larger than the unadjusted results, as we expected). For the TIME effect the adjusted results are 0.7229 and 0.7426.

The program also gives the Greenhouse–Geisser and Huynh–Feldt estimates of ε. These are, respectively, 0.8318 and 0.9239. Perhaps not apparently very far from 1.0, but enough to make the sphericity test above significant and to induce some fairly substantial changes in the tests involving time.

Plots of group means, shown in Fig. 3.3, show no apparent regular pattern of differences between the groups and support our conclusion of no difference.

Example 3.2

The data for this example are described in Andersen, Jensen and Schou (1981) and reproduced in Table A. 11. They show measurements of plasma citrate concentrations in micromoles per litre on ten subjects measured at five times. A plot of the profiles of the ten subjects is given in Fig. 3.4. This figure gives no suggestion of skewness, so we have not transformed the data before analysis. Since there is only

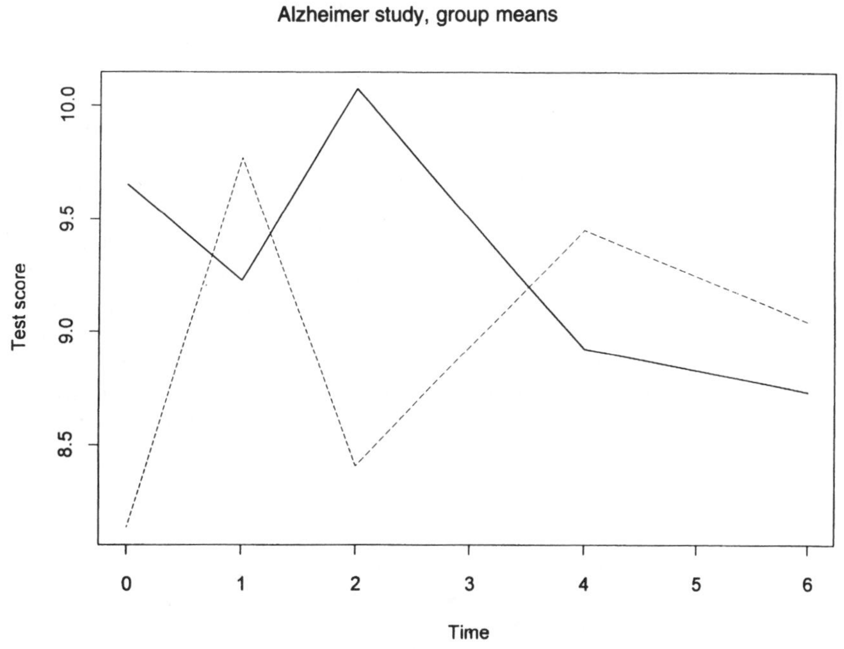

Figure 3.3. *Alzheimer study group means by time. The solid line shows the placebo group and the broken line the lecithin group.*

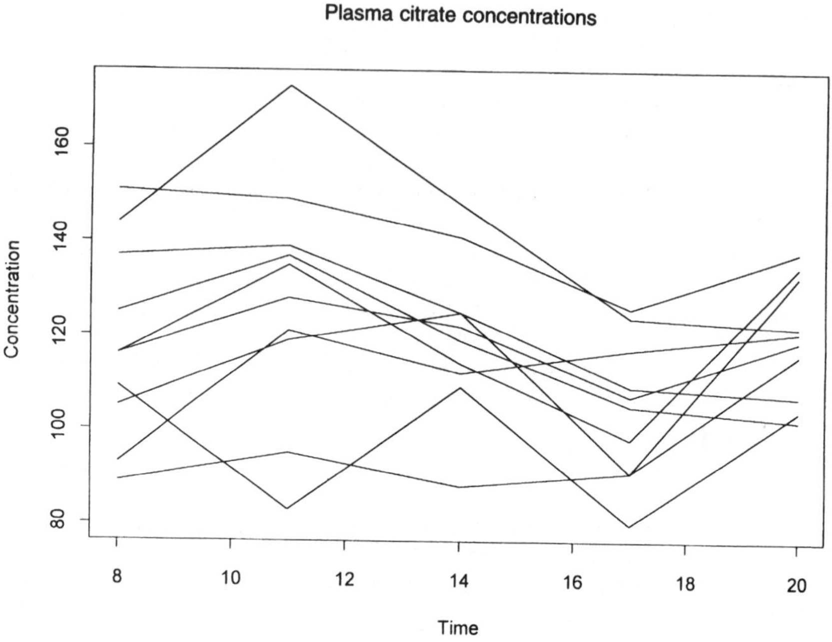

Figure 3.4. *Profiles of plasma citrate concentrations.*

a single group of subjects, there is no between-subjects aspect to the study, other than possible comparisons of means for this single group with zero (or some other constant).

For this example, we used the MANOVA routine in SPSSPC+. The basic MANOVA commands are illustrated in Table 3.4. The MANOVA command merely says that the analysis will use five response variables, called TIME1 to TIME5. The WSFACTORS subcommand specifies that there is a single within-subjects factor, TIME, with five levels. The WSDESIGN subcommand specifies the within subjects design – in this case simply a saturated design on the single within-subjects factor. Since the observations are taken at five points in time, and since our aim is to model the time evolution, it is natural to describe the profiles in terms of their polynomial components: the CONTRAST subcommand specifies that TIME

```
MANOVA TIME1 TO TIME5
    /WSFACTORS = TIME(5)
    /CONTRAST(TIME) = POLYNOMIAL
    /RENAME = CONS, LIN, QUAD, CUB, QUAR
    /WSDESIGN
    /PRINT TRANSFORM.
```

Table 3.4. *The SPSSPC+ MANOVA commands for Example 3.2*

is to be transformed into constant, linear, quadratic, cubic and quartic polynomial components. The RENAME command simply renames the resulting derived variables so that the output is easier to interpret. The PRINT TRANSFORM subcommand prints the transformation matrix so that we can see that it is really doing what we want it to do. The matrix is shown in Table 3.5, and it is easy to see that the first column will generate a derived variable proportional to the overall mean of the five scores, the second will be proportional to the linear trend, and so on. The last four derived variables will together span the space of differences between the five scores.

The first analysis results from MANOVA (Table 3.6) are for the first of these derived variables. It simply compares the average score of the means of the five times for each subject with zero. Since the raw scores are necessarily positive, one would expect this test to produce a significant result.

Table 3.7 is more interest. This shows a multivariate test of the hypothesis that the four components of change over time are zero. Significance at the 0.025% level is

Orthonormalized Transformation Matrix (Transposed)

	CONS	LIN	QUAD	CUB	QUAR
TIME1	.447	−.632	.535	−.316	.120
TIME2	.447	−.316	−.267	.632	−.478
TIME3	.447	.000	−.535	.000	.717
TIME4	.447	.316	−.267	−.632	−.478
TIME5	.447	.632	.535	.316	.120

Table 3.5. *The transformation matrix for Example 3.2*

Tests of Significance for CONS using UNIQUE sums of squares

Source of Variation	SS	DF	MS	F	Sig. of F
WITHIN CELLS	10592.72	9	1176.97		
CONSTANT	699034.88	1	699034.88	593.93	.000

Table 3.6. *First anova table output by MANOVA – a test of the hypothesis that the overall mean of the five scores for each subject is zero*

EFFECT.. TIME

Multivariate Tests of Significance (S = 1, M = 1, N = 2)

Test Name	Value	Approx. F	Hypoth. DF	Error DF	Sig. of F
Pillais	.80560	6.21605	4.00	6.00	0.025
Hotellings	4.14403	6.21605	4.00	6.00	0.025
Wilks	.19440	6.21605	4.00	6.00	0.025
Roys	.80560				

Table 3.7. *Multivariate statistics to test the hypothesis of no change over time*

achieved, suggesting that there is change over time. Note that these tests involve a single degree of freedom 'between groups' since there is only one possible group comparison – that of the single group mean with a constant (in this case, zero). Because of this, the various test statistics yield the same F value and significance level.

But what about the univariate approach? Table 3.8 shows the results of the Mauchly test for sphericity. This is seen to be non-significant – we have no evidence to suppose the sphericity assumption is wrong. Even so, the estimates of ε are not very close to 1. Table 3.9 shows the unadjusted univariate tests. This shows a highly significant result. If we follow the strategy outlined in section 2.3, we should now carry out a conservative test, adjusting the degrees of freedom by a factor $1/(p-1)$. This is the lower bound on ε and Table 3.8 shows it to have value 0.25. This adjustment means we have to compare the F-statistic of 4.75 with an F-distribution on $4 \times 0.25 = 1$ and $36 \times 0.25 = 9$ degrees of freedom. The 5% level of such a distribution is 5.12, so our result fails to achieve significance at that level (though it does achieve significance at that 10% level, 3.36). Under these circumstances, the strategy of section 2.3 suggests we should now estimate ε from the sample covariance matrix and conduct an ε-adjusted test.

Mauchly sphericity test, W =	.26038
Chi-square approx. =	9.98003 with 9 D.F.
Significance =	.352
Greenhouse-Geisser Epsilon =	.60711
Huynh-Feldt Epsilon =	.84776
Lower-bound Epsilon =	.25000

Table 3.8. *The sphericity test and estimated ε's for the plasma citrate data*

AVERAGED Tests of Significance for TIME using UNIQUE sums of squares

Source of Variation	SS	DF	MS	F	Sig. of F
WITHIN CELLS	5308.48	36	147.46		
TIME	2803.92	4	700.98	4.75	.003

Table 3.9. *The unadjusted univariate F-tests for the plasma citrate data*

Taking the Greenhouse–Geisser estimate of ε gives degrees of freedom of $4 \times 0.61 = 2.44$ and $36 \times 0.61 = 21.96$. Interpolating in F tables shows that the obtained statistic of 4.75 is significant at the 5% level. The Huynh–Feldt estimate of 0.85 yields larger adjusted degrees of freedom and so also leads to a significant result. For variety, Table 3.10 shows a segment of results produced by BMDP4V with these adjustments being made automatically.

Having established that there is some pattern to the change over time, to complete this example, we would want to model that pattern. We can do this by identifying what order of polynomial should be used, and using a model of that, order. Table 3.11 shows separate univariate tests of the four polynomial components (obtained by including a SIGNIF(UNIV) in the MANOVA PRINT command). The quartic term can be safely neglected, but the cubic seems to be highly significant. A glance back at Fig. 3.4 shows that this is reasonable.

EFFECT	VARIATE	STATISTIC		F	DF		P
T							
	DEP_VAR						
		TSQ =	37.2963	6.22	4,	6	0.0251
		WCP SS =	2803.920000				
		WCP MS =	700.980000	4.75	4,	36	0.0035
	GREENHOUSE-GEISSER ADJ. DF			4.75	2.43,	21.86	0.0147
	HUYNH-FELDT ADJUSTED DF			4.75	3.39,	30.52	0.0061
ERROR							
	DEP_VAR						
		WCP SS =	5308.48000000				
		WCP MS =	147.45777778				
		GGI EPSILON	0.60711				
		H-F EPSILON	0.84776				

Table 3.10. *Adjusted F-tests using BMDP4V*

Univariate F-tests with (1, 9) D.F.

Variable	Hypoth. SS	Error SS	Hypoth MS	Error MS	F	Sig. of F
LIN	428.49000	2733.21000	428.29000	303.69000	1.41095	.265
QUAD	6.86429	881.77857	6.86429	97.97540	.07006	.797
CUB	2246.76000	723.04000	2246.76000	80.33778	27.96642	.001
QUAR	121.80571	970.45143	121.80571	107.82794	1.12963	.316

Table 3.11. *Separate univariate tests of the polynomial components*

3.5 Further reading

The univariate analysis of variance approach to repeated measures data has been very popular in psychology, and several expositional papers and papers comparing it with the multivariate approach have appeared in the literature of that discipline. Examples are Rouanet and Lepine (1970), McCall and Appelbaum (1973), O'Brien and Kaiser (1985), Hertzog and Rovine (1985) and Anderson (1991).

The Mauchly test for sphericity appears to be the only one which is widely given in software packages, but John (1971;1972), Sugiura (1972) and Harris (1984) have developed alternative methods. Grieve (1984) has related the Mauchly test to ε.

CHAPTER 4

Regression methods

4.1 The model

We remarked, in Chapter 3, that in some scientific disciplines analysis of variance was very widely used. Given this, it is practically advantageous to scientists working in those areas to use methods for analysing repeated measures data which adopt an anova approach. To be valid, however, the covariance matrix of the repeated measures had to satisfy certain criteria ($p-1$ variables derived using orthogonal contrasts of the p measured variables had to satisfy sphericity). Admittedly, non-spherical distributions could still yield approximately valid F-tests by means of adjustments to the degrees of freedom, but one might prefer to seek a more natural model fit, rather than make an *ad hoc* adjustment.

Similarly, the manova method, described in Chapter 2, with its anova basis, also has appeal in those disciplines which make heavy use of anova. This approach had the advantage that it imposed no restrictions on the covariance matrix of the repeated measures. But it required that the data be balanced, in the sense that each individual was measured on the same p occasions.

In this chapter we relax the sphericity assumption and allow the covariance matrix to be arbitrary. Moreover, we permit individuals to be measured on different numbers of occasions. In particular, this has the merits that individuals with an incomplete set of measurements can be straightforwardly included in the analysis, and that individuals can have different covariates.

Let $\mathbf{y}_i = [y_{i1}, \ldots, y_{ip_i}]^{\mathrm{T}}$ be the $p_i \times 1$ vector of responses for the ith individual. We can model the distribution of this vector by

$$\mathbf{y}_i = \mathbf{X}_i\boldsymbol{\beta} + \mathbf{u}_i \tag{4.1}$$

Here $\boldsymbol{\beta}$ is a $q \times 1$ vector of parameters, common for all individuals, and including param-eters describing both differences between individuals and differences between measurement occasions (i.e. differences 'within' individuals). $\mathbf{X}_i$ is a $p_i \times q$ design matrix for the ith individual. $\mathbf{u}_i$ is a vector of random values assumed to follow a multivariate normal distribution, $\mathbf{u}_i \sim N(\mathbf{0}, \boldsymbol{\Sigma}_{pi})$. $\boldsymbol{\Sigma}_{pi}$ is taken to be an appropriate submatrix of a matrix $\boldsymbol{\Sigma}_p$ of covariances for all p possible measurement occasions.

The model for the univariate anova approach, described in Chapter 3, has the same parameter vector, $\boldsymbol{\beta}$, but concatenates the $\mathbf{y}_i$ to form a single $np \times 1$ observation vector, the $\mathbf{X}_i$ to form a single $np \times q$ design matrix, and the $\boldsymbol{\Sigma}_{pi}$ to form a single covariance matrix $\boldsymbol{\Sigma}$.

■ *Illustration 4.1* Suppose there are two groups of individuals, each potentially measured at times 1, 2 and 3 (cf. Illustration 3.1). Suppose that we model the expected response of the kth group by

$$[\mu_{k1} + \mu_{k2}\ \mu_{k1} + \mu_{k3}\ \mu_{k1} - \mu_{k2} - \mu_{k3}]^{\mathrm{T}}$$

Then for an individual in the first group who is measured at all three times we have

$$E\begin{bmatrix} y_{i1} \\ y_{i2} \\ y_{i3} \end{bmatrix} = \begin{bmatrix} 1 & 1 & 0 & 0 & 0 & 0 \\ 1 & 0 & 1 & 0 & 0 & 0 \\ 1 & -1 & -1 & 0 & 0 & 0 \end{bmatrix} \begin{bmatrix} \mu_{11} \\ \mu_{12} \\ \mu_{13} \\ \mu_{21} \\ \mu_{22} \\ \mu_{23} \end{bmatrix}$$

and for an individual in the second group measured at all three times we have

$$E\begin{bmatrix} y_{i1} \\ y_{i2} \\ y_{i3} \end{bmatrix} = \begin{bmatrix} 0 & 0 & 0 & 1 & 1 & 0 \\ 0 & 0 & 0 & 1 & 0 & 1 \\ 0 & 0 & 0 & 1 & -1 & -1 \end{bmatrix} \begin{bmatrix} \mu_{11} \\ \mu_{12} \\ \mu_{13} \\ \mu_{21} \\ \mu_{22} \\ \mu_{23} \end{bmatrix}$$

For an individual from the first group who is measured only at times 1 and 3, we have

$$E\begin{bmatrix} y_{i1} \\ y_{i2} \end{bmatrix} = \begin{bmatrix} 1 & 1 & 0 & 0 & 0 & 0 \\ 1 & -1 & -1 & 0 & 0 & 0 \end{bmatrix} \begin{bmatrix} \mu_{11} \\ \mu_{12} \\ \mu_{13} \\ \mu_{21} \\ \mu_{22} \\ \mu_{23} \end{bmatrix}$$

■

4.2 Estimation and testing

From (4.1) the normal log-liklihood is given by

$$L = -\frac{1}{2}\left\{\sum_i \log|\boldsymbol{\Sigma}_{pi}| - \sum_i (\mathbf{y}_i - \mathbf{X}_i\boldsymbol{\beta})^{\mathrm{T}}\boldsymbol{\Sigma}_{pi}^{-1}(\mathbf{y}_i - \mathbf{X}_i\boldsymbol{\beta})\right\}$$

which yields the maximum likelihood estimate of $\boldsymbol{\beta}$ as

$$\hat{\boldsymbol{\beta}} = \left(\sum_{i=1}^{n} \mathbf{X}_i^{\mathrm{T}} \hat{\boldsymbol{\Sigma}}_{pi}^{-1} \mathbf{X}_i\right)^{-1} \left(\sum_{i=1}^{n} \mathbf{X}_i^{\mathrm{T}} \hat{\boldsymbol{\Sigma}}_{pi}^{-1} \mathbf{y}_i\right) \tag{4.2}$$

(where $\hat{\boldsymbol{\Sigma}}_{pi}$ is the ML estimator of $\boldsymbol{\Sigma}_{pi}$), with estimated covariance matrix

$$\hat{\mathbf{V}}_{\boldsymbol{\beta}} = \left(\sum_{i=1}^{n} \mathbf{X}_i^{\mathrm{T}} \hat{\boldsymbol{\Sigma}}_{pi}^{-1} \mathbf{X}_i\right)^{-1}$$

A test statistic for the hypothesis $\underset{r\times q}{\mathbf{H}}\ \underset{q\times 1}{\boldsymbol{\beta}} = \underset{r\times 1}{\mathbf{h}}$ is given by

$$(\mathbf{H}\hat{\boldsymbol{\beta}} - \mathbf{h})^{\mathrm{T}} (\mathbf{H}\hat{\mathbf{V}}_{\boldsymbol{\beta}}\mathbf{H}^{\mathrm{T}})^{-1} (\mathbf{H}\hat{\boldsymbol{\beta}} - \mathbf{h})$$

which is approximately χ_r^2-distributed in large samples.

Of course, to evaluate (4.2) one needs an estimate $\hat{\boldsymbol{\Sigma}}_p$ from which submatrices $\hat{\boldsymbol{\Sigma}}_{pi}$ can be extracted. This is defined in terms of $\hat{\boldsymbol{\beta}}$, so yielding an iterative solution, cycling alternately through $\hat{\boldsymbol{\beta}} = \hat{\boldsymbol{\beta}}(\hat{\boldsymbol{\Sigma}}_p)$ and $\hat{\boldsymbol{\Sigma}}_p = \hat{\boldsymbol{\Sigma}}_p(\hat{\boldsymbol{\beta}})$. If there were no missing values, we would have $\hat{\boldsymbol{\Sigma}}_p = \frac{1}{n}\sum_{i=1}^{n}(\mathbf{y}_i - \mathbf{X}_i\hat{\boldsymbol{\beta}})(\mathbf{y}_i - \mathbf{X}_i\hat{\boldsymbol{\beta}})^{\mathrm{T}}$. When there are missing values, the (j, k)th element of $\boldsymbol{\Sigma}_p$ will be based on an average of those values which have been measured.

Thus far we have imposed on constraints on the form of the covariance matrix $\boldsymbol{\Sigma}_p$. Frequently, however, this matrix is described in terms of a small parameter set. This can lead to improved estimates if the number of parameters is substantially less than the number of distinct terms in $\boldsymbol{\Sigma}_p$ and situations frequently arise which lead to natural structures for $\boldsymbol{\Sigma}_p$. Detailed discussion is given in Chapter 6, but examples of important structures are:

1. $\boldsymbol{\Sigma}_p = \sigma^2\mathbf{I}$, arising if all the observations can be taken to be independent, with identical variances and a single parameter.
2. The compound symmetry model, $\boldsymbol{\Sigma}_p = \sigma_a^2\,\mathbf{1}\mathbf{1}^{\mathrm{T}} + \sigma^2\mathrm{I}$, with two parameters, described in Chapter 3.
3. The random effects model already introduced in Chapter 3 and outlined at length because of its importance in Chapter 5, with a number of parameters which depends on the number of random effects.
4. Models which parametrize correlations between successive error terms, beyond any correlation induced by random between-individual effects, such as the *antedependence* model of Kenward (1987).
5. Models which combine both (3) and (4) above, having the general form $\boldsymbol{\Sigma}_{pi} = \mathbf{Z}_i\mathbf{B}\mathbf{Z}_i^{\mathrm{T}} + \mathbf{E}$, such as that suggested by Diggle (1988), which has $\boldsymbol{\Sigma}_p = (\sigma_a^2\,\mathbf{1}\mathbf{1}^{\mathrm{T}} + \sigma^2\mathbf{I}) + \sigma_b^2\mathbf{R}$.

In general we take $\boldsymbol{\Sigma}_{pi} = \boldsymbol{\Sigma}_{pi}(\boldsymbol{\phi})$, where $\boldsymbol{\phi}$ denotes the vector of parameters governing $\boldsymbol{\Sigma}_p$ and is assumed to be the same for all individuals. The log-likelihood is then a function of $\boldsymbol{\beta}$ and $\boldsymbol{\phi}$. The maximum likelihood estimate of

$\boldsymbol{\beta}$ is then, of course (via the $\boldsymbol{\Sigma}_{pi}$) a function of $\boldsymbol{\phi}$. If this estimate is substituted in the overall likelihood we can find the solution by maximizing over $\boldsymbol{\phi}$. Details of this maximization are given in Crowder and Hand (1990).

4.3 Examples

Example 4.1
This example explores data on the effect of diet on the growth of guinea pigs given in Crowder and Hand (1990, Table 3.1). There are three groups of five animals each, and each animal was given a growth-inhibiting substance four weeks before the start of the measurements. One week before the first measurement, group 1 was given no vitamin E, group 2 was given a low dose, and group 3 was given a high dose. The body weights (in grams) one week, two weeks and three weeks after administration of the vitamin E were measured and are shown as the last three columns of Table A.12. The profiles of responses are shown in Fig. 4.1.

Interest centres on whether the three groups grow in different ways. In particular, it might be natural to be interested in growth *rate* here. We therefore fitted a regression model with a separate regression intercept and slope for each group. The BMDP5V commands for this are shown in Table 4.1. GROUP is the name we have given to the grouping factor, and WT1 to WT3 are the three weight measurements.

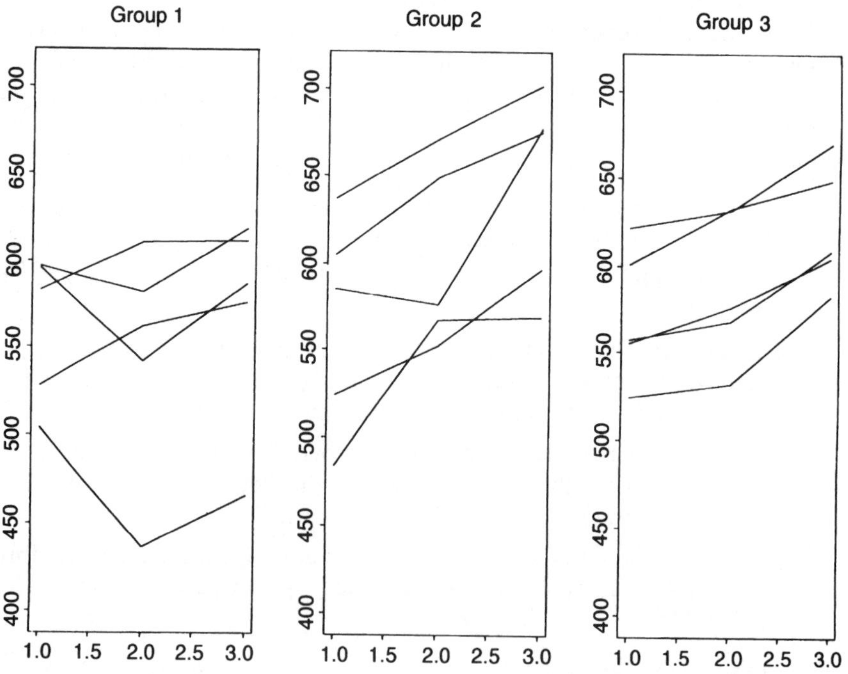

Figure 4.1. *Weights of guinea pigs on three occasions after commencement of administration of vitamin E supplements.*

The DESIGN paragraph specifies the between-subjects grouping structure and names the repeated factor as TIME, specifying that it has three levels. The CONTRAST command defines a contrast X on time and the line below shows that this is a linear contrast on the three measurements (which are equally spaced). DPNAME gives the name WEIGHT to the dependent variable, and DPVAR identifies the three weight measurements WT1, WT2 and WT3 as being repeated measurements of WEIGHT. The MODEL paragraph defines the model to be fitted–in this case it has a main group effect, a linear time effect, and the group by linear time interaction. Finally, the STRUCTURE paragraph says that we have not imposed any particular form (such as compound symmetry) on the covariance matrix. (In the next chapter we extend this model to the case when each guinea pig can have its own regression line. This is equivalent to assuming a particular structure for the covariance matrix of observations. BMDP5V gives the observed covariance and correlation matrices for the data as well as estimates obtained using the particular form specified. This permits a comparison to be made between the observed and hypothesized forms of the matrix. In the present example, of course, where we have made no restrictive assumptions about the form of the matrix, the observed matrices and the estimates obtained from the model are the same.)

This routine gives the Akaike information criterion (AIC), which can be used to compare models. (This criterion takes into account the number of parameters in a model, and hence makes an adjustment for the better fit of models with more parameters). For this model it is -218.563.

The parameter estimates given by BMDP5V are shown in Table 4.2. From these estimates the regression slopes for the three groups are given by

$$\beta_1 = X + G1.X = 24.87 - 16.38 = 8.49$$

$$\beta_2 = X + G2.X = 24.87 - 14.04 = 38.91$$

$$\beta_3 = X - (G1.X + G2.X) = 24.87 - (-16.38 + 14.04) = 27.21$$

Table 4.3 shows the results of Wald tests on the main effects and group by linear slope interaction. Both the interaction and the overall linear slope are highly significant, as might have been expected from the plots.

```
/VARIABLE    NAMES = GROUP, WT1, WT2, WT3.
/DESIGN      GROUPING = GROUP.
             REPEATED = TIME.
             LEVEL = 3.
             CONTRAST (TIME) = X.
             X = 1, 2, 3.
             DPNAME = WEIGHT.
             DPVAR = WT1, WT2, WT3.
/MODEL       WEIGHT = 'GROUP + X + GROUP.X'.
/STRUCTURE   TYPE = UNSTRUC.
```

Table 4.1. *BMDP5V commands for fitting a separate regression line to each of the three guinea pig groups*

PARAMETER		ESTIMATE	ASYMPTOTIC SE	Z	TWO-SIDED P-VALUE
1	CONST.	538.94837	11.2 9540	47.714	0.0000
2	G1	8.51635	15.97411	0.533	0.5939
3	G2	−11.55422	15.97411	−0.723	0.4695
4	X	24.87188	2.16757	11.475	0.0000
5	G1.X	−16.37951	3.06541	−5.343	0.0000
6	G2.X	14.03559	3.06541	4.579	0.0000

Table 4.2. *Parameter estimates from BMDP5V*

TEST	DF	CHI-SQUARE	P-VALUE
G	2	0.56	0.755
X	1	131.66	0.000
G.X	2	33.40	0.000

Table 4.3. *Wald test results*

Example 4.2

The data for this example come from Jones (1993, Table 1.1) and are reproduced in Table A.13. Ten individuals, forming two groups of five, are measured on seven occasions. However, the final observation for the ninth patient is missing. The profiles of the groups are shown in Fig. 4.2.

Table 4.4 shows the BMDP5V commands for fitting a linear regression model to each group, assuming a compound symmetric covariance matrix. The X contrast specifies that a simple linear regression will be fitted (later we extend this to a quadratic fit) and the STRUCTURE TYPE = CS command indicates that a compound symmetry form will be used.

```
/VARIABLE     NAMES = GROUP, T1, T2, T3, T4, T5, T6, T7.
/DESIGN       GROUPING = GROUP.
              REPEATED = TIME.
              LEVEL = 7.
              CONTRAST (TIME) = X.
              X = 0, 5, 10, 20, 30, 45, 60.
              DPNAME = SCORE.
              DPVAR = T1, T2, T3, T4, T5, T6, T7.
/MODEL        SCORE = 'GROUP*X'.
/STRUCTURE    TYPE = CS.
```

Table 4.4. *BMDP5V commands for fitting a linear regression model to each group, assuming a compound symmetric covariance matrix*

a)

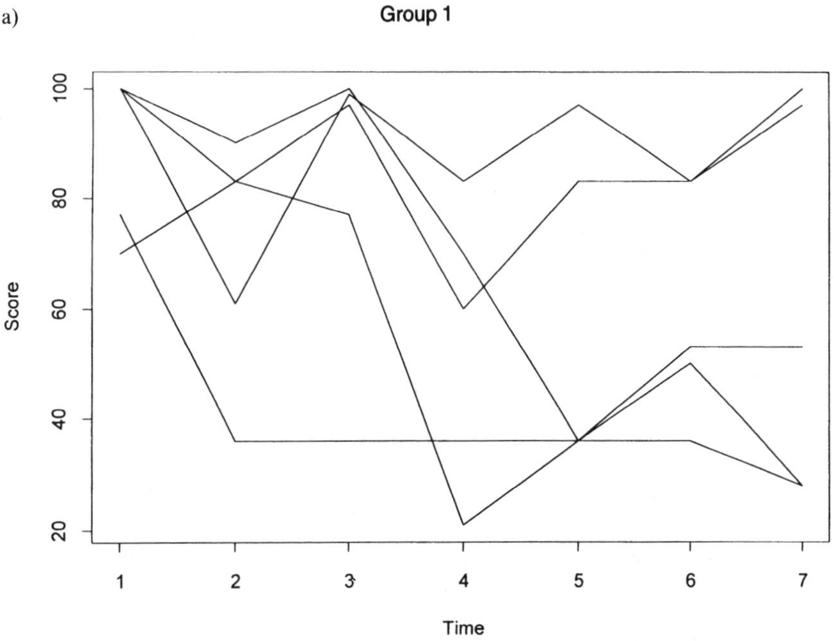

b)

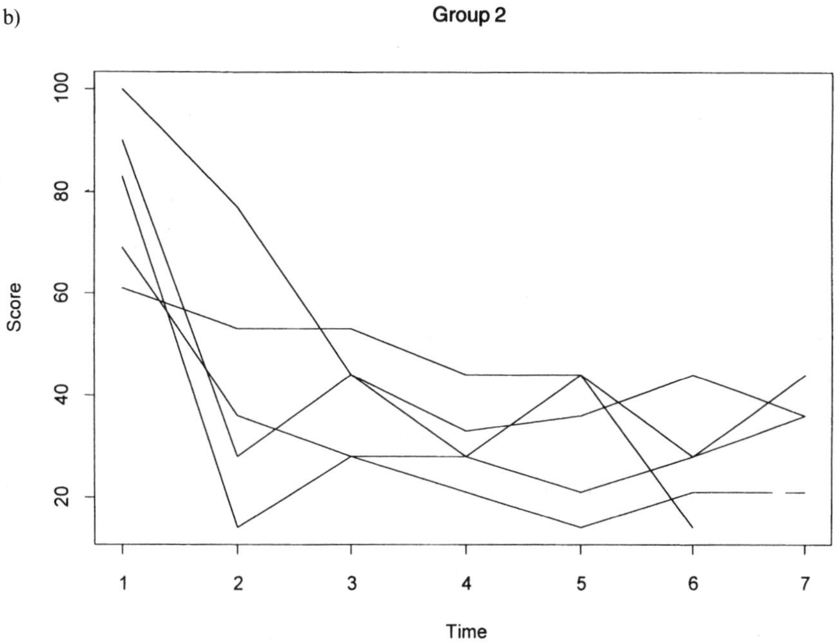

Figure 4.2. *Profiles of the scores in the medical experiment in Example 4.2: (a) group 1; (b) group 2.*

The two parameters of the covariance matrix are estimated to be 340.86 and 143.49, leading to the estimated compound symmetric covariance matrix shown in the top half of Table 4.5. The estimate obtained directly from the data (using all available pairs in the calculations) appears in the lower half of the table. Although the variances do not differ substantially, the covariances do, so perhaps compound symmetry is not a good model. The Akaike information criterion for this fit is − 307.9.

It is obvious from Fig. 4.2 that using linear regression slopes is not ideal. A better model would also use quadratic terms. The DESIGN and MODEL paragraphs of the above BMDP5V commands are modified as shown in Table 4.6 to introduce

MODEL: WITHIN-SUBJECT COVARIANCE MATRIX

	1	2	3	4	5	6	7
1	484.34399						
2	143.48865	484.34399					
3	143.48865	143.48865	484.34399				
4	143.48865	143.48865	143.48865	484.34399			
5	143.48865	143.48865	143.48865	143.48865	484.34399		
6	143.48865	143.48865	143.48865	143.48865	143.48865	484.34399	
7	143.48865	143.48865	143.48865	143.48865	143.48865	143.48865	484.34399

ALL-PAIRS WITHIN-SUBJECT COVARIANCE MATRIX

	1	2	3	4	5	6	7
1	626.20284						
2	−78.13157	487.62186					
3	46.99699	273.02121	418.27430				
4	−134.61506	141.22343	206.65667	478.34476			
5	9.95207	137.28611	224.98079	337.03712	483.86029		
6	48.91713	3.00743	195.10722	151.29184	242.28713	231.66099	
7	283.33330	−55.21335	249.71996	87.84758	405.95166	344.32721	692.70776

Table 4.5. *Top: estimated compound symmetric covariance matrix. Bottom: unconstrained covariance matrix estimated directly from residuals*

```
/DESIGN     GROUPING = GROUP.
            REPEATED = TIME.
            LEVEL = 7.
            CONTRAST (TIME) = X, Q.
            X = 0, 5, 10, 20, 30, 45, 60.
            Q = 0, 25, 100, 400, 900, 2025, 3600.
            DPNAME = SCORE.
            DPVAR = T1, T2, T3, T4, T5, T6, T7.
/MODEL      SCORE = 'GROUP + X + Q + GROUP.X + GROUP.Q'.
```

Table 4.6. *How the DESIGN and MODEL paragraphs are modified to introduce quadratic terms for each group*

MODEL: WITHIN-SUBJECT COVARIANCE MATRIX

	1	2	3	4	5	6	7
1	410.91027						
2	161.05276	410.91027					
3	161.05276	161.05276	410.91027				
4	161.05276	161.05276	161.05276	410.91027			
5	161.05276	161.05276	161.05276	161.05276	410.91027		
6	161.05276	161.05276	161.05276	161.05276	161.05276	410.91027	
7	161.05276	161.05276	161.05276	161.05276	161.05276	161.05276	410.91027

ALL-PAIRS WITHIN-SUBJECT COVARIANCE MATRIX

	1	2	3	4	5	6	7
1	297.27681						
2	−36.78527	558.16470					
3	85.62195	259.55156	413.24462				
4	98.17126	90.23451	181.85262	320.93538			
5	190.80342	37.97146	213.03024	231.19828	461.66964		
6	101.80206	−41.53628	196.80696	121.22213	249.48878	243.56503	
7	83.39283	127.82307	275.78786	227.75551	477.18516	316.40248	608.04502

Table 4.7. *Covariance matrices arising when linear and quadratic terms are used to model the time evolution of each group*

	PARAMETER	ESTIMATE	ASYMPTOTIC SE	Z	TWO-SIDED P-VALUE
1	CONST.	76.02470	5.50202	13.818	0.0000
2	G1	9.97502	5.50202	1.813	0.0698
3	X	−1.95023	0.34259	−05.693	0.0000
4	Q	0.02570	0.00566	4.537	0.0000
5	G1.X	0.37714	0.34259	1.101	0.2710
6	G1.Q	−0.00570	0.00566	−1.005	0.3147

Table 4.8. *Estimated regression parameters of the quadratic model*

	Group 1	*Group 2*
Intercept	$76.025 + 9.975 = 86.000$	$76.025 - 9.975 = 66.050$
Linear	$-1.950 + 0.377 = -1.573$	$-1.950 - 0.377 = -2.327$
Quadratic	$0.026 - 0.006 = 0.020$	$0.026 + 0.006 = 0.032$

Table 4.9. *Estimated parameters for the regression curves of each group*

separate quadratic terms for each group. Now X provides the linear term and Q the quadratic term. The resulting covariance matrix is shown in Table 4.7.

The Akaike information criterion of this new model is -298.9, an improvement over the model using only linear terms, as we expected. The estimates of the regression parameters of this new model are shown in Table 4.8 and the resulting parameters for the estimated mean curves of each group in Table 4.9. The tests in Table 4.8 suggest that the interaction terms might be dropped from the model.

CHAPTER 5

Random effects models

5.1 The models

In Chapter 4 we modelled the expected response vector for an individual as a linear function of a set of parameters via a regression model, $E(\mathbf{y}_i) = \mathbf{X}_i\boldsymbol{\beta}$, and, initially, imposed no restriction on the form of the covariance matrix of the $\mathbf{y}_i$. This can be advantageous, in that it implies that the model is very flexible – but, on the other hand, it can be disadvantageous because a potentially large number of covariance parameters must be estimated. As we pointed out in Chapter 4, one way to overcome the latter problem is to restrict the form of the covariance matrix, letting $\boldsymbol{\Sigma}_p = \boldsymbol{\Sigma}_p(\boldsymbol{\phi})$, a function of a vector of parameters. In fact, we saw precisely this approach to modelling Σ_p in Chapter 3, when we assumed compound symmetry. Then $\boldsymbol{\phi} = [\sigma^2, \sigma_a^2]$ and $\boldsymbol{\Sigma}_p = \sigma^2\mathbf{I} + \sigma_\alpha^2\mathbf{1}\mathbf{1}^{\mathrm{T}}$. This was a natural form for $\boldsymbol{\Sigma}_p$ in certain circumstances. In this chapter we generalize this approach.

The basic model we described in Chapter 3 had the form $y_{ij} = \alpha + a_i + \beta_j + e_{ij}$, with a_i a random individual effect and β_j a fixed occasion effect. We can generalize this to (in matrix form)

$$\mathbf{y}_i = \mathbf{Z}_i\mathbf{b}_i + \mathbf{X}_i\boldsymbol{\beta} + \mathbf{e}_i \tag{5.1}$$

Here the $\mathbf{b}_i$ are random effects, yielding a combined contribution to $\mathbf{y}_i$ via the design matrix $\mathbf{Z}_i$, and $\boldsymbol{\beta}$ is a vector of fixed effects. Such models are sometimes called **two-stage models** because the $\mathbf{e}_i$ refer to within-individual variation (stage 1) and the $\mathbf{b}_i$ refer to between-individual variation (stage 2). Models with this general form have now become very popular. An early exposition is Laird and Ware (1982). For a Bayesian perspective see Lindley and Smith (1972).

■ *Illustration 5.1* Figure 5.1, reproducing Fig. 1.2, shows the body weights of 20 chicks on a normal diet measured on alternate days. The data are given in Crowder and Hand (1990 Table 5.1), and are reproduced in Table A.2.

The shapes of these curves suggest that a reasonable model might involved random linear and quadratic components. ■

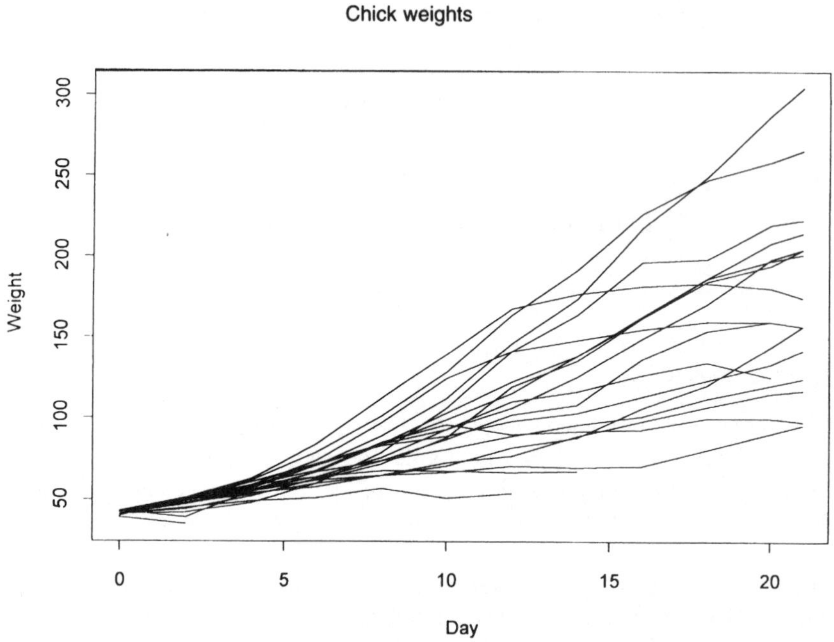

Figure 5.1. *Body weights of chicks over time.*

In Chapter 3 we saw that the special case dealt with there led to a particularly convenient form for the covariance matrix of the $\mathbf{y}_i$. We can generalize this for model (5.1). Suppose that $E(\mathbf{b}_i) = \mathbf{0}$, $E(\mathbf{e}_i) = \mathbf{0}$, $\text{var}(\mathbf{b}_i) = \mathbf{B}_i$, $\text{var}(\mathbf{e}_i) = \mathbf{E}_i$ and that $\mathbf{b}_i$ and $\mathbf{e}_i$ are independent. Then

$$\mathbf{\Sigma}_{pi} = \text{var}(\mathbf{y}_i) = \text{var}(\mathbf{Z}_i\mathbf{b}_i) + \text{var}(\mathbf{e}_i) = \mathbf{Z}_i\mathbf{B}_i\mathbf{Z}_i^{\mathrm{T}} + \mathbf{E}_i$$

Also, as elsewhere in this first part of the book, we in fact assume that the distributions of $\mathbf{b}_i$ and $\mathbf{e}_i$ are multivariate normal.

The matrix $\mathbf{\Sigma}_{pi}$ above is now described in terms of a set of parameters, some defining $\mathbf{B}_i$ and some defining $\mathbf{E}_i$. Normally a common form is assumed for the $\mathbf{B}_i$ and $\mathbf{E}_i$. The most common form for $\mathbf{E}_i$ is $\mathbf{E}_i = \sigma^2\mathbf{I}$ (the **conditional-independence model**). Parameter estimates are obtained, as in Chapter 4, from maximizing the log-likelihood, yielding

$$\hat{\boldsymbol{\beta}} = \left(\sum \mathbf{X}_i^{\mathrm{T}}\hat{\mathbf{\Sigma}}_{pi}^{-1}\mathbf{X}_i\right)^{-1}\left(\sum \mathbf{X}_i^{\mathrm{T}}\hat{\mathbf{\Sigma}}_{pi}^{-1}\mathbf{y}_i\right)$$

Here, $\hat{\mathbf{\Sigma}}_{pi}$ is the ML estimate of the appropriate submatrix (if there are missing values) of $\hat{\mathbf{\Sigma}}_p$ obtained from the ML estimates of the parameter vector $\boldsymbol{\phi}$.

5.2 Residual maximum likelihood

As is well known, ML estimation produces biased estimators of variance components. For example, for n observations from a linear model $\mathbf{y} = \mathbf{X}\boldsymbol{\beta} + \mathbf{u}$, with independent errors $\mathbf{u}$, $\operatorname{var}(\mathbf{u}) = \sigma^2 \mathbf{I}$, the ML estimate of σ^2 is given by dividing the residual sum of squares by n. An unbiased estimator, however, is given by dividing by $n - p$, where there are p parameters in $\boldsymbol{\beta}$. In effect, the estimate has not taken account of the fact that $\boldsymbol{\beta}$, also, is estimated. Residual or restricted maximum likelihood (REML: Patterson and Thompson, 1971; Harville 1977) is a modified approach which does take this into account, and so yields unbiased estimators.

Rather than estimating the variance components from the overall likelihood

$$L = -\frac{1}{2}\left\{\sum_i \log|\boldsymbol{\Sigma}_{pi}| + \sum_i (\mathbf{y}_i - \mathbf{X}_i\boldsymbol{\beta})^{\mathrm{T}} \boldsymbol{\Sigma}_{pi}^{-1} (\mathbf{y}_i - \mathbf{X}_i\boldsymbol{\beta})\right\}$$

REML maximizes the part of the likelihood which is invariant to $\boldsymbol{\beta}$. Equivalently, it maximizes the likelihood of a vector of linear combinations of observations which are invariant to $\mathbf{X}\boldsymbol{\beta}$. In terms of our model in (5.1), a linear combination of the observations $\mathbf{L}\mathbf{y}_i = \mathbf{L}\mathbf{Z}_i\mathbf{b}_i + \mathbf{L}\mathbf{X}_i\boldsymbol{\beta} + \mathbf{L}\mathbf{e}_i$ will only be invariant to $\mathbf{X}_i\boldsymbol{\beta}$ if $\mathbf{L}\mathbf{X}_i = 0$. This will be true when $\mathbf{L}$ is of the form $\mathbf{L} = \mathbf{I} - \mathbf{X}_i(\mathbf{X}_i^{\mathrm{T}}\mathbf{X}_i)\mathbf{X}_i^{\mathrm{T}}$. Then the $\mathbf{L}\mathbf{y}_i$ are residuals after estimating $\boldsymbol{\beta}$ by ordinary least squares. The elements $\mathbf{L}\mathbf{y}_i$ are sometimes called 'error contrasts'. Basing the estimation only on the residuals is equivalent to choosing the parameters of $\Sigma = \Sigma(\boldsymbol{\phi})$ which maximize the slightly modified form

$$L = -\frac{1}{2}\left\{\sum_i \log|\boldsymbol{\Sigma}_{pi}| + \sum_i (\mathbf{y}_i - \mathbf{X}_i\boldsymbol{\beta})^{\mathrm{T}} \boldsymbol{\Sigma}_{pi}^{-1} (\mathbf{y}_i - \mathbf{X}_i\boldsymbol{\beta}) + \sum_i \log|\mathbf{X}_i^{\mathrm{T}}\boldsymbol{\Sigma}_{pi}^{-1}\mathbf{X}_i|\right\}$$

Many packages will give the user a choice of ordinary ML estimation or REML estimation. Note that, if interest is primarily in inference about $\boldsymbol{\beta}$ rather than the variance components, REML should not make much difference – if it does, n is probably too small!

In any case, perhaps too much weight should not be placed on unbiasedness. In a random sample from $N(\mu, \sigma^2)$, for example, the sample variance s^2 is unbiased for σ^2, but s is not unbiased for σ, and it is σ, not σ^2, which occurs in the formulae for confidence intervals for μ. Further, it is well known that $\tilde{\sigma}^2 = (n-1)s^2/(n+1)$ has smaller mean square error than s^2, so the ML estimator $\hat{\sigma}^2 = (n-1)s^2/n$ can be regarded as a compromise between s^2 and $\tilde{\sigma}^2$. Some useful points are made by Harville (1977, section 4.3 and discussion).

5.3 Examples

Example 5.1

In Example 1 of Chapter 4 we analysed the growth of guniea pigs by fitting a separate regression line to each of the three groups involved. Here we extend this analysis to fit a separate regression line to each of the animals, where the parameters of the regressions are assumed to be randomly chosen from a suitable bivariate normal distribution (bivariate because there are two parameters involved – an intercept and a slope – for each animal).

BMDP5V commands for analysing these data are shown in Table 5.1. These are basically the same as those of Example 4.1, except that the STRUCTURE paragraph indicates that a random effects model is to be fitted. ZCOL1 is the first column of matrix **Z** above and ZCOL2 is the second column. In this example all the $\mathbf{Z}_i$ are the same since each guinea pig is measured at the same three times and there are no missing values. Thus

$$\mathbf{Z} = \begin{bmatrix} 1 & 1 \\ 1 & 2 \\ 1 & 3 \end{bmatrix}$$

The random effects structure of BMDP5V generated by the RANDOM option automatically assumes that the common $\mathbf{E} = \sigma^2\mathbf{I}$. This means that our covariance matrix has four parameters (three in the assumed common $\mathbf{B}_i$ and one from σ^2). This is in contrast to the six that would be required to fit a completely unrestricted covariance matrix. (The reduction from six to four is not great – but then this is a simple and small example. Had more measurements been taken, and had they been fitted by just a few random components, then the effect would be much more marked.)

```
/VARIABLE      NAMES = GROUP, WT1, WT2, WT3.
/DESIGN        GROUPING = GROUP.
               REPEATED = TIME.
               LEVEL = 3.
               CONTRAST (TIME) = X.
               X = 1, 2, 3.
               DPNAME = WEIGHT.
               DPVAR = WT1, WT2, WT3.
/MODEL         WEIGHT = 'GROUP + X + GROUP.X'.
/STRUCTURE     TYPE = RANDOM.
               ZCOLI = 1, 1, 1.
               ZCOL2 = 1, 2, 3.
```

Table 5.1. *BMDP5V commands for analysing the guinea pig data using random regression coefficients*

The output from this analysis shows that the Akaike information criterion has value −218.106. The analysis of Chapter 4, in which an unrestricted covariance matrix was used, yielded a value of −218.563. Thus, although the extra flexibility in the model resulting from allowing the covariance matrix to take any form might be expected to lead to a better fit to the data, when the reduced number of parameters in the random effects model is taken into account a better overall fit results.

MODEL: WITHIN-SUBJECT COVARIANCE MATRIX

	1	2	3
1	1967.4215		
2	1733.3504	2358.7526	
3	2014.5459	1953.6215	2407.9637

ALL-PAIRS WITHIN-SUBJECT COVARIANCE MATRIX

	1	2	3
1	1945.6141		
2	1711.3319	2534.0563	
3	1926.8941	2063.2919	2254.4674

Table 5.2. *Top: the covariance matrix as predicted by the model. Bottom: the covariance matrix as estimated directly from the residuals*

IN THE TABLE BELOW EACH FIXED EFFECT PART OF THE MODEL IS DECOMPOSED INTO SINGLE DEGREE OF FREEDOM REGRESSION TERMS AND COVARIATES

PARAMETER		ESTIMATE	ASYMPTOTIC SE	Z	TWO-SIDED P-VALUE
1	CONST.	540.11111	11.31663	47.727	0.0000
2	G1	9.55556	16.00414	0.597	0.5505
3	G2	−12.44444	16.00414	−0.778	0.4368
4	X	23.13333	2.40241	9.629	0.0000
5	G1.X	−17.93333	3.39752	−5.278	0.0000
6	G2.X	15.36667	3.39752	4.523	0.0000

Table 5.3. *Estimates of the regression parameters* $\boldsymbol{\beta}$

The estimated covariance matrix $\hat{\Sigma}$, arising from the model is as shown in the top half of Table 5.2. The lower half of Table 5.2. shows the covariance matrix as estimated directly from the data. Clearly the model does provide quite a good fit to the residuals.

Table 5.3 shows the estimates of the fixed effects regression parameters, and Table 5.4 significance tests. There is clearly a time effect, and it differs between the groups. Glancing back at the plots of the profiles in Fig 4.1, one might wonder how much of the interaction effect is due to the possibly anomalously low-scoring individual in group 1. A reanalysis without this apparent outlier might be revealing. The results of such an analysis are presented in Table 5.5. In fact, it seems that dropping this individual does not substantially change the results.

Further analyses of these data are given in Wilson (1991).

Example 5.2

The data for this example are described in Andersen, Jensen and Schou (1981) and are reproduced in Table A.11. They show measurements of plasma citrate

WALD TESTS OF SIGNIFICANCE OF FIXED EFFECTS AND COVARIATES

TEST	DF	CHI-SQUARE	P-VALUE
G	2	0.66	0.718
X	1	92.72	0.000
G.X	2	32.59	0.000

Table 5.4. *Significance tests for each factor*

PARAMETER	ESTIMATE	ASYMPTOTIC SE	Z	TWO-SIDED P-VALUE
1 CONST.	543.69444	11.69295	46.498	0.0000
2 G1	16.72222	17.16056	0.974	0.3298
3 G2	−16.02778	16.21521	−0.988	0.3229
4 X	25.15000	1.70958	14.711	0.0000
5 G1.X	−13.90000	2.50898	−5.540	0.0000
6 G2.X	13.35000	2.37077	5.631	0.0000

WALD TESTS OF SIGNIFICANCE OF FIXED EFFECTS AND COVARIATES

TEST	DF	CHI-SQUARE	P-VALUE
G	2	1.26	0.533
X	1	216.42	0.000
G.X	2	40.81	0.000

Table 5.5. *Estimates of the regression parameters* $\boldsymbol{\beta}$ *and significance tests for each factor, without the low scoring first guinea pig in group 1*

concentrations in micromoles per litre on ten subjects measured at five equally spaced times. They were previously analysed in Example 2 of Chapter 3.

Here we fit a linear model for each subject, allowing the slopes and intercepts to be random effects and AR(1) errors. We have used the very flexible SAS PROC MIXED procedure to do this. Table 5.6 shows the SAS commands.

The DATA step loads the data into a SAS data set named A with variables Y (the response), SUB (the subject), and TIME (the time period). PROC MIXED requires all of the response values to be in one variable, and so A has 50 observations.

The PROC MIXED code fits a model with a random intercept and slope along with AR(1) errors. The CLASS statement indicates that SUB is to be regarded as a categorical variable.

The MODEL statement specifies the dependent variable and the fixed effects. Since an intercept is included by default, these fixed effects represent a simple linear regression. Hence the $\mathbf{X}_i$ matrices are 5×2, with first columns consisting entirely of 1's and the second the values of the TIME variable.

The RANDOM statement sets up the $\mathbf{Z}_i$ and $\mathbf{B}_i$ matrices. It indicates that both INTERCEPT (a keyword) and TIME are to be regarded as random effects (so the $\mathbf{Z}_i$

```
DATA A;
    SUB = _N_;
    DO TIME = 1 TO 5;
        INPUT Y @@;
        OUTPUT;
    END;
    DATALINES;
-----------DATA GO HERE -----------------
RUN;
PROC MIXED DATA = A;
    CLASS SUB;
    MODEL Y = TIME;
    RANDOM INTERCEPT TIME / TYPE = UN SUBJECT = SUB;
    REPEATED / TYPE = AR (1) SUBJECT = SUB;
RUN;
```

Table 5.6. *SAS PROC MIXED commands for analysing the plasma citrate data*

are 5 × 2 with the same structure as the $\mathbf{X}_i$ matrices), and the TYPE = UN option specifies that a correlation is to be modelled between them. The SUBJECT = SUB option produces a 50 × 20 global **Z** matrix which is block diagonal with blocks the 5 × 2 $\mathbf{Z}_i$ matrices and also produces a corresponding 20 × 20 global block diagonal **B** matrix.

The REPEATED statement sets up a 50 × 50 global **E** matrix (block diagonal with ten 5 × 5 identical $\mathbf{E}_i$ matrices), with the SUBJECT = option inducing blocking and the TYPE option specifying that each $\mathbf{E}_i$ matrix has the same AR(1) structure, with elements parametrized as $\sigma^2 \rho^k$, where k is the 'distance' between the two measurements.

Table 5.7 shows the REML estimates of the covariance parameters. The first three rows represent the elements of the common $\mathbf{B}_i$, so that

$$\mathbf{B}_i = \begin{bmatrix} 789.05 & -124.49 \\ -124.49 & 21.27 \end{bmatrix}$$

789.05 is the estimated variance of the intercept, and 21.27 is that of the slope. They have a strong negative correlation between them of −0.96.

Covariance Parameter Estimates (REML)

Cov. Parm		Ratio	Estimate	Std Drror	Z	Pr > |Z|
INTERCEPT	UN (1, 1)	5.77629972	789.05027390	425.39497006	1.85	0.0636
	UN (2, 1)	−0.91131829	−124.4872989	76.66593768	−1.62	0.1044
	UN (2, 2)	0.15572656	21.27245718	15.13959971	1.41	0.1600
DIAG AR (1)		−0.00258361	−0.35292515	0.27051381	−1.30	0.1920
Residual		1.00000000	136.60133870	35.39157082	3.86	0.0001

Table 5.7. *Estimated covariance parameters from SAS PROC MIXED*

The final two rows give the elements of the assumed common $\mathbf{E}_i$. The estimated AR(1) parameter ρ is -0.353, its negative sign modelling an alternation about the random regression line, and the estimated σ^2 is 136.60.

SAS also gives the results of global model fitting tests and of tests of the fixed effects. In this example the test that the overall mean slope is zero yields a p value of 0.1176, not rejecting the hypothesis.

CHAPTER 6

Covariance structures

6.1 Regression models with structured covariance

In Chapters 2 and 4 the multivariate approach to repeated measures and longitudinal data was described in which no prior constraints were imposed on the covariance matrix $\mathbf{\Sigma}$. Linear models for $\boldsymbol{\mu}$ were formulated and fitted, and $\mathbf{\Sigma}$ was estimated without prejudice, i.e. without prior judgements such as constraint relationships between its elements. Now, while it is a good thing to be flexible and robust, not making assumptions about $\mathbf{\Sigma}$ which might be untrue, it can also be costly. Thus, with ten repeated measures there are 55 independent parameters in $\mathbf{\Sigma}$ to be estimated. (A $p \times p$ matrix, constrained only to be symmetric, as $p(p+1)/2$ independent elements.) With a limited amount of data, this might leave too little information available for the aspects of main interest, i.e. usually those concerning $\boldsymbol{\mu}$. In Chapters 3 and 5 some structure was assumed for $\mathbf{\Sigma}$ which was meant to reflect the situation under which the observations were generated. As always, a balance has to be struck: with too little structure too many parameters might be left to estimate, leading to weaker inferences; with too much structure there is a risk of model misspecification, leading to apparently stronger, but biased, inferences.

■ *Illustration 6.1* Consider a random sample of n vector observations, $\mathbf{y}_1, \ldots, \mathbf{y}_n$, from $N_p(\boldsymbol{\mu}, \mathbf{\Sigma})$. The maximum likelihood estimator of $\boldsymbol{\mu}$ is $\bar{\mathbf{y}}$, the sample mean vector, and its sampling variance is estimated by n^{-1} $\mathbf{S}$, $\mathbf{S}$ being the sample covariance matrix. The standard test of the hypothesis that $\boldsymbol{\mu} = \boldsymbol{\mu}_0$ is based on Hotelling's T^2-statistic, $T^2 = n(\bar{\mathbf{y}} - \boldsymbol{\mu}_0)^{\mathrm{T}}$ $\mathbf{S}^{-1}$ $(\bar{\mathbf{y}} - \boldsymbol{\mu}_0)$: under the hypothesis, $(n-p)T^2/\ \{(n-1)\ p\}$ has distribution $F(p, n-p)$. As p increases, with fixed sample size n, the denominator degrees of freedom, $n-p$, decreases. It reaches 0 when $p = n$, at which point $\mathbf{S}$ becomes singular. This is an extreme example of using up too many degrees of freedom for estimating the covariance matrix. At the other extreme, if $\mathbf{\Sigma}$ were assumed to be completely known, and so did not have to be estimated, the statistic $(\bar{\mathbf{y}} - \boldsymbol{\mu}_0)^{\mathrm{T}} \mathbf{\Sigma}^{-1} (\bar{\mathbf{y}} - \boldsymbol{\mu}_0)$ could be referred to χ_p^2 for any n, even as small as 1. (Of course, the power of the test would not be great with small n, but that is another issue.) However, if the assumed $\mathbf{\Sigma}$ is wrong, the test statistic will not have the χ_p^2 distribution and the inference will be based on incorrect distribution theory. ■

Consider a general framework in which data vectors $\mathbf{y}_1, \ldots, \mathbf{y}_n$ are available, together with model specifications $E(\mathbf{y}_i) = \boldsymbol{\mu}_i$ and $\text{var}(\mathbf{y}_i) = \boldsymbol{\Sigma}_i$. In general, $\boldsymbol{\mu}_i$ will be a function of explanatory variables, or covariates, and of regression parameters. For instance, the all-pervasive linear model has $\boldsymbol{\mu}_i = \mathbf{X}_i\boldsymbol{\beta}$ or, in component form, $y_{ij} = \mathbf{x}_{ij}^{\mathrm{T}}\boldsymbol{\beta}$; here, $\mathbf{x}_{ij}^{\mathrm{T}}$, the jth row of $\mathbf{X}_i$, comprises the covariates for y_{ij}, and $\boldsymbol{\beta}$ is the vector of linear regression coefficients. Likewise, $\boldsymbol{\Sigma}_i$ will generally depend on covariates and parameters, the latter comprising $\boldsymbol{\beta}$, possibly, and a further set, say $\boldsymbol{\tau}$. In practice, $\boldsymbol{\Sigma}_i$ is often rather more restricted: it might not involve covariates at all, or only in some limited way, and it might depend only on $\boldsymbol{\tau}$, a parameter vector distinct from $\boldsymbol{\beta}$. Having said that, 'often' here just means 'in most applications'; whole classes of exceptions, which owe nothing to artificial contrivance, can be found in this book.

In order to get started, let us consider a particular case, the Laird–Ware two-stage linear model as introduced in section 5.1:

$$\mathbf{y}_i = \mathbf{X}_i\boldsymbol{\beta} + \mathbf{Z}_i\mathbf{b}_i + \mathbf{e}_i \tag{6.1}$$

The resulting forms for the mean $\boldsymbol{\mu}_i$ and covariance matrix $\boldsymbol{\Sigma}_i$ of $\mathbf{y}_i$ are

$$\boldsymbol{\mu}_i = \mathbf{X}_i\boldsymbol{\beta} \quad \text{and} \quad \boldsymbol{\Sigma}_i = \mathbf{Z}_i\mathbf{B}\mathbf{Z}_i^{\mathrm{T}} + \mathbf{E}_i \tag{6.2}$$

where $\mathbf{B} = \text{var}(\mathbf{b}_i)$, $\mathbf{E}_i = \text{var}(\mathbf{e}_i)$, and $\mathbf{b}_i$ and $\mathbf{e}_i$ are uncorrelated. Now, the restriction of $\boldsymbol{\Sigma}_i$ to this form, rather than placing no constraint upon it, reflects the attempt to model the data with more insight, i.e. expressing the idea that the mean profile for the observation $\mathbf{y}_i$ is composed of a population-averaged term, $\mathbf{X}_i\boldsymbol{\beta}$, and an individual-specific term, $\mathbf{Z}_i\mathbf{b}_i$. (The terminology is borrowed from Zeger, Liang and Albert (1988): they used the descriptions 'population-averaged' and 'subject-specific' in the context of generalized linear models, which are the subject of Chapter 9 below.) Over the populations of individuals, $\boldsymbol{\beta}$ is constant but $\mathbf{b}_i$ varies with mean $\mathbf{0}$ and covariance matrix $\mathbf{B}$; in applications it is common practice to assume no particular structure for $\mathbf{B}$, and so to estimate it without constraint. Given that it is individual i who is being watched, variation about $\mathbf{X}_i\boldsymbol{\beta} + \mathbf{Z}_i\mathbf{b}_i$, his expected profile, is governed by the covariance structure $\mathbf{E}_i$. In the so-called conditional-independence model, often assumed in applications, $\mathbf{E}_i$ is taken to be $\sigma_e^2\mathbf{I}$, i.e. the random departures e_{ij} $(j = 1, \ldots, p)$, the components of $\mathbf{e}_i$, are uncorrelated and all have the same variance σ_e^2.

In the previous paragraph it was seen how, in a particular area, considerations of the context might be introduced to constrain $\boldsymbol{\Sigma}_i$ to have a certain form. One can go further down this path and look critically at $\mathbf{Z}_i\mathbf{B}\mathbf{Z}_i^{\mathrm{T}}$ and $\mathbf{E}_i$ in order to propose more appropriate forms for them. This is what this chapter is about: we will look at various situations in which there are good grounds for adopting a structured form for $\boldsymbol{\Sigma}_i$, rather than allowing it to be entirely undisciplined.

6.2 The error matrix $\mathbf{E}_i$

In this section some basic models for $\mathbf{E}_i$ are described. In applications these can be used as they are, or as building blocks for more complex situations. At the risk of being repetitive, we emphasize that $\mathbf{E}_i = \text{var}(\mathbf{e}_i)$ describes the covariances among the observations when we follow exclusively the progress of individual i, forsaking all others; it is the covariance matrix of the ith individual's deviations from his personal mean profile $\mathbf{X}_i\boldsymbol{\beta} + \mathbf{Z}_i\mathbf{b}_i$.

6.2.1 Conditionally uncorrelated errors

In some situations it might be reasonable to believe that the successive observations on an individual might be uncorrelated. For instance, one would not expect to detect much relationship between pulse rates taken ten days apart after removing any assignable causes of variation such as athletic exertions or late-night movies. The corresponding form for $\mathbf{E}_i$ is diagonal. If, in addition, there is some reason to trust in homogeneity of variance, the form $\sigma_e^2\mathbf{I}_p$ for $\mathbf{E}_i$ would be appropriate; $\mathbf{I}_p$ here denotes the $p \times p$ unit matrix. This form is commonly applied but should not be taken too lightly. For instance, to use the athletic example again, it might well be that higher pulse rates are associated with larger variances. In such a case it would be reasonable to entertain a model in which E_{ijj}, the jth diagonal element of $\mathbf{E}_i$, were related to the expected value of y_{ij}, i.e. to $\mathbf{x}_{ij}$. Thus, we might take E_{ijj} as $\sigma_e^2 w_{ij}$, where w_{ij} is a measure of the work rate on occasion j and σ_e^2 is constant over j and i. In the context of normally distributed observations, where correlation rules dependence, models in which $\mathbf{E}_i$ is diagonal are known as conditional-independence models (section 5.1).

6.2.2 Equicorrelated errors

A step up from uncorrelated errors is to equicorrelated errors. In this model $\mathbf{E}_i$ has diagonal elements σ_j^2 and off-diagonal elements $\rho\sigma_j\sigma_k$; each pair of components of the error vector $\mathbf{e}_i$ has the same correlation, ρ. In terms of $\mathbf{I}_p$ and $\boldsymbol{J}_p$, the $p \times p$ matrix of 1's, $\mathbf{E}_i = \mathbf{s}\{(1-\rho)\mathbf{I}_p + \rho\boldsymbol{J}_p\}\mathbf{s}$, where $\mathbf{s} = \text{diag}[\sigma_1 \cdots \sigma_p]$. The kind of situation for which this structure might be appropriate is where the measurements are all made at the same time in the same place, with no reason for stronger interaction between any one pair than any other. This is the set-up known in the trade as **split plots**. For measurements of the same type made in the same way it is usual to take the σ_j^2 equal for all j, i.e. to assume variance homogeneity. The resulting form of covariance matrix, $a\mathbf{I}_p + b\mathbf{J}_p$ for some a and b, is known as **compound symmetric** (section 3.1). We will not emphasize this structure for $\mathbf{E}_i$ because, in the kinds of applications addressed in this book, it is more relevant to $\boldsymbol{\Sigma}_i$ than to $\mathbf{E}_i$. In such contexts, to be covered below, it arises from a natural random effects model which shows where the correlation comes from rather than just adopting it without explanation (section 3.1).

6.2.3 Autoregressive errors

When the measurements on an individual have been made in sequence over time the errors may be serially correlated. Thus, over hypothetical repetitions of the process with the ith individual, one could discern some consistent relationship between the components e_{ij} of the error vectors in single runs. A widely used time-series model is **autoregressive process**: dropping the subscript i, this has

$$e_j = \rho e_{j-1} + u_j \tag{6.3}$$

for $j \geqslant 2$, where ρ is the regression parameter and the u_j are innovation errors, usually assumed to be $N(0, \sigma_u^2)$ variates each independent of the past. Repeated application of (6.3) for $e_{j-1}, e_{j-2}, \dots,$ yields, for $k > 0$,

$$e_j = u_j + \rho u_{j-1} + \rho^2 u_{j-2} + \cdots + \rho^{k-1} u_{j-k+1} + \rho^k e_{j-k} \tag{6.4}$$

There are two associated submodels. In the first the process (6.3) is assumed to have been running long before observation started, i.e. since $j = -\infty$, and to be 'stable': this means that $|\rho| < 1$ and e_j all have the same distribution, which is therefore normal with mean zero and a common variance. From (6.4),

$$\operatorname{var}(e_j) = \sigma_u^2(1 + \rho^2 + \rho^4 + \cdots + \rho^{2(k-1)}) + \rho^{2k} \operatorname{var}(e_{j-k})$$

and so this common variance is $\operatorname{var}(e_j) = \sigma_u^2/(1 - \rho^2)$. Also from (6.4),

$$E(e_j e_{j-k}) = \rho^k E(e_{j-k}^2) = \sigma_u^2 \rho^k/(1 - \rho^2)$$

and it follows that

$$E_{jk} = \sigma_u^2 \rho^{|j-k|}/(1 - \rho^2)$$

The inverse matrix $\mathbf{E}^{-1}$ is tridiagonal, with diagonal

$$\sigma_u^{-2}[1, 1 + \rho^2, 1 + \rho^2, \dots, 1 + \rho^2, 1]$$

and subdiagonal $-\sigma_u^{-2}[\rho \dots \rho]$; also, det $\mathbf{E} = \sigma_u^{2p}/(1 - \rho^2)$. The tridiagonal form of $\mathbf{E}^{-1}$ corresponds to the Markov property of the error process. To explain, note that the element E^{jk} of the inverse is the partial covariance of e_j and e_k after conditioning on all the other e's. The fact that E^{jk} is zero when j and k differ by more than 1 means that the correlation between e_j and e_k, given the values of intervening e's, is zero, i.e. a Markov-type property. The full Markov property, with conditioning on just one intervening e, can be argued similarly.

In the second submodel the process (6.3) is deemed to have started up with the first observation, i.e. at $j = 1$, with $u_0 = 0$ and $e_1 = u_1$. From (6.4) one can obtain the covariance structure now as

$$E_{jk} = \sigma_u^2 \rho^{j-k}(1-\rho^{2k})/(1-\rho^2)$$

for $j \geqslant k$. In contrast to the stable submodel, the e_j now have unequal variances and there is no restriction to $|\rho| < 1$. The inverse matrix $\mathbf{E}^{-1}$ is tridiagonal, with diagonal

$$\sigma_u^{-2}[1, 1+\rho^2, 1+\rho^2, \ldots, 1+\rho^2, 1]$$

and subdiagonal $-\sigma_u^{-2}[\rho \ldots \rho]$, and det $\mathbf{E} = \sigma_u^{2p}$. Thus the inverse matrices $\mathbf{E}^{-1}$ for the two submodels differ only in the (1, 1)th entry.

As $\rho \to 1$ in the unstable submodel, $E_{jk} \to \sigma_u^2 \min(j, k)$. This represents a **cumulative error process**: setting $\rho = 1$ in (6.4) shows that the innovation errors u_j are straightforwardly accumulated as time goes on. An alternative description is that, for $\rho = 1$, (6.3) is a **random walk** model.

6.2.4 Markov correlation structure

The stable autoregressive submodel in section 6.2.3, which is a homogeneous Markov chain, has correlations $\rho^{|j-k|}$. Thus the correlation between components e_j decreases geometrically with their separation in time. When the data are unequally spaced over time, the measurement occasions being at times t_j, say, the geometrically decreasing correlation form can be sensibly incorporated by taking $E_{jk} = \sigma_e^2 \rho^{|t_j - t_k|}/(1-\rho^2)$. This form is invariant under linear transformation of the time scale: replacement of t_j by $a + bt_j$ leads to $E_{jk} = \sigma_1^2 \rho_1^{|t_j - t_k|}/(1-\rho_1^2)$ with $\rho_1 = \rho^b$ and $\sigma_1^2 = \sigma_e^2(1-\rho_1^2)/(1-\rho_1^{2/b})$. Usually $0 \leqslant \rho < 1$, but in some cases, for example when the $|t_j - t_k|$ are integers, one can allow $-1 < \rho < 1$. Here $\mathbf{E}^{-1}$ has tridiagonal form, with diagonal

$$\sigma_e^{-2}(1-\rho^2)[f_1, f_1 + f_2 - 1, \ldots, f_{p-2} + f_{p-1} - 1, f_{p-1}]$$

and subdiagonal

$$-\sigma_e^{-2}(1-\rho^2)[f_1 \rho^{d_1}, \ldots, f_{p-1}\rho^{d_{p-1}}]$$

where $d_j = t_{j+1} - t_j$ and $f_j = (1-\rho^{2d_j})^{-1}$; also

$$\det \mathbf{E} = \sigma_e^{2p}(1-\rho^2)^{-p}(f_1 f_2 \cdots f_{p-1})^{-1}$$

The structure corresponding to the unstable submodel of section 6.2.3 is $E_{jk} = \sigma_e^2 \rho^{|t_j - t_k|}(1-\rho^{2t_k})/(1-\rho^2)$ for $j \geqslant k$. As $\rho \to 1$, $E_{jk} \to \sigma_e^2 \min(t_j, t_k)$.

6.2.5 Antedependence

The error covariance structures described in section 6.2.3 are capable of various generalizations. Thus, the autoregression (6.3) can be extended to higher orders than 1, and the error variances may be unequal, as in the second submodel of that section. Also, the time intervals between successive observations can be unequal as in section 6.2.4. A structure covering all these extensions is the so-called **antedependence** form of Gabriel (1961; 1962). Needless to say, the number of parameters required to formulate the extended structures is often considerably more than the former two, σ_u and ρ.

When the errors e_j are independent their covariance matrix **E**, and hence also $\mathbf{E}^{-1}$, is diagonal. In the case of first-order Markov dependence, as in sections 6.2.3 and 6.2.4, $\mathbf{E}^{-1}$ is tridiagonal, i.e. it has one additional non-zero entry on either side of the diagonal. In the rth-order case $\mathbf{E}^{-1}$ has r non-zero entries on either side of the diagonal, i.e. it has a diagonal band of width $2r+1$ with zeros filling the remaining top right and bottom left corners. Gabriel (1962) showed that this structure is equivalent to the following condition on the partial correlations $\rho_{js} = \text{corr}(e_j, e_{j-s-1} | e_{j-1}, \ldots, e_{j-s})$: that $\rho_{js} = 0$ for all j and all $s > r$. In words, the partial correlation between two e's more than r epochs apart, given the intermediate e's, is zero. The extreme cases, $r = 0$ and $r = p - 1$, yield respectively complete independence (strictly speaking, zero correlation) and correlation of arbitrary pattern. In terms of autoregression, e_j is expressed as a linear function of $e_{j-1}, \ldots, e_{j-r}$ plus an independent error u_j, like equation (6.3). However, unlike the usual model, the regression coefficients and error variances are not constrained to be constant over j.

In the framework of multivariate normal distributions correlation is synonymous with dependence, and Gabriel (1962) defines **r-antedependence** as above for this case.

6.2.6 Diggle's covariance structures

Diggle (1988; 1990, Chapter 5) has suggested adoption of the form $\sigma_1^2\mathbf{I}_p + \sigma_2^2\mathbf{R}$ for $\mathbf{E}_i$, where the correlation matrix **R** has elements $R_{jk} = r(|t_j - t_k|)$, $r(\cdot)$ being a suitably specified function. In particular, Diggle suggested $r(u) = \exp(-\rho u^c)$ as a useful form. Here, ρ is a parameter and c is taken to be 1 or 2, the choice $c = 1$ giving a structure equivalent to the stable model of section 6.2.4. Diggle (1988) outlined a practical approach to the choice and validation of the covariance structure using the empirical semi-variogram.

For measures equispaced over time, $|t_j - t_k|$ depends only on $|j - k|$. Then, for general $r(\cdot)$, **R** takes the Toeplitz form familiar to students of stationary time series in which all the elements at the same distance from the main diagonal are equal.

6.3 The matrices $\mathbf{Z}_i$ and B

In the mixed linear model (6.1), $\mathbf{Z}_i$ is the multiplier for $\mathbf{b}_i$ just as $\mathbf{X}_i$ is for $\boldsymbol{\beta}$. Correspondingly, a $\mathbf{Z}_i$ matrix is constructed from explanatory variables in exactly the same way as an $\mathbf{X}_i$ matrix. In the extreme, $\mathbf{Z}_i$ might even be equal to

$\mathbf{X}_i$, as in the pure random coefficients, or growth curve, model $\mathbf{y}_i = \mathbf{X}_i(\boldsymbol{\beta} + \mathbf{b}_i) + \mathbf{e}_i$. However, more usually $\mathbf{Z}_i$ is different from $\mathbf{X}_i$ because it has a different purpose: $\mathbf{X}_i$ is used to model the mean $\boldsymbol{\mu}_i$, whereas $\mathbf{Z}_i$ is used to model the covariance $\boldsymbol{\Sigma}_i$.

There follow now some examples of $\mathbf{Z}_i$ matrices covering some of the more commonly applied models. These will go some way towards illustrating the potential flexibility of the contribution $\mathbf{Z}_i\mathbf{B}\mathbf{Z}_i^{\mathrm{T}}$ to the overall covariance structure (6.2). In some cases $\mathbf{B}$ will be unstructured, and in others prior considerations will suggest a particular form for it. In the former case, it is convenient for computation to set $\mathbf{B}$ up as $\mathbf{A}\mathbf{A}^{\mathrm{T}}$, where $\mathbf{A}$ is an $r \times r$ lower triangular matrix: this ensures that $\mathbf{B}$ is automatically positive semi-definite.

6.3.1 *Random individual levels*

The basic model can be represented as

$$y_{ij} = \mu_{ij} + a_i^{\mathrm{I}} + e_{ij}$$

where μ_{ij} is the mean for individual i on measurement j, and a_i^{I} is the **random individual level**. For example, μ_{ij} might be the jth component of the mean profile vector for the treatment group containing individual i, or, more generally, of a linear model $\mathbf{X}_i\boldsymbol{\beta}$ or even of a nonlinear one; see section 3.1 for a fuller discussion of these models, particularly in the anova context. In vector format,

$$\mathbf{y}_i = \boldsymbol{\mu}_i + a_i^{\mathrm{I}}\mathbf{1}_p + \mathbf{e}_i$$

where $\mathbf{1}_p$ is the $p \times 1$ vector of 1's. In terms of the general representation $\mathbf{Z}_i\mathbf{b}_i$ for the random effects we have here $\mathbf{Z}_i = \mathbf{1}_p$ and $\mathbf{b}_i = (a_i^{\mathrm{I}})$. The resulting covariance matrix is obtained from (6.2) as $\boldsymbol{\Sigma}_i = \sigma_e^2\mathbf{I}_p + \sigma_{\mathrm{I}}^2\mathbf{J}_p$, where $\sigma_{\mathrm{I}}^2 = \mathrm{var}(a_i^{\mathrm{I}})$, $\mathbf{I}_p$ is the $p \times p$ unit matrix, $\mathbf{J}_p$ is the $p \times p$ matrix of 1's, and the usual assumption in this context, that $\mathbf{E} = \sigma_e^2\mathbf{I}_p$ (section 6.2.1), has been made. Note that $\boldsymbol{\Sigma}_i$ here has the compound symmetric form referred to in sections 3.1 and 6.2.2.

A more elaborate model contains random interaction terms:

$$\mathbf{y}_i = \boldsymbol{\mu}_i + a_i^{\mathrm{I}}\mathbf{1}_p + \mathbf{a}_i^{\mathrm{IO}} + \mathbf{e}_i$$

where $\mathbf{a}_i^{\mathrm{IO}} = [a_{i1}^{\mathrm{IO}} \cdots a_{ip}^{\mathrm{IO}}]^{\mathrm{T}}$ contains as components the random individuals by occasions interactions. The usual identifiability constraint for a two-way interaction term in anova is $a_{i+}^{\mathrm{IO}} = 0$, the $+$ denoting summation over $j = 1, \ldots, p$. The other possible constraint, $a_{+j}^{\mathrm{IO}} = 0$, is not appropriate in this context: it would entail collusion between the individuals in the sample, so compromising their independence. The random effects $a_i^{\mathrm{I}}\mathbf{1}_p + a_i^{\mathrm{IO}}$ can now be represented as $\mathbf{Z}_i\mathbf{b}_i$ by taking $r = p + 1$, $\mathbf{b}_i = [a_i^{\mathrm{I}}\ \mathbf{a}_i^{\mathrm{IO}}]^{\mathrm{T}}$ and $\mathbf{Z}_i = [\mathbf{1}_p\ \mathbf{I}_p]$. We will assume that a_i^{I} and $\mathbf{a}_i^{\mathrm{IO}}$ are uncorrelated, and that the components of $\mathbf{a}_i^{\mathrm{IO}}$ have equal variances and are equicorrelated. These are, more or less, the most basic assumptions that can be made here. It is not difficult to think of situations where a_i^{I} and $\mathbf{a}_i^{\mathrm{IO}}$ might be correlated, or where symmetry in the correlations among the a_{ij}^{IO} might not obtain. Nevertheless, pursuing this basic situation, it follows from these two assumptions that

$$\mathbf{B} = \mathrm{var}\,(\mathbf{b}_i) = \begin{bmatrix} \sigma_{\mathrm{I}}^2 & \mathbf{0} \\ \mathbf{0} & \sigma_{\mathrm{IO}}^2 \mathbf{K}_p \end{bmatrix}$$

$$\mathbf{Z}_i \mathbf{B} \mathbf{Z}_i^{\mathrm{T}} = \sigma_{\mathrm{I}}^2 \mathbf{J}_p + \sigma_{\mathrm{IO}}^2 \mathbf{K}_p$$

where $\mathbf{0} = [0 \cdots 0]^{\mathrm{T}}$ and $\mathbf{K}_p = (p\mathbf{I}_p - \mathbf{J}_p)/(p-1)$ has (j, k)th element 1 for $j = k$ and $-1/(p-1)$ for $j \neq k$. The form of $\mathbf{K}_p$, the correlation matrix of $\mathbf{a}_i^{\mathrm{IO}}$, results from the constraint assumed for the a_{ij}^{IO}: if $a_{i+}^{\mathrm{IO}} = 0$ then, assuming that $\rho = \mathrm{corr}\,(a_{ij}^{\mathrm{IO}}, a_{ik}^{\mathrm{IO}})$ is the same for all $j \neq k$,

$$0 = \mathrm{var}(a_{i+}^{\mathrm{IO}}) = \sum_{j=1}^{p} \mathrm{var}\,(a_{ij}^{\mathrm{IO}}) + \sum_{j \neq k} \mathrm{cov}(a_{ij}^{\mathrm{IO}}, a_{ik}^{\mathrm{IO}}) = p\sigma_{\mathrm{IO}}^2 + p(p-1)\sigma_{\mathrm{IO}}^2 \rho$$

so $\rho = -1/(p-1)$. The matrix $\mathbf{K}_p$, and therefore $\mathbf{B}$, is singular, To avoid this, one could instead incorporate the constraint automatically by taking $r = p$, $\mathbf{b}_i = [a_i^{\mathrm{I}}\ a_{i1}^{\mathrm{IO}} \ldots a_{i,\,p-1}^{\mathrm{IO}}]^{\mathrm{T}}$, and modifying $\mathbf{Z}_i$ by deleting the last column and changing the last row to $(1, -1, \ldots, -1)$. However, $\mathbf{Z}_i \mathbf{b}_i$ and $\mathbf{Z}_i \mathbf{B} \mathbf{Z}_i^{\mathrm{T}}$, the quantities actually appearing in the $\mathbf{y}_i$ model, are then the same as before.

With $\mathbf{E} = \sigma_e^2 \mathbf{I}_p$, $\mathbf{\Sigma}_i$ is of compound symmetric form, as for the basic model. In fact, the three parameters σ_e^2, σ_{I}^2 and σ_{IO}^2 are not separately identifiable in this covariance structure since it can be expressed as

$$\mathbf{\Sigma}_i = \{\sigma_e^2 + \sigma_{\mathrm{IO}}^2 p/(p-1)\} \mathbf{I}_p + \{\sigma_{\mathrm{I}}^2 - \sigma_{\mathrm{IO}}^2/(p-1)\} \mathbf{J}_p$$

in terms of the two parametric combinations in the curly brackets. (This accords with Example 3.1 in Crowder and Hand (1990), where it was found that only $\sigma_e^2 + \sigma_{\mathrm{IO}}^2 p/(p-1)$ and $\sigma_e^2 + p\sigma_{\mathrm{I}}^2$ were estimable from the mean squares in the corresponding anova table.) Thus, in a Gaussian likelihood, the three parameters would be unidentifiable unless different individuals contributed different numbers of repeated measures. In the latter case p, and hence the two linear combinations of parameters, would vary over individuals. This is one of those odd cases where suitable deletion of some of the data would yield 'more information'.

An alternative assumption for $\mathbf{B}$ in common use is that $\mathrm{var}\,(\mathbf{a}_i^{\mathrm{IO}}) = \sigma_{\mathrm{IO}}^2 \mathbf{I}_p$, rather than $\sigma_{\mathrm{IO}}^2 \mathbf{K}_p$, corresponding to an unconstrained, uncorrelated set of components. In this case, $\mathbf{Z}_i \mathbf{B} \mathbf{Z}_i^{\mathrm{T}}$ comes out as $\sigma_{\mathrm{I}}^2 \mathbf{J}_p + \sigma_{\mathrm{IO}}^2 \mathbf{I}_p$, and $\mathbf{\Sigma}_i$ as $(\sigma_e^2 + \sigma_{\mathrm{IO}}^2)\,\mathbf{I}_p + \sigma_{\mathrm{I}}^2 \mathbf{J}_p$. This has the same compound symmetric form as before but with different combinations of the parameters σ_e^2, σ_{I}^2 and σ_{IO}^2. However, any method of inference which depends on these parameters only through $\mathbf{\Sigma}_i$ will be unaffected: the estimates for them will be different under the alternative assumptions, but that for $\mathbf{\Sigma}_i$ will be the same. In particular, this applies to normal likelihood methodology (section 4.2).

6.3.2 A two-way anova model

Suppose that each subject is observed over a complete $a \times b$ factorial design. The full set of random effects comprises

$$a_i^{\mathrm{I}} + a_{ij}^{\mathrm{IA}} + a_{ik}^{\mathrm{IB}} + a_{ijk}^{\mathrm{IAB}}$$

where the three interaction terms are subject to suitable identifiability constraints. To begin with, let us take $r = 1 + a + b$ with

$$\mathbf{b}_i = [a_i^{\mathrm{I}} | a_{i1}^{\mathrm{IA}} \cdots a_{ia}^{\mathrm{IA}} | a_{i1}^{\mathrm{IB}} \cdots a_{ib}^{\mathrm{IB}}]^{\mathrm{T}}$$

just involving terms up to the two-way random interactions. Let us make the basic assumptions that there is no correlation between the three sets of variates $\{a_i^{\mathrm{I}}\}$, $\{a_{ij}^{\mathrm{IA}}\}$ and $\{a_{ik}^{\mathrm{IB}}\}$, and that the a_{ij}^{IA} $(j = 1, \ldots, a)$ have equal variances and are equicorrelated, with a similar assumption for the a_{ik}^{IB} $(k = 1, \ldots, b)$. It follows that $\mathbf{B} = \mathrm{diag}[\sigma_{\mathrm{I}}^2 \; \sigma_{\mathrm{A}}^2 \mathbf{K}_a \; \sigma_B^2 \mathbf{K}_b]$, where $\mathbf{K}_a$ $(a \times a)$ and $\mathbf{K}_b$ $(b \times b)$ here have the form of $\mathbf{K}_p$ in section 6.3.1, as induced by the constraints $a_{i+}^{\mathrm{IA}} = 0$ and $a_{i+}^{\mathrm{IB}} = 0$. Also, ordering the components of $\mathbf{y}_i$ with the B-factor varying faster than the A-factor, the matrix $\mathbf{Z}_i$ has the form

$$\mathbf{Z}_i = \begin{bmatrix} \mathbf{1}_b & \mathbf{L}_1 & \mathbf{I}_b \\ \mathbf{1}_b & \mathbf{L}_2 & \mathbf{I}_b \\ \cdots & \cdots & \cdots \\ \mathbf{1}_b & \mathbf{L}_a & \mathbf{I}_b \end{bmatrix}$$

where $\mathbf{L}_j$ is $b \times a$ with jth column $\mathbf{1}_b$ and all other elements zero. The resulting covariance contribution $\mathbf{Z}_i\mathbf{B}\mathbf{Z}_i^{\mathrm{T}}$ comprises an $a \times a$ arrangement of $b \times b$ blocks, the (j, k)th block being

$$\sigma_{\mathrm{I}}^2 \mathbf{J}_b + \sigma_{\mathrm{A}}^2 \mathbf{L}_j \mathbf{K}_a \mathbf{L}_k^{\mathrm{T}} + \sigma_{\mathrm{B}}^2 \mathbf{K}_b = \sigma_{\mathrm{B}}^2 \{b/(b-1)\} \mathbf{I}_b + \{\sigma_{\mathrm{I}}^2 + \sigma_{\mathrm{A}}^2 \kappa_{jk} - \sigma_{\mathrm{B}}^2/(b-1)\} \mathbf{J}_b$$

where $\kappa_{jk} = (a\delta_{jk} - 1)/(a - 1)$ and δ_{jk} is the Kronecker delta. Note that $\sigma_{\mathrm{I}}^2, \sigma_{\mathrm{A}}^2$ and σ_{B}^2 are all identifiable in this covariance structure which is assembled from two distinct compound symmetric blocks.

The alternative assumption $\mathbf{B} = \mathrm{diag}[\sigma_{\mathrm{I}}^2 \; \sigma_{\mathrm{A}}^2 \mathbf{I}_a \; \sigma_{\mathrm{B}}^2 \mathbf{I}_b]$ leads to the (j, k)th block

$$\sigma_{\mathrm{B}}^2 \mathbf{I}_b + (\sigma_{\mathrm{I}}^2 + \sigma_{\mathrm{A}}^2 \delta_{jk}) \mathbf{J}_b$$

In terms of a statistical model, this is indistinguishable from the previous version.

Let us consider now the inclusion of the full set of random effects, adding the three-way interactions vector

$$\mathbf{a}_i^{\mathrm{IAB}} = \left[a_{i11}^{\mathrm{IAB}} \dots a_{i1b}^{\mathrm{IAB}} | \dots | a_{ia1}^{\mathrm{IAB}} \dots a_{iab}^{\mathrm{IAB}}\right]^{\mathrm{T}}$$

of length ab, to the former $\mathbf{b}_i$. The corresponding $\mathbf{Z}_i$ matrix, say $\mathbf{Z}_i^+$ is related to the former $\mathbf{Z}_i$ by $\mathbf{Z}_i^+ = (\mathbf{Z}_i | \mathbf{I}_{ab})$, and the corresponding augmented $\mathbf{B}$ matrix is

$$\mathbf{B}^+ = \mathrm{diag}[\sigma_{\mathrm{I}}^2 \; \sigma_{\mathrm{A}}^2 \mathbf{K}_a \; \sigma_{\mathbf{B}}^2 \mathbf{K}_b \; \sigma_{\mathbf{AB}}^2 \mathbf{R}]$$

where $\mathbf{R}(ab \times ab)$ is to be determined as follows. The usual anova constraints for the interaction terms are $a_{i+k}^{\mathrm{IAB}} = 0$ for each k and $a_{ij+}^{\mathrm{IAB}} = 0$ for each j. Let

$$\rho_{jk,j'k'} = \mathrm{corr}(a_{ijk}^{\mathrm{IAB}}, a_{ij'k'}^{\mathrm{IAB}})$$

By the same equicorrelation argument used for the a_{ij}^{IA} the constraint $a_{i+k}^{\mathrm{IAB}} = 0$ induces

$$\rho_{jk,j'k} = -1/(a-1)$$

for $j \neq j'$, and $a_{ij+}^{\mathrm{IAB}} = 0$ induces

$$\rho_{jk,jk'} = -1/(b-1)$$

for $k \neq k'$. Remaining to be determined are the $\rho_{jk,j'k'}$ for $j \neq j'$ and $k \neq k'$. It turns out that, without further assumption, these correlations are all equal to $1/\{(a-1)(b-1)\}$. Hence, $\mathbf{R}$ is an $a \times a$ arrangement of $b \times b$ blocks, the (j, k)th of which is $\mathbf{K}_b$ for $j = k$, and $-\mathbf{K}_b/(a-1)$ for $j \neq k$; more succinctly, the (j, k)th block of $\mathbf{R}$ is $\kappa_{jk}\mathbf{K}_b$, where κ_{jk} is defined above; even more succinctly $\mathbf{R} = \mathbf{K}_a \otimes \mathbf{K}_b$, using the Kronecker product. The (j, k)th block in the resulting covariance contribution is

$$\begin{aligned}(\mathbf{Z}_i^+ \mathbf{B}^+ \mathbf{Z}_i^{+\mathrm{T}})_{jk} &= \sigma_{\mathrm{I}}^2 \mathbf{J}_b + \sigma_{\mathrm{A}}^2 \mathbf{L}_j \mathbf{K}_a \mathbf{L}_k^{\mathrm{T}} + \sigma_{\mathrm{B}}^2 \mathbf{K}_b + \sigma_{\mathbf{AB}}^2 a_{jk} \mathbf{K}_b \\ &= \sigma_{\mathbf{B}}^2 \{b/(b-1)\} \mathbf{I}_b + \{\sigma_{\mathrm{I}}^2 + \sigma_{\mathrm{A}}^2 \kappa_{jk} - \sigma_{\mathbf{B}}^2/(b-1) \\ &\quad - \sigma_{\mathrm{AB}}^2 \kappa_{jk}/(b-1)\} \mathbf{J}_b\end{aligned}$$

The introduction of non-zero σ_{AB}^2 into $\mathbf{B}$ has evidently led to non-identifiability: only the parametric combinations

$$\sigma_{\mathrm{B}}^2, \; \sigma_{\mathrm{I}}^2 + \sigma_{\mathrm{A}}^2 - \sigma_{\mathrm{AB}}^2/(b-1) \;\text{ and }\; \sigma_{\mathrm{I}}^2 - \sigma_{\mathrm{A}}^2/(a-1) + \sigma_{\mathrm{AB}}^2/\{(a-1)(b-1)\}$$

are now separately identified in $\mathbf{Z}_i^+ \mathbf{B}^+ \mathbf{Z}_i^{+\mathrm{T}}$. This is in accordance with Crowder and Hand's (1990) Example 3.2, where only four mean squares were available in the anova table to estimate five variance components, the fifth being σ_e^2.

Had we started with $\mathbf{B}^+ = \text{diag}[\sigma_{\text{I}}^2 \; \sigma_{\text{A}}^2\mathbf{I}_a \; \sigma_{\text{B}}^2\mathbf{I}_b \; \sigma_{\text{AB}}^2\mathbf{I}_{ab}]$, we would have ended up with

$$(\mathbf{Z}_i^+ \mathbf{B}^+ \mathbf{Z}_i^{+\text{T}})_{jk} = \sigma_{\text{I}}^2 \mathbf{J}_b + \sigma_{\text{A}}^2 \mathbf{L}_j \mathbf{L}_k^{\text{T}} + \sigma_{\text{B}}^2 \mathbf{I}_b + \sigma_{\text{AB}}^2 \delta_{jk} \mathbf{I}_b$$
$$= (\sigma_{\text{B}}^2 + \sigma_{\text{AB}}^2 \delta_{jk})\mathbf{I}_b + (\sigma_{\text{I}}^2 + (\sigma_{\text{A}}^2 \delta_{jk})\mathbf{J}_b$$

Thus, again, inferences through $\mathbf{B}$ are essentially unaffected.

6.3.3 *Multilevel models*

Mixed-model anova for nested or hierarchical designs (see, for example, Scheffé, 1959, section 7.6) has been a focus of attention in education statistics for some years, as seen in the promotion of **multilevel models** (Goldstein, 1987). Consider, for instance, children in classes in schools. Let y_{ij} be the examination score of child j who is in class $c(j)$ in school i. Then, in the general model (6.1), $\mathbf{X}_i$ may contain child, class/teacher and school covariates, and the jth component of $\mathbf{Z}_i\mathbf{b}_i$ may take the form $a_i + a_{i,c(j)}$ in the multilevel context, where a_i is the random individual school effect, and $a_{i,c(j)}$ is the random class effect for class $c(j)$. In multilevel terms, the set-up has three levels: the child (level 1), the class (level 2), and the school (level 3). In our terms, the 'individual' here is the school and the 'repeated measures' arise from the children, one observation per child in the simple case. The data vector $\mathbf{y}_i$ contains the scores for all the tested children in school i, with subvectors representing different classes within the school. The vector $\mathbf{b}_i = (a_i, a_{i1}, \ldots, a_{ic})$ when there are c classes, and the jth row of $\mathbf{Z}_i$ has 1's in the first and $c(j)$th positions and 0's elsewhere. The usual assumption made in this context is that

$$\mathbf{B} = \begin{bmatrix} \sigma_{\text{S}}^2 & \mathbf{0}^{\text{T}} \\ \mathbf{0} & \sigma_{\text{C}}^2 \mathbf{I}_{\text{C}} \end{bmatrix}$$

i.e. that the a's are all uncorrelated with the indicated variances. Then, in the covariance structure (6.2), $\mathbf{Z}_i\mathbf{B}\mathbf{Z}_i^{\text{T}}$ has the form

$$\begin{bmatrix} (\sigma_{\text{S}}^2 + \sigma_{\text{C}}^2)\mathbf{J}_{11} & \sigma_{\text{S}}^2 \mathbf{J}_{12} & \sigma_{\text{S}}^2 \mathbf{J}_{12} & \cdots \\ \sigma_{\text{S}}^2 \mathbf{J}_{21} & (\sigma_{\text{S}}^2 + \sigma_{\text{C}}^2)\mathbf{J}_{22} & \sigma_{\text{S}}^2 \mathbf{J}_{23} & \cdots \\ \cdots & \cdots & \cdots & \cdots \end{bmatrix}$$

where $\mathbf{J}_{jk}$ is a $p_j \times p_k$ matrix of 1's, p_j being the number of children in class j. Hence, the correlation between classmates is $(\sigma_{\text{S}}^2 + \sigma_{\text{C}}^2)/(\sigma_{\text{S}}^2 + \sigma_{\text{C}}^2 + \sigma^2)$, and that between children in different classes in the same school is $\sigma_{\text{S}}^2/(\sigma_{\text{S}}^2 + \sigma_{\text{C}}^2 + \sigma^2)$.

The multilevel framework is obviously applicable in other contexts. Many systems have a similar hierarchical structure, e.g. patients < doctors < health areas and leaves < trees < fields < farms < regions. The basic framework needs to be extended to allow for designs in which children are crossed with teachers, say, as well as being nested within them, though this is straight forward.

6.3.4 *Inter-laboratory comparisons*

Education is not the only rich source of random effects models. A less well-developed area is that of comparison trials between different laboratories. To take a specific example (Crowder, 1992), suppose that certain chemical and physical properties of alloys are to be studied. Samples of the alloys are prepared and rolled into bars which are cut into pieces to be sent out to a number of laboratories for analysis. The bars can reasonably be regarded as a random sample from a hypothetical population, and so be represented by random effects. On the other hand, different alloy specifications would be a fixed effect. Again, the laboratories could be a few randomly selected from a population of such (random effects), or particular ones of special interest such as national institutes (fixed effects). In addition, the two factors, bars and laboratories, could be crossed or nested or some combination of these.

Suppose that b bars are made from an alloy and that each is cut into l pieces for distribution to l laboratories. Thus, each laboratory gets one piece of each bar, and the determination made by laboratory i on bar j is y_{ij}. The linear model, with laboratories as a fixed effect, is

$$y_{ij} = \mu + \alpha_i^{\mathrm{L}} + a_j^{\mathrm{B}} + e_{ij} \qquad (i = 1, \ldots, l;\; j = 1, \ldots, b)$$

Here, μ is the 'true' value of the property under study, α_i^{L} is the fixed laboratory effect and a_j^{B} is the random bar effect; we take the a_j^{B} and e_{ij} to have zero means, variances σ_{B}^2 and σ_e^2, and all $b + lb$ of them to be uncorrelated. The covariance structure for the data follows from

$$\begin{aligned} \operatorname{cov}(y_{ij}, y_{i'j'}) &= \operatorname{cov}(a_j^{\mathrm{B}} + e_{ij}, a_{j'}^{\mathrm{B}} + e_{i'j'}) \\ &= \delta_{jj'}(\sigma_{\mathrm{B}}^2 + \delta_{ii'}\sigma_e^2) \end{aligned}$$

Let $\mathbf{y}_j = (y_{1j}, \ldots, y_{1j})^{\mathrm{T}}$ be the set of determinations from bar j. Hence, the $\mathbf{y}_j$ ($j = 1, \ldots, b$) are uncorrelated and each has covariance matrix $\sigma_e^2\mathbf{I}_l + \sigma_{\mathrm{B}}^2\mathbf{J}_l$, of familiar compound symmetric form. It is the bars which are the 'individuals' here and, for once in this book, they are labelled by subscript j instead of i, though this will be but a temporary aberration. In the general two-stage framework the model can be written as

$$\mathbf{y}_j = \mathbf{X}_j\boldsymbol{\beta} + \mathbf{Z}_j\mathbf{b}_j + \mathbf{e}_j$$

with $\boldsymbol{\beta} = [\mu\ \alpha_1^{\mathrm{L}} \cdots \alpha_l^{\mathrm{L}})^{\mathrm{T}}$, $\mathbf{X}_j = [\mathbf{1}_l\ \mathbf{I}_l]$, $\mathbf{b}_j = (a_j^{\mathrm{B}})$ and $\mathbf{Z}_j = \mathbf{1}_l$. The general formula $\mathbf{Z}_j\mathbf{B}\mathbf{Z}_j^{\mathrm{T}} + \mathbf{E}_j$, with $\mathbf{B} = \sigma_{\mathrm{B}}^2\mathbf{I}_l$ and $\mathbf{E}_j = \sigma_e^2\mathbf{I}_l$, then produces the covariance matrix as given.

Now suppose that the l laboratories are a random selection of such. The model becomes

$$y_{ij} = \mu + a_i^{\mathrm{L}} + a_j^{\mathrm{B}} + e_{ij} \quad (i = 1, \ldots, l; j = 1, \ldots, b)$$

where now a_i^{L} is the random laboratory effect. With assumptions and notation similar to the previous ones, we have

$$\begin{aligned}\operatorname{cov}(y_{ij}, y_{i'j'}) &= \operatorname{cov}(a_i^{\mathrm{L}} + a_j^{\mathrm{B}} + e_{ij}, a_{i'}^{\mathrm{L}} + a_{j'}^{\mathrm{B}} + e_{i'j'}) \\ &= \delta_{ii'}\sigma_{\mathrm{L}}^2 + \delta_{jj'}\sigma_{\mathrm{B}}^2 + \delta_{ii'}\delta_{jj'}\sigma_e^2\end{aligned}$$

The resulting covariance matrix for the whole $lb \times 1$ data vector $\mathbf{y} = [\mathbf{y}_1^{\mathrm{T}} \dots \mathbf{y}_b^{\mathrm{T}}]^{\mathrm{T}}$ has block structure: we have var $(\mathbf{y}_j) = (\sigma_e^2 + \sigma_{\mathrm{L}}^2)\mathbf{I}_l + \sigma_{\mathrm{B}}^2\mathbf{J}_l$ for the diagonal blocks, and $\operatorname{cov}(\mathbf{y}_j, \mathbf{y}_{j'}) = \sigma_{\mathrm{L}}^2\mathbf{I}_l$ for the off-diagonal ones with $j \neq j'$. In short,

$$\operatorname{var}(\mathbf{y}) = \mathbf{I}_b \otimes (\sigma_e^2\mathbf{I}_l + \sigma_{\mathrm{B}}^2\mathbf{J}_l) + \mathbf{J}_b \otimes (\sigma_{\mathrm{L}}^2\mathbf{I}_l)$$

reminiscent of the form in section 6.3.2. Note that there is only one 'individual' here, it not being possible to split the data vector tidily into uncorrelated subvectors. In terms of the two-stage linear model we have

$$\mathbf{y} = \mathbf{X}\boldsymbol{\beta} + \mathbf{Z}_1\mathbf{a}^{\mathrm{L}} + \mathbf{Z}_2\mathbf{a}^{\mathrm{B}} + \mathbf{e}$$

where $\mathbf{X} = \mathbf{1}_{lb}$, $\boldsymbol{\beta} = (\mu)$, and the usual term $\mathbf{Z}_i\mathbf{b}_i$ has been written in expanded form as $\mathbf{Z}_1\mathbf{a}^{\mathrm{L}} + \mathbf{Z}_2\mathbf{a}^{\mathrm{B}}$ with $\mathbf{a}^{\mathrm{L}} = [a_1^{\mathrm{L}} \dots a_l^{\mathrm{L}}]^{\mathrm{T}}, \mathbf{a}^{\mathrm{B}} = [a_1^{\mathrm{B}} \dots a_b^{\mathrm{B}}]^{\mathrm{T}}$, $\mathbf{Z}_1 = \mathbf{1}_b \otimes \mathbf{I}_l$ and $\mathbf{Z}_2 = \mathbf{I}_b \otimes \mathbf{1}_l$. The resulting covariance matrix is

$$\begin{aligned}\mathbf{Z}_1\operatorname{var}(\mathbf{a}^{\mathrm{L}})\mathbf{Z}_1^{\mathrm{T}} + \mathbf{Z}_2\operatorname{var}(\mathbf{a}^{\mathrm{B}})\mathbf{Z}_2^{\mathrm{T}} + \mathbf{E} &= \sigma_{\mathrm{L}}^2\mathbf{Z}_1\mathbf{Z}_1^{\mathrm{T}} + \sigma_{\mathrm{B}}^2\mathbf{Z}_2\mathbf{Z}_2^{\mathrm{T}} + \sigma_e^2\mathbf{I}_{lb} \\ &= \sigma_{\mathrm{L}}^2\mathbf{J}_b \otimes \mathbf{I}_l + \sigma_{\mathrm{B}}^2\mathbf{I}_b \otimes \mathbf{J}_l + \sigma_e^2\mathbf{I}_{lb}\end{aligned}$$

agreeing with the formula given above.

For a final example from this field, let us allow for different methods of measurement of the property under investigation. Consider a full $l \times b \times m$ factorial layout, in which laboratory i has used method k on bar j to produce the determination y_{ijk}. With methods as a fixed effect, and laboratories, bars and the interaction laboratories by methods as random effects, the linear model is

$$y_{ijk} = \mu + a_i^{\mathrm{L}} + a_j^{\mathrm{B}} + \beta_k^{\mathrm{M}} + a_{ik}^{\mathrm{LM}} + e_{ijk} \quad (i = 1, \dots, l;\ j = 1, \dots, b; k = 1, \dots, m)$$

With assumptions and notation similar to those above, we have

$$\operatorname{cov}(y_{ijk}, y_{i'j'k'}) = \delta_{ii'}\sigma_{\mathrm{L}}^2 + \delta_{jj'}\sigma_{\mathrm{B}}^2 + \delta_{ii'}\delta_{kk'}\sigma_{kk'} + \delta_{ii'}\delta_{jj'}\delta_{kk'}\sigma_e^2)$$

where $\sigma_{kk'} = \operatorname{cov}(a_{ik}^{\mathrm{LM}}, a_{ik'}^{\mathrm{LM}})$. Let

$$\mathbf{y}_i = [y_{i11} \dots y_{i1m} | y_{i21} \dots y_{i2m} | \dots | y_{ib1} \dots y_{ibm}]^{\mathrm{T}}$$

be the data vector from laboratory i. Then the two-stage linear model for the whole $lbm \times 1$ data vector $\mathbf{y} = [\mathbf{y}_1^{\mathrm{T}} \dots \mathbf{y}_l^{\mathrm{T}}]^{\mathrm{T}}$ can be written as

$$\mathbf{y} = \mathbf{X}\boldsymbol{\beta} + \mathbf{Z}_1\mathbf{a}^{\mathrm{L}} + \mathbf{Z}_2\mathbf{a}^{\mathrm{B}} + \mathbf{Z}_3\mathbf{a}^{\mathrm{LM}} + \mathbf{e}$$

Here, $\mathbf{X}\boldsymbol{\beta}$ represents the fixed effects (methods) and the random effects are $\mathbf{a}^{\mathrm{L}}$ and $\mathbf{a}^{\mathrm{B}}$ as given above, and

$$\mathbf{a}^{\mathrm{LM}} = [a_{11}^{\mathrm{LM}} \ldots a_{1m}^{\mathrm{LM}} | \ldots | a_{l1}^{\mathrm{LM}} \ldots a_{lm}^{\mathrm{LM}})^{\mathrm{T}}$$

with corresponding $\mathbf{Z}$ matrices

$$\mathbf{Z}_1(lbm \times 1) = \mathbf{I}_l \otimes \mathbf{1}_b \otimes \mathbf{1}_m$$

$$\mathbf{Z}_2(lbm \times b) = \mathbf{1}_l \otimes \mathbf{I}_b \otimes \mathbf{1}_m$$

$$\mathbf{Z}_3(lbm \times lm) = \mathbf{I}_l \otimes \mathbf{1}_b \otimes \mathbf{I}_m$$

With assumptions analogous to those used above, and with $\mathbf{C}_m = \mathrm{var}(\mathbf{a}^{\mathrm{LM}})$, we find that

$$\begin{aligned}\boldsymbol{\Sigma} = \mathrm{var}(\mathbf{y}) &= \mathbf{Z}_1(\sigma_{\mathrm{L}}^2 \mathbf{I}_l)\mathbf{Z}_1^{\mathrm{T}} + \mathbf{Z}_2(\sigma_{\mathrm{B}}^2 \mathbf{I}_b)\mathbf{Z}_2^{\mathrm{T}} + \mathbf{Z}_3(\mathbf{I}_l \otimes \mathbf{C}_m)\mathbf{Z}_3^{\mathrm{T}} + \sigma_e^2 \mathbf{I}_{lbm'} \\ &= \sigma_{\mathrm{L}}^2 \mathbf{I}_l \otimes \mathbf{J}_b \otimes \mathbf{J}_m + \sigma_{\mathrm{L}}^2 \mathbf{J}_l \otimes \mathbf{I}_b \otimes \mathbf{J}_m + \mathbf{I}_l \otimes \mathbf{J}_b \otimes \mathbf{C}_m + \sigma_e^2 \mathbf{I}_{lbm}\end{aligned}$$

Although fairly involved, the covariance structures derived above are still only of random effects or unstructured type: they do not, as yet, contain any genuine longitudinal component. This is easily rectified. Suppose that the study is ongoing, with regular reruns at intervals designed to monitor the laboratories or the production of the alloy. Then, the covariance matrix for the whole set of data over time would have diagonal blocks like $\boldsymbol{\Sigma}$ given above. The off-diagonal blocks would need to reflect covariation over time, the jth run being performed at time t_j, say. One simple suggestion would be to represent the (j, k)th block as $\rho^{|t_j - t_k|}\boldsymbol{\Sigma}$, by analogy with the Markov structure of section 6.2.4. Thus, the complete covariance matrix would be $\mathbf{R} \otimes \boldsymbol{\Sigma}$, where $R_{jk} = \rho^{|t_j - t_k|}$.

Another, quite different source of longitudinal variation in this context is the position along a bar. We have seen some studies in which such an effect has been detected. Then, for random variation with distance along the bar, one could adopt a covariance structure of Markov type (section 6.2.4). Combination of this with the ongoing monitoring described above would produce repeated measurements over two dimensions.

One of the issues which arises in dealing with such data is the size of the covariance matrices, often too large for routine numerical work. On the other hand, the regular structures which crop up are often amenable to algebraic reduction, though this can be tedious to do by hand. It would be useful to develop software for such operations, perhaps employing one of the popular computer algebra packages.

6.3.5 *Growth curves*

Much of the classic work on two-stage models concerns the fitting of polynomial curves to animal growth measurements over time. Here $\mathbf{X}_i$ has jth row $[1\, x_{ij}\, x_{ij}^2 \ldots x_{ij}^{q-1}]$, where x_{ij} is the age of the animal, or the 'orthogonal polynomial' version of this. Also, in that literature, $\mathbf{X}_i$ is usually equal to $\mathbf{Z}_i$ or, more generally, to $\mathbf{Z}_i\mathbf{A}_i$ for some $\mathbf{A}_i$. This arises from $\mathbf{y}_i = \mathbf{Z}_i\boldsymbol{\beta}_i + \mathbf{e}_i$ at the first stage, and then $\boldsymbol{\beta}_i$ having distribution $N_q(\mathbf{A}_i\boldsymbol{\beta}, \mathbf{B})$ at the second, where $\mathbf{A}_i$ incorporates covariates and design indicators and $\mathbf{B}$ is normally left unstructured. Thus $E(\mathbf{y}_i) = \mathbf{X}_i\boldsymbol{\beta}$, with $\mathbf{X}_i = \mathbf{Z}_i\mathbf{A}_i$, and $\text{var}(\mathbf{y}_i) = \mathbf{Z}_i\mathbf{B}\mathbf{Z}_i^{\mathrm{T}} + \mathbf{E}_i$.

For polynomial growth curves many authors have noted the possibility that individual curves may be of higher degree than the population mean curve. This occurs when the later components of $\boldsymbol{\beta}_i$ have zero means.

Laird and Ware (1982) drew the following distinction between growth curves and repeated measures. In the former, the natural development or ageing process of the individual tends to be monitored without intervention, and comparisons are made between different groups. In the latter, individual characteristics tend to be constant over the period of study while the individual undergoes different treatments or conditions. Of course, the distinction is not rigid, and one does not have to look far for exceptions.

6.4 Further reading

Dempster's (1972) method of 'covariance selection' could be regarded as a generalization of Gabriel's (1962) antedependence structures. In both models certain elements of $\mathbf{E}^{-1}$ are set to zero: in Gabriel's these are strictly the elements off the diagonal band, whereas in Dempster's the choice is open to wider considerations. Byrne and Arnold (1983) used an antedependence structure in testing for a specified μ-profile in the single-sample case. Kenward (1987) showed how the antedependence structure could be exploited for routine data analysis. He re-expressed certain likelihood ratio tests of fit of the model in the form of analyses of covariance and applied his methods to some cattle growth data.

Scheffé (1959, Chapters 7 and 8) and Searle (1971, Chapters 9 and 10) contain detailed discussion and analysis for random effects models in anova and regression. On the question of different assumptions for the covariance structure in mixed-model anova, relevant work includes Hocking (1973) and Hartley and Searle (1969).

Goldstein (1979; 1986a; 1986b; 1987) and others have been applying multi-level models to education data for some years. Goldstein's (1987) book gives an extended account, with thorough discussion of all aspects including more elaborate models.

Srivastava (1984) and Srivastava and Keen (1988) considered inference about the parameters of the model $\mathbf{y}_i \sim N(\boldsymbol{\mu}_i, \boldsymbol{\Sigma}_i)$ with $\boldsymbol{\mu}_i^{\mathrm{T}} = [\mu_{\mathrm{m}}\, \mu_{\mathrm{c}}\mathbf{1}_p^{\mathrm{T}}]$, representing a mother and a number of children, and

$$\boldsymbol{\Sigma}_i = \begin{bmatrix} \sigma_{\mathrm{m}}^2 & \sigma_{\mathrm{mc}}\mathbf{1}_p^{\mathrm{T}} \\ \sigma_{\mathrm{mc}}\mathbf{1}_p & \sigma_{\mathrm{c}}^2\boldsymbol{\Phi}_i \end{bmatrix}$$

with $\boldsymbol{\Phi}_i = (1 - \rho_{\mathrm{cc}})\mathbf{I}_p + \rho_{\mathrm{cc}}\mathbf{J}_p$ (an equicorrelated structure).

An important class of covariance forms is that of 'linear structure' introduced by Anderson (1969; 1970; 1973). Here $\boldsymbol{\Sigma}$ or $\boldsymbol{\Sigma}^{-1}$ is expressible as a sum of terms $\phi_j\mathbf{G}_j$ where the matrices $\mathbf{G}_j$ are known but the scalars ϕ_j are not. Rogers and Young (1977) give some further results for this set-up, and Andrade and Helms (1984) suggest computation based on the EM algorithm.

Longford's (1993) book is concerned with random effects of the types described in section 6.3, particularly as occurring in multi-level models.

Part Two: Non-normal Error Distributions

CHAPTER 7

Continuous non-normal measures: Gaussian estimation

This chapter contains some developments of the material of Chapters 4 and 5. The linear models and covariance structures covered there are wide-ranging and applicable in many different contexts. However, it is fair to ask how critical is the third ingredient, i.e. the assumption of normal distributions, and whether this is really needed. The approach described in this chapter, Gaussian estimation (due to Whittle, 1961), is to perform the same computations as before, obtaining parameter estimates from a normal likelihood, but then to see what modifications are necessary to make the inferences valid when normality might not obtain. Put like that, one might wonder why this is not done routinely. Perhaps it should be done more often, but it does not give an entirely free ride. The modifications themselves depend on higher-order properties of the observations than the means and covariances: they depend on skewness and kurtosis, and these usually have to be estimated by reusing the sample. This process naturally introduces more random error, the effects of which could wipe out any advantage gained. As usual, it comes down to statistical judgement whether such modifications are likely to be beneficial for the problem at hand. It would be useful if some case studies were to appear in the literature so that more experience of Gaussian estimation in a variety of practical applications could be accumulated.

7.1 Gaussian estimation: theory

7.1.1 Generalities

Suppose that we have a random sample $\mathbf{y} = [y_1, \ldots, y_n]$ of observations believed to arise from some distribution with density $f(y; \boldsymbol{\theta})$, where $\boldsymbol{\theta}$ is the parameter of the distribution. We can use the form of $f(y; \boldsymbol{\theta})$, say by maximizing the corresponding likelihood function or by using some other characteristic of the distribution, to obtain an estimator $\tilde{\boldsymbol{\theta}}$. This estimator is defined by a 'numerical recipe', an explicit formula or an iterative procedure which specifies how the estimate is obtained from the sample.

Study of $\tilde{\boldsymbol{\theta}}$, as a function of the sample elements, may show that it has some attractive properties independent of the true distribution from which the sample arose. The estimator might, for example, be unbiased or consistent for the corresponding parameter of the true source distribution of the y_i, irrespective of the precise form of this distribution, i.e. whether or not it has density of form $f(y;\boldsymbol{\theta})$. If this does turn out to be the case, then it seems natural to use $\tilde{\boldsymbol{\theta}}$, the numerical recipe, without worrying about the true form of the source distribution.

The role of $f(y;\boldsymbol{\theta})$ here has been solely to lead to an estimator which has attractive properties. Once this estimator has been found, $f(y;\boldsymbol{\theta})$ can be discarded. The numerical recipe obtained has, with luck, yielded an estimator with those generally attractive properties, general in the sense that they are achieved regardless of the true underlying distribution from which the data arose.

To make parametric inferences we need to estimate such things as the variance of $\tilde{\boldsymbol{\theta}}$. We could derive such estimates from calculations based on the original $f(y;\boldsymbol{\theta})$, but this would negate the advantage that we have just won. Instead, we can make use of the fact that $\tilde{\boldsymbol{\theta}}$ is simply a function of the sample data. Then, such estimates can be derived directly from the form of this function, and will usually involve sample moments such as s^2, the sample variance. The papers by White (1982) and Royall (1986) are concerned with these ideas.

■ *Illustration 7.1* Suppose that we believe the data to have arisen from an exponential distribution with density $f(y;\theta) = \theta^{-1} e^{-y/\theta}$ on $(0, \infty)$, where θ is the mean of the distribution. The log-likelihood function is then

$$Q(\mathbf{y};\theta) = \log \prod_{i=1}^{n} (\theta^{-1} e^{-y_i/\theta}) = -n \log \theta - n\bar{y}/\theta$$

where $\bar{y} = n^{-1} \sum y_i$ is the sample mean. The corresponding score function is

$$q(\mathbf{y};\theta) = \partial Q(y;\theta) / \partial \theta = n\theta^{-2} (\bar{y} - \theta)$$

and so the likelihood estimating equation $q(\mathbf{y};\theta) = 0$ yields the maximum likelihood estimator $\hat{\theta} = \bar{y}$. Now, $\bar{y}$ has some attractive properties as an estimate of the mean of a distribution, whatever the distribution concerned, exponential or otherwise. It is, for example, unbiased. This means that $\bar{y}$ would be a 'good' estimate of the mean even if we were mistaken in supposing that the source distribution was exponential.

To make inferences from $\hat{\theta}$ we will probably need an estimate of var $(\hat{\theta})$, its variance. For exponential distributions var $(y) = \theta^2$, so we could estimate var$(\bar{y})$ by $\bar{y}^2/n$. However, this relies upon the exponential assumptions being correct. What we really want is a variance estimate which is not tied to this assumption.

We pointed out above that we were happy to use $\bar{y}$ as our estimate whatever the true distribution which might have produced the data. Now, $\bar{y}$ is defined solely as a function of the data, by the formula $\bar{y} = \sum y_i/n$. From this formula alone, $\text{var}(\bar{y}) = \sigma^2/n$, where $\sigma^2 = \text{var}(y_i)$, and the corresponding estimate s^2/n for $\text{var}(\hat{\theta})$ does not depend on the exponential distributional assumption. ■

To summarize, the strategy is as follows:

1. Use some estimating function to yield an estimator $\tilde{\boldsymbol{\theta}}$. This estimating function might be obtained from a log-likelihood based on some assumed distribution, but it need not be. The resulting estimator will simply be a numerical recipe, which can be applied to any sample, whatever the true source distribution.
2. To make inferences, estimate the variance (or whatever is required) of $\tilde{\boldsymbol{\theta}}$ by basing the calculations directly on the numerical recipe specified in 1.

The following asymptotic (large-sample) results from the general theory of estimating functions (e.g. Crowder, 1986) will be used repeatdly throughout this and the subsequent chapters. Let the estimator $\tilde{\boldsymbol{\theta}}$ of the vector parameter $\boldsymbol{\theta}$ be defined as a solution of the vector estimating equation $\mathbf{q}(\mathbf{y};\boldsymbol{\theta}) = \mathbf{0}$. Then, provided that $E\{\mathbf{q}(\mathbf{y};\boldsymbol{\theta})\} \to \mathbf{0}$ for all $\boldsymbol{\theta}$ as $n \to \infty$, i.e. the estimating equation is asymptotically unbiased, and provided that certain regularity conditions hold, $\tilde{\boldsymbol{\theta}}$ is unique, consistent, and has asymptotic distribution $N(\boldsymbol{\theta}, \mathbf{V}_\theta)$, where

$$\mathbf{V}_\theta = \mathbf{D}^{-1}\mathbf{C}(\mathbf{D}^{-1})^{\mathrm{T}}, \quad \mathbf{C} = \text{cov}\{\mathbf{q}(\mathbf{y};\boldsymbol{\theta}), \mathbf{D} = E\{-\mathbf{q}'(\mathbf{y};\boldsymbol{\theta})\}$$

and $\mathbf{q}'(\mathbf{y};\boldsymbol{\theta})$ is the matrix of derivatives with (j,k)th element $\partial q_j/\partial\theta_k$. If $\mathbf{q}(\mathbf{y};\boldsymbol{\theta})$ is the true score function, then, in the 'regular' case, $E\{\mathbf{q}(\mathbf{y};\boldsymbol{\theta})\} = \mathbf{0}$ for all n, and $\mathbf{C} = \mathbf{D}$, the Fisher information matrix. The form $\mathbf{D}^{-1}\mathbf{C}(\mathbf{D}^{-1})^{\mathrm{T}}$ has come to be known as the **information sandwich**; an early serving of it in a general setting occurs in Huber (1967).

The asymptotic covariance matrix $\mathbf{V}_\theta$ of $\tilde{\boldsymbol{\theta}}$, in estimated form as shown below, can be used to test parametric hypotheses and generate confidence regions for parametric functions in the usual ways. For instance, suppose that the linear parametric function $\mathbf{a}^{\mathrm{T}}\boldsymbol{\theta}$ is of interest, where $\mathbf{a}$ is some given vector. The corresponding estimate is $\mathbf{a}^{\mathrm{T}}\tilde{\boldsymbol{\theta}}$, with variance $\mathbf{a}^{\mathrm{T}}\mathbf{V}_\theta\mathbf{a}$ and standard error the square root of this. Again, to test a linear hypothesis $\mathbf{H}\boldsymbol{\theta} = \mathbf{h}$, where $\mathbf{H}$ is a given $r \times q$ matrix of rank $r \leqslant q$ and $\mathbf{h}$ a given vector, the so-called **Wald test statistic**

$$W = (\mathbf{H}\tilde{\boldsymbol{\theta}} - \mathbf{h})^{\mathrm{T}}(\mathbf{H}\mathbf{V}_\theta\mathbf{H}^{\mathrm{T}})^{-1}(\mathbf{H}\tilde{\boldsymbol{\theta}} - \mathbf{h})$$

would be referred to χ_r^2 (Silvey, 1970, section 7.1).

Alternative procedures to the Wald test exist. One which is not strictly valid in this context is a **likelihood ratio** analogue based on $Q(\mathbf{y};\boldsymbol{\theta})$: When $Q(\mathbf{y};\boldsymbol{\theta})$ is not the true log-likelihood the asymptotic distribution does not reduce to a simple chi-square. (In any case, sample sizes are not usually infinite.) However, the test is invariant to parameter transformation, i.e. not dependent on the way in which the model is formulated in terms of parameters. We will not, therefore, eschew comparisons of Gaussian log-likelihoods as a vehicle for inference altogether, but just drive with due care and attention.

Throughout the present chapter $Q(\mathbf{y};\boldsymbol{\theta})$ will be taken to be a Gaussian log-likelihood, and $\mathbf{q}(\mathbf{y};\boldsymbol{\theta})$ to be the corresponding Gaussian score function. These would be 'correct' if the observations were known to arise from a normal distribution as in Chapter 4. However, it is not now assumed that the observations are normally distributed. In fact, we do not make any particular distributional assumption, the observations could even arise from a discrete distribution. (In Crowder (1985) Gaussian estimation was seen to work surprisingly well for correlated binomial data.) However, it is natural to suppose that Gaussian estimation is likely to be more efficient the closer the true distribution is to normal.

7.1.2 Univariate observations

Suppose that data are available comprising univariate observations y_i $(i=1,\ldots,n)$ with means $\mu_i = E(y_i)$ and variances $\sigma_i^2 = \text{var}\,(y_i)$ depending on a $q \times 1$ parameter vector $\boldsymbol{\theta}$.

■ *Illustration 7.2* For a basic straight-line regression model we have $\mu_i = \alpha + \beta x_i$ and, in the homoscedastic case, $\sigma_i^2 = \sigma^2$. Thus, in the general scheme, $\boldsymbol{\theta} = [\alpha\,\beta\,\sigma]$ with $q = 3$ components. If, instead of homoscedasticity, there were constant coefficients of variation $\phi^{1/2}$, we would have $\sigma_i^2 = \phi\mu_i^2$, and then $\boldsymbol{\theta} = [\alpha\,\beta\,\phi]$. In a more elaborate model σ_i^2 might comprise both types of variability: $\sigma_i^2 = \sigma^2 + \phi\mu_i^2$. Then μ_i depends on α and β only, whereas σ_i^2 depends on all four components of $\boldsymbol{\theta} = [\alpha\,\beta\,\sigma\,\phi]$. ■

To calculate the Gaussian estimate for $\boldsymbol{\theta}$ we maximize the normal distribution log-likelihood

$$Q(\mathbf{y};\boldsymbol{\theta}) = \log \prod_{i=1}^{n} [(2\pi\sigma_i^2)^{-1/2} \exp\{-(y_i - \mu_i)^2/2\sigma_i{}^2\}] \tag{7.1}$$

Differentiation of Q with respect to $\boldsymbol{\theta}$ produces the Gaussian score function

$$\mathbf{q}(\mathbf{y};\boldsymbol{\theta}) = \sum_{i=1}^{n} [\sigma_i{}^{-2}(y_i - \mu_i)\boldsymbol{\mu}_i' + \sigma_i{}^{-3}\{(y_i - \mu_i)^2 - \sigma_i^2\}\boldsymbol{\sigma}_i'] \tag{7.2}$$

where $\boldsymbol{\mu}_i'(q \times 1) = \partial\mu_i/\partial\boldsymbol{\theta}$ and $\boldsymbol{\sigma}_i'(q \times 1) = \partial\sigma_i/\partial\boldsymbol{\theta}$. Under some standard conditions (section 7.1.1), in particular that the specified parametric forms of μ_i and σ_i are correct, the resulting estimator $\tilde{\boldsymbol{\theta}}$ is consistent and asymptotically normal with asymptotic covariance matrix $\mathbf{V}_\theta = \mathbf{D}^{-1}\mathbf{C}(\mathbf{D}^{-1})^{\mathrm{T}}$, where

$$\mathbf{D} = E\{-\mathbf{q}'(\mathbf{y};\boldsymbol{\theta})\} = \sum \sigma_i{}^{-2}\,(\boldsymbol{\mu}_i'\boldsymbol{\mu}_i'^{\mathrm{T}} + 2\boldsymbol{\sigma}_i'\boldsymbol{\sigma}_i'^{\mathrm{T}}),$$

$$\mathbf{C} = \text{cov}\{\mathbf{q}(\mathbf{y};\boldsymbol{\theta})\} = \sum \sigma_i^{-2}\{\boldsymbol{\mu}_i'\boldsymbol{\mu}_i'^{\mathrm{T}} + \gamma_{1i}(\boldsymbol{\mu}_i'\boldsymbol{\sigma}_i'^{\mathrm{T}} + \boldsymbol{\sigma}_i'\boldsymbol{\mu}_i'^{\mathrm{T}}) + (\gamma_{2i}+2)\boldsymbol{\sigma}_i'\boldsymbol{\sigma}_i'^{\mathrm{T}}\} \tag{7.3}$$

γ_{1i} and γ_{2i} are respectively the skewness and kurtosis of y_i. The matrix $\mathbf{D}$ will be estimable from the sample by inserting $\tilde{\boldsymbol{\theta}}$ for $\boldsymbol{\theta}$ in the formulae for σ_i, $\boldsymbol{\mu}_i'$ and $\boldsymbol{\sigma}_i'$. This estimate is model-based to the extent that it depends on the model specified for μ_i and σ_i but not on a normal distribution model for the observations. If the forms of γ_{1i} and γ_{2i} are known in terms of $\boldsymbol{\theta}$ these can be used for $\mathbf{C}$ with $\boldsymbol{\theta}$ replaced by $\tilde{\boldsymbol{\theta}}$. Otherwise, more usually, $\mathbf{C}$ can be estimated by inserting the sample estimates, $(y_i - \tilde{\mu}_i)^3/\tilde{\sigma}_i^3$ and $(y_i - \tilde{\mu}_i)^4/\tilde{\sigma}_i^4 - 3$, for γ_{1i} and γ_{2i}. Then the estimate for $\mathbf{C}$ will be partly model-based, through μ_i and σ_i, and partly sample-based, through the y_i.

If the y_i were actually normally distributed, γ_{1i} and γ_{2i} would both be zero and then $\mathbf{D} = \mathbf{C}$ in (7.3), giving $\mathbf{V}_{\boldsymbol{\theta}} = \mathbf{D}^{-1}$. This is the form used in previous chapters and is the one produced by a standard normal-models computer package. The use of the form $\mathbf{D}^{-1}\mathbf{C}(\mathbf{D}^{-1})^{\mathrm{T}}$ for $\mathbf{V}_{\boldsymbol{\theta}}$, instead of just $\mathbf{D}^{-1}$, effectively gives some protection against a false assumption of normality; it gives a more robust estimator for $\mathbf{V}_{\boldsymbol{\theta}}$. However, the sample size must be quite large to give a reliable estimate of $\mathbf{C}$ because it involves third and fourth moments of y_i.

■ *Illustration 7.2 (cont'd)* Take $\mu_i = \alpha + \beta x_i$ and $\sigma_i^2 = \sigma^2 + \phi\mu_i^2$, with $\boldsymbol{\theta} = [\alpha\ \beta\ \sigma\ \phi]$. Then we have $\boldsymbol{\mu}_i' = [1\ x_i\ 0\ 0]^{\mathrm{T}}$, $\boldsymbol{\sigma}_i' = [2\phi\mu_i\ 2\phi\mu_i x_i\ 2\sigma\ \mu_i^2]^{\mathrm{T}}$, and

$$\boldsymbol{\mu}_i'\boldsymbol{\mu}_i'^{\mathrm{T}} = \begin{bmatrix} 1 & x_i & 0 & 0 \\ x_i & x_i^2 & 0 & 0 \\ 0 & 0 & 0 & 0 \\ 0 & 0 & 0 & 0 \end{bmatrix}, \boldsymbol{\sigma}_i'\boldsymbol{\sigma}_i'^{\mathrm{T}} = \begin{bmatrix} 4\phi^2\mu_i^2 & 4\phi^2\mu_i^2 x_i & 4\phi\sigma\mu_i & 2\phi\mu_i^3 \\ 4\phi^2\mu_i^2 x_i & 4\phi^2\mu_i^2 x_i^2 & 4\phi\sigma\mu_i x_i & 2\phi\mu_i^3 x_i \\ 4\phi\sigma\mu_i & 4\phi\sigma\mu_i x_i & 4\sigma^2 & 2\sigma\mu_i^2 \\ 2\phi\mu_i^3 & 2\phi\mu_i^3 x_i & 2\sigma\mu_i^2 & \mu_i^4 \end{bmatrix},$$

$$\boldsymbol{\mu}_i'\boldsymbol{\sigma}_i'^{\mathrm{T}} = \begin{bmatrix} 2\phi\mu_i & 2\phi\mu_i x_i & 2\sigma & \mu_i^2 \\ 2\phi\mu_i x_i & 2\phi\mu_i x_i^2 & 2\sigma x_i & \mu_i^2 x_i \\ 0 & 0 & 0 & 0 \\ 0 & 0 & 0 & 0 \end{bmatrix}$$ ■

For many models used in practice the parameters governing μ_i and σ_i are distinct, as in Chapter 4. Thus, $\boldsymbol{\theta} = [\boldsymbol{\beta}\ \boldsymbol{\tau}]$ with $\mu_i = \mu_i(\boldsymbol{\beta})$ and $\sigma_i = \sigma_i(\boldsymbol{\tau})$, where the $\boldsymbol{\beta}$ components are the regression parameters or coefficients, and the $\boldsymbol{\tau}$ components govern the variances. Suppose that $\boldsymbol{\beta}$ is a $q_{\boldsymbol{\beta}} \times 1$ vector and $\boldsymbol{\tau}$ is a $q_{\boldsymbol{\tau}} \times 1$ vector, with $q_{\boldsymbol{\beta}} + q_{\boldsymbol{\tau}} = q$. Then $\boldsymbol{\mu}_i'^{\mathrm{T}} = [\boldsymbol{\mu}_{i\beta}'^{\mathrm{T}}\ \mathbf{0}]$ and $\boldsymbol{\sigma}_i'^{\mathrm{T}} = [\mathbf{0}\ \ \boldsymbol{\sigma}_{i\tau}'^{\mathrm{T}}]$, where $\boldsymbol{\mu}_{i\beta} = \partial\mu_i/\partial\boldsymbol{\beta}$ and $\boldsymbol{\sigma}_{i\tau}' = \partial\sigma_i/\partial\boldsymbol{\tau}$. In consequence, inspection of (7.2) shows that $\mathbf{q}(\mathbf{y};\boldsymbol{\theta})$ decouples into component vectors $\mathbf{q}_{\boldsymbol{\beta}}(\mathbf{y};\boldsymbol{\theta})$ and $\mathbf{q}_{\boldsymbol{\tau}}(\mathbf{y};\boldsymbol{\theta})$:

$$\mathbf{q}_{\boldsymbol{\beta}}(\mathbf{y};\boldsymbol{\theta}) = \sum \sigma_i^{-2}(y_i - \mu_i)\boldsymbol{\mu}_{i\boldsymbol{\beta}}'$$

$$\mathbf{q}_{\boldsymbol{\tau}}(\mathbf{y};\boldsymbol{\theta}) = \sum \sigma_i^{-3}\{(y_i - \mu_i)^2 - \sigma_i^2\}\boldsymbol{\sigma}_{i\tau}' \tag{7.4}$$

The $\boldsymbol{\beta}$ component, $\mathbf{q}_\beta(\mathbf{y};\boldsymbol{\theta})$, is essentially the quasi-likelihood score function (section 9.1). The use of $\mathbf{q}_\beta(\mathbf{y};\boldsymbol{\theta})$ even in the non-decoupled case, i.e. where $\mu_i = \mu_i(\boldsymbol{\beta})$ and $\sigma_i = \sigma_i(\boldsymbol{\beta}, \boldsymbol{\tau})$, is a useful variant when there is some uncertainty in the parametric specification of σ_i^2. The reason for this is that for (7.2) to provide an unbiased estimating equation we need both $E(y_i - \mu_i) = 0$ and $E\{(y_i - \mu_i)^2 - \sigma_i^2 = 0$, whereas only the former condition is needed for $\mathbf{q}_\beta(\mathbf{y};\boldsymbol{\theta})$. This use of $\mathbf{q}_\beta(\mathbf{y};\boldsymbol{\theta})$ was suggested by Williams (1959, section 4.5), though for cases in which the term involving $\partial\sigma/\partial\boldsymbol{\beta}$ makes a negligible contribution in (7.2): to quote him, 'In general, this quantity is small, and so the information may be ignored', and 'For this reason, and also for computational simplicity, the least-squares equations rather than the maximum likelihood equations are generally used'.

The matrices $\mathbf{C}$ and $\mathbf{D}$ associated with the estimating functions (7.4) can be partitioned as follows:

$$\mathbf{D} = \begin{bmatrix} \mathbf{D}_{\beta\beta} & \mathbf{D}_{\beta\tau} \\ \mathbf{D}_{\tau\beta} & \mathbf{D}_{\tau\tau} \end{bmatrix} \quad \text{and} \quad \mathbf{C} = \begin{bmatrix} \mathbf{C}_{\beta\beta} & \mathbf{C}_{\beta\tau} \\ \mathbf{C}_{\tau\beta} & \mathbf{C}_{\tau\tau} \end{bmatrix} \tag{7.5}$$

with

$$\mathbf{D}_{\beta\beta} = \sum \sigma_i^{-2} \boldsymbol{\mu}'_{i\beta} \boldsymbol{\mu}'^{\mathrm{T}}_{i\beta} \qquad \mathbf{D}_{\tau\tau} = 2\sum \sigma_i^{-2} \boldsymbol{\sigma}'_{i\tau} \boldsymbol{\sigma}'^{\mathrm{T}}_{i\tau} \quad \mathbf{D}_{\beta\tau} = \mathbf{0} \quad \mathbf{D}_{\tau\beta} = 2\sum \sigma_i^{-2} \boldsymbol{\sigma}'_{i\tau} \boldsymbol{\sigma}'^{\mathrm{T}}_{i\beta}$$

$$\mathbf{C}_{\beta\beta} = \sum \sigma_i^{-2} \boldsymbol{\mu}'_{i\beta} \boldsymbol{\mu}'^{\mathrm{T}}_{i\beta} \quad \mathbf{C}_{\tau\tau} = \sum \sigma_i^{-2} (\gamma_{2i} + 2) \boldsymbol{\sigma}'_{i\tau} \boldsymbol{\sigma}'^{\mathrm{T}}_{i\tau} \quad \mathbf{C}_{\beta\tau} = \mathbf{C}^{\mathrm{T}}_{\tau\beta} = \sum \sigma_i^{-2} \gamma_{1i} \boldsymbol{\mu}'_{i\beta} \boldsymbol{\sigma}'^{\mathrm{T}}_{i\tau}$$

In the decoupled case $\boldsymbol{\sigma}'_{i\beta} = \mathbf{0}$, and then $\mathbf{D}_{\tau\beta} = \mathbf{0}$ and

$$\mathbf{V}_\theta = \mathbf{D}^{-1}\mathbf{C}(\mathbf{D}^{-1})^{\mathrm{T}} = \begin{bmatrix} \mathbf{D}_{\beta\beta}^{-1} & \mathbf{D}_{\beta\beta}^{-1}\mathbf{C}_{\beta\tau}(\mathbf{D}_{\tau\tau}^{-1})^{\mathrm{T}} \\ \mathbf{D}_{\tau\tau}^{-1}\mathbf{C}_{\tau\beta}(\mathbf{D}_{\beta\beta}^{-1})^{\mathrm{T}} & \mathbf{D}_{\tau\tau}^{-1}\mathbf{C}_{\tau\tau}(\mathbf{D}_{\tau\tau}^{-1})^{\mathrm{T}} \end{bmatrix}$$

In particular, the asymptotic covariance matrix $\mathbf{V}_\beta$ of $\hat{\boldsymbol{\beta}}$ is just $\mathbf{D}_{\beta\beta}^{-1}$, the same as if $\boldsymbol{\tau}$ were known. This occurs even though the parameters $\boldsymbol{\beta}$ and $\boldsymbol{\tau}$ are asymptotically correlated, i.e. are not 'orthogonal' (Cox and Reid, 1987); in this non-likelihood setting it is a consequence of $\mathbf{D}_{\beta\tau} = \mathbf{0}$. An additional bonus is that $\mathbf{D}_{\beta\beta}$ does not involve the unknown γ_{1i} and γ_{2i}.

In summary, the case of decoupled parameters is quite common in practical applications and yields a neat separation of the Gaussian score function into two parts, $\mathbf{q}_\beta(\mathbf{y};\boldsymbol{\theta})$, associated with $\boldsymbol{\mu}_i(\boldsymbol{\beta})$, and $\mathbf{q}_\tau(\mathbf{y};\boldsymbol{\theta})$, associated with $\sigma_i(\boldsymbol{\tau})$. Also in this case, the covariance matrix of $\tilde{\boldsymbol{\beta}}$ reduces to $\mathbf{V}_\beta = \mathbf{D}_{\beta\beta}^{-1}$, not involving the $\boldsymbol{\gamma}_{1i}$ and $\boldsymbol{\gamma}_{2i}$. For the special case of a linear model, $\boldsymbol{\mu}_i = \mathbf{x}_i^{\mathrm{T}}\boldsymbol{\beta}$, $\mathbf{D}_{\beta\beta}^{-1}$ would be produced by a standard normal-linear-models package.

7.1.3 Multivariate observations

Consider now the multivariate case: vector observations $\mathbf{y}_i (i = 1, \ldots, n)$ are available, $\mathbf{y}_i$ being $p_i \times 1$ with mean $\boldsymbol{\mu}_i$ and covariance matrix $\boldsymbol{\Sigma}_i$. This is the situation for repeated measures or longitudinal data. The Gaussian log-likelihood for the complete data vector $\mathbf{y} = [y_1 \ldots y_n]^\mathrm{T}$ is

$$Q(\mathbf{y};\boldsymbol{\theta}) = \log \prod_{i=1}^{n} \left[\{\det(2\pi\boldsymbol{\Sigma}_i)\}^{-1/2} \exp\left\{ -\frac{1}{2}(\mathbf{y}_i - \boldsymbol{\mu}_i)^\mathrm{T} \boldsymbol{\Sigma}_i^{-1} (\mathbf{y}_i - \boldsymbol{\mu}_i) \right\} \right]. \quad (7.7)$$

The corresponding Gaussian score vector $\mathbf{q}(\mathbf{y};\boldsymbol{\theta}) = \partial Q/\partial\boldsymbol{\theta}$ can be expressed as $\sum \mathbf{q}_i(\mathbf{y};\boldsymbol{\theta})$, where $\mathbf{q}_i(\mathbf{y};\boldsymbol{\theta})$ has lth component

$$q_{il}(\mathbf{y};\boldsymbol{\theta}) = -\frac{1}{2}\mathrm{tr}\left(\boldsymbol{\Sigma}_i^{-1} \frac{\partial \boldsymbol{\Sigma}_i}{\partial \boldsymbol{\theta}_l} \right) + \left(\frac{\partial \mu_i}{\partial \theta_l} \right)^\mathrm{T} \boldsymbol{\Sigma}_i^{-1} (\mathbf{y}_i - \boldsymbol{\mu}_i)$$
$$+\frac{1}{2}(\mathbf{y}_i - \boldsymbol{\mu}_i)^\mathrm{T} \boldsymbol{\Sigma}_i^{-1} \left(\frac{\partial \boldsymbol{\Sigma}_i}{\partial \theta_l} \right) \boldsymbol{\Sigma}_i^{-1} (\mathbf{y}_i - \boldsymbol{\mu}_i) \quad (7.8)$$

Note that $E\{q_{il}(\mathbf{y};\boldsymbol{\theta})\} = 0$, by applying the identity $E\{(\mathbf{y}_i - \boldsymbol{\mu}_i)^\mathrm{T}\mathbf{A}(\mathbf{y}_i - \boldsymbol{\mu}_i)\} = \mathrm{tr}(\mathbf{A}\boldsymbol{\Sigma})$; this confirms that the estimating equation $\mathbf{q}(\mathbf{y};\boldsymbol{\theta}) = \mathbf{0}$ is unbiased. Under regularity conditions, maximization of (7.7) leads to a consistent, asymptotically normal estimator for $\boldsymbol{\theta}$ with asymptotic covariance matrix $\mathbf{V}_\theta = \mathbf{D}^{-1}\mathbf{C}\,(\mathbf{D}^{-1})^\mathrm{T}$, where $\mathbf{C}$ and $\mathbf{D}$ have (l,m)th entries given by

$$D_{lm} = E(-\partial^2 Q/\partial\theta_l \partial\theta_m) = \sum_{i=1}^{n} \left\{ \frac{1}{2}\mathrm{tr}\left(\boldsymbol{\Sigma}_i^{-1} \frac{\partial \boldsymbol{\Sigma}_i}{\partial \theta_l} \boldsymbol{\Sigma}_i^{-1} \frac{\partial \boldsymbol{\Sigma}_i}{\partial \theta_m} \right) \right.$$
$$\left. + \left(\frac{\partial \mu_i}{\partial \theta_l} \right)^\mathrm{T} \boldsymbol{\Sigma}_i^{-1} \frac{\partial \mu_i}{\partial \theta_m} \right\} \quad (7.9)$$

$$C_{lm} = \mathrm{cov}(\partial Q/\partial\theta_l, \partial Q/\partial\theta_m) = \Sigma E\{q_{il}(\mathbf{y};\boldsymbol{\theta})\, q_{im}(\mathbf{y};\boldsymbol{\theta})\}$$

Model-based estimates can be used for the contributions to D_{lm}, i.e. by inserting $\tilde{\boldsymbol{\theta}}$ for $\boldsymbol{\theta}$ in the various functional forms in (7.9). However, experience with the computations has suggested that there is no guarantee that $\mathbf{D}$ will be positive definite when calculated in this way. The problem can be sidestepped by adopting the alternative, actually simpler, technique of computing $\mathbf{D}$ directly as $-\partial^2 Q/\partial\boldsymbol{\theta}^2$: in the computer program used for the examples in this chapter, the routine which supplies $Q(\mathbf{y};\boldsymbol{\theta})$ is called with incremented values of $\boldsymbol{\theta}$ to obtain a numerical, second-differences estimate of $\mathbf{D}$. Provided that a genuine local maximum of $Q(\mathbf{y};\boldsymbol{\theta})$ has been located, $\mathbf{D}$ will automatically be positive semi-definite. The use of the sample version, rather than the expected value of $-\partial^2 Q/\partial\boldsymbol{\theta}^2$, is analogous to the use of the sample information matrix in likelihood methodology (Efron and Hinkley, 1978). The mixed model- and sample-based form $\sum q_{il}(\mathbf{y};\tilde{\boldsymbol{\theta}})\, q_{im}(\mathbf{y};\tilde{\boldsymbol{\theta}})$ provides a convenient estimate for C_{lm}; this also has the virtue of being automatically positive semi-definite.

In the decoupled case, where $\boldsymbol{\theta} = [\boldsymbol{\beta}\ \boldsymbol{\tau}]$ with $\boldsymbol{\mu}_i = \boldsymbol{\mu}_i(\boldsymbol{\beta})$ and $\boldsymbol{\Sigma}_i = \boldsymbol{\Sigma}_i(\boldsymbol{\tau})$, the formulae simplify. The components of $\mathbf{q}(\mathbf{y};\boldsymbol{\theta})$ are given by

$$\{\mathbf{q}_\beta(\mathbf{y};\boldsymbol{\theta})\}_l = \sum \left(\frac{\partial \boldsymbol{\mu}_i}{\partial \beta_l} \right)^\mathrm{T} \boldsymbol{\Sigma}_i^{-1} (\mathbf{y}_i - \boldsymbol{\mu}_i)$$

$$\{\mathbf{q}_\tau(\mathbf{y};\boldsymbol{\theta})\}_l = -\frac{1}{2}\sum\left\{\mathrm{tr}\left(\boldsymbol{\Sigma}_i^{-1}\frac{\partial\boldsymbol{\Sigma}_i}{\partial\tau_l}\right) - (\mathbf{y}_i - \boldsymbol{\mu}_i)^{\mathrm{T}}\boldsymbol{\Sigma}_i^{-1}\frac{\partial\boldsymbol{\Sigma}_i}{\partial\tau_l}\boldsymbol{\Sigma}_i^{-1}(\mathbf{y}_i - \boldsymbol{\mu}_i)\right\}$$

$$= \frac{1}{2}\mathrm{tr}\sum\left[\boldsymbol{\Sigma}_i^{-1}\frac{\partial\boldsymbol{\Sigma}_i}{\partial\tau_l}\boldsymbol{\Sigma}_i^{-1}\{(\mathbf{y}_i - \boldsymbol{\mu}_i)(\mathbf{y}_i - \boldsymbol{\mu}_i)^{\mathrm{T}} - \boldsymbol{\Sigma}_i\}\right].$$

The submatrices of **C** and **D** in the partition (7.5) are given by

$$(\mathbf{C}_{\boldsymbol{\beta\beta}})_{lm} = (\mathbf{D}_{\boldsymbol{\beta\beta}})_{lm} = \sum\left(\frac{\partial\boldsymbol{\mu}_i}{\partial\beta_l}\right)^{\mathrm{T}}\boldsymbol{\Sigma}_i^{-1}\frac{\partial\boldsymbol{\mu}^i}{\partial\beta_m}, \quad \mathbf{D}_{\beta\tau} = \mathbf{D}_{\tau\beta} = \mathbf{0},$$

$$(\mathbf{D}_{\tau\tau})_{lm} = \frac{1}{2}\sum\mathrm{tr}\left(\boldsymbol{\Sigma}_i^{-1}\frac{\partial\boldsymbol{\Sigma}_i}{\partial\tau_l}\boldsymbol{\Sigma}_i^{-1}\frac{\partial\boldsymbol{\Sigma}_i}{\partial\tau_m}\right)$$

in this case, $\mathbf{C}_{\boldsymbol{\beta}\tau}$ and $\mathbf{C}_{\tau\tau}$ being easier to estimate via partly sample-based quantities as in section 7.1.2. The asymptotic covariance matrix $\mathbf{V}_{\boldsymbol{\beta}}$ of $\tilde{\boldsymbol{\beta}}$ conveniently reduces to $\mathbf{D}_{\boldsymbol{\beta\beta}}^{-1}$, as in the univariate decoupled case; now, $\mathbf{D}_{\boldsymbol{\beta\beta}} = \Sigma\,\boldsymbol{\mu}_{i\boldsymbol{\beta}}'^{\mathrm{T}}\boldsymbol{\Sigma}_i^{-1}\boldsymbol{\mu}_{i\boldsymbol{\beta}}'$, where $\boldsymbol{\mu}_{i\boldsymbol{\beta}}'$ is $p_i \times \dim(\boldsymbol{\beta})$ with (j,k)th element $\partial\mu_{ij}/\partial\beta_k$, is automatically positive semi-definite.

7.1.4 Goodness of fit and parametric tests

A vector of Gaussian residuals for the ith case can be computed as $(\mathbf{y}_i - \boldsymbol{\mu}_i)$ or, in standardized form, as $\mathbf{T}_i^{\mathrm{T}}(\mathbf{y}_i - \boldsymbol{\mu}_i)$, where $\mathbf{T}_i$ is any matrix satisfying $\mathbf{T}_i^{\mathrm{T}}\boldsymbol{\Sigma}_i\mathbf{T}_i = \mathbf{I}$. If the $\boldsymbol{\mu}_i$ and $\boldsymbol{\Sigma}_i$ were precisely correct, the components of the standardized vector would have zero mean, unit variance and be uncorrelated. In practice, $\boldsymbol{\mu}_i(\tilde{\boldsymbol{\beta}})$ will be used for $\boldsymbol{\mu}_i$, and $\mathbf{T}_i$ can conveniently be computed as a lower triangular matrix satisfying $\mathbf{T}_i\mathbf{T}_i^{\mathrm{T}} = \tilde{\boldsymbol{\Sigma}}_i^{-1}$. The fitted values and resulting residuals, both standardized and unstandardized, are all of use in model checking. For instance, the sample covariance matrix **S** of the unstandardized residuals should provide some approximation to the real covariance structure. To avoid the effect of any consistent bias in $\boldsymbol{\mu}_i(\tilde{\boldsymbol{\beta}})$, i.e. consistent over i, the cross-product contributions to **S** should be corrected for the sample mean. In the very special case where $\boldsymbol{\mu}_i$ is the same for all i, i.e. the data form a random sample, **S** is equal to the sample covariance matrix of the observations $\mathbf{y}_1, \ldots, \mathbf{y}_n$. If the pattern of **S** conflicts with that of $\tilde{\boldsymbol{\Sigma}}_i$, a rethink is called for. A more formal comparison can be made by computing the sample covariance matrix of the standardized residuals: this is equal to $\mathbf{T}_i^{\mathrm{T}}\mathbf{S}_i\mathbf{T}_i$ and should resemble a unit matrix.

It is important to assess the fit of the assumed covariance structure, not least because it will be used to provide error estimates for the quantities of primary interest. At the same time, such error estimates, and tests based on them, will often be quite robust to misspecification. To be very safe, a range of covariance structures can be fitted to see what effect this has on the main inferences. However, from the lofty heights of Mount Academia, it is ever easy to be righteous in recommendation to others. In real life, working statisticians under time constraints have to learn how to be flexible and robust without spending an eternity on the job.

The standardized residuals can be plotted in various ways: to check for outliers, versus x-variables included and not included in the fitted model, versus the fitted values, and perhaps even to hazard some tentative judgement of their distributional shape with a view to constructing a better model.

Regarding parametric hypothesis tests, there will be two main strands here. These concern (a) the regression coefficients which govern $\boldsymbol{\mu}_i$, and (b) the covariance structure, $\boldsymbol{\Sigma}_i$. Often, the major concern will be (a), with (b) acting as an enabling stage. Thus, a common approach will be to fit a full model, incorporating all the $\boldsymbol{\mu}_i$ features that the statistician judges the sample size will stand, together with a first stab at a likely covariance structure. Then, one might try to refine the covariance structure to optimize the goodness of fit, for example in terms of Gaussian log-likelihoods and residuals. Finally, the regression coefficients governing $\boldsymbol{\mu}_i$ can be attacked, keeping the selected covariance structure intact.

Formal tests on regression coefficients can be made as described in section 7.1.1, using the Wald statistic. For instance, consider a linear hypothesis $\mathbf{H}\boldsymbol{\beta}_1 = \mathbf{0}$, where $\boldsymbol{\beta}_1$ is a subset of regression coefficients and $\mathbf{H}$ is a specified matrix. Let $\tilde{\mathbf{V}}_1$ be the estimated covariance matrix of $\tilde{\boldsymbol{\beta}}_1$ obtained as the appropriate submatrix of $\mathbf{D}^{-1}\mathbf{C}(\mathbf{D}^{-1})^{\mathrm{T}}$. Then the test statistic

$$W = (\mathbf{H}\tilde{\boldsymbol{\beta}}_1)^{\mathrm{T}}(\mathbf{H}\tilde{\mathbf{V}}_1\mathbf{H}^{\mathrm{T}})^{-1}(\mathbf{H}\tilde{\boldsymbol{\beta}}_1)$$

is asymptotically distributed as χ^2_v, with degrees of freedom $v = \dim(\beta_1)$, under the hypothesis. Standard errors and Wald tests are less useful for covariance parameters which are usually range-restricted: their profile log-likelihoods are often extremely non-quardratic in shape, causing normal-theory confidence intervals and tests to be very wide of the mark.

7.2 Gaussian estimation: some examples

In this section some straightforward linear regression examples, using small but manageable data sets taken from Crowder and Hand (1990), will be worked through. The purpose is to illustrate the construction and fitting of typical models for the mean vector and covariance matrix, not to present deep scientific investigations. A wide range of techniques is available nowadays. These include a variety of ways of looking at residuals, diagnostics, influence, tests for specific aspects of the model, etc. In order to maintain the gripping, fast-paced, edge-of-the-seat, couldn't-put-it-down style of this narrative, not every single technique is applied to every single example below. In practice, of course, one would pay more respect to one's data.

Example 7.1 Reactions of 12 patients to treatment
The data, displayed in Table A.5 and plotted in Fig.1.5, are repeated measures of plasma ascorbic acid. There is a treatment-phase partition of the seven occasions, the three phases spanning respectively the first two, middle three, and last two observation times.

Since we have a random sample here, a single group with no between-individuals factors, the sample covariance matrix can be inspected as a guide to the underlying covariance structure. More generally, with sufficiently sizeable groups of individuals, we could examine the sample covariance matrices within groups. The sample variances for the seven measures are

$$0.125,\ 0.222,\ 0.084,\ 0.116,\ 0.066,\ 0.072,\ 0.054$$

and the sample correlation matrix is

$$\mathbf{R} = \begin{bmatrix} 1.000 & 0.680 & 0.586 & 0.462 & 0.140 & 0.219 & 0.395 \\ 0.680 & 1.000 & 0.733 & 0.553 & 0.354 & 0.270 & 0.109 \\ 0.586 & 0.733 & 1.000 & 0.527 & 0.281 & 0.314 & -0.060 \\ 0.462 & 0.553 & 0.527 & 1.000 & 0.792 & 0.635 & -0.188 \\ 0.140 & 0.354 & 0.281 & 0.792 & 1.000 & 0.318 & -0.296 \\ 0.219 & 0.270 & 0.314 & 0.635 & 0.318 & 1.000 & 0.306 \\ 0.395 & 0.109 & -0.060 & -0.188 & -0.296 & 0.306 & 1.000 \end{bmatrix}$$

The sample variances vary from 0.054 to 0.222, and in $\mathbf{R}$ the correlations fall off with separation in time for the most part. Also, from the raw data plot, not to mention everyday experience, we should allow for random individual levels. This all suggests that we might not go too far wrong in taking $\mathbf{Z}_i$ as $\mathbf{1}$ (section 6.3.1) and $\mathbf{E}_i$ as the error covariance matrix for the Markov form with unequal time spacing (section 6.2.4).

The previous mixed-model anova gives way to regression here but the analysis concerns the same parametric questions. A model specifying constant mean level within each phase will be fitted. This is supported by a test, based on Hotelling's T^2, for differences within phases: $F(4, 8) = 2.69\ (p > 0.10)$; see Crowder and Hand (1990, Example 4.2); we remind the reader that the exact distribution theory for T^2 is based on normality. The rows of $\mathbf{X}_i$ for this model comprise a constant term 1 followed by an indicator vector for phase:

$$\mathbf{X}_i = \begin{bmatrix} 1 & 1 & 0 \\ 1 & 1 & 0 \\ 1 & 0 & 1 \\ 1 & 0 & 1 \\ 1 & 0 & 1 \\ 1 & -1 & -1 \\ 1 & -1 & -1 \end{bmatrix}$$

The numerical results from fitting the model specified are

$$\tilde{\boldsymbol{\beta}} = [0.919 \quad -0.371 \quad 0.312], \quad \tilde{\boldsymbol{\tau}} = [\tilde{\sigma}_e \; \tilde{\rho} \; \tilde{\sigma}_I] = [0.252 \; 0.00 \; 0.187]$$

and

$$\mathbf{V}_\beta = \mathbf{D}_{\beta\beta}^{-1} = 10^{-3}\begin{bmatrix} 3.22 & 2.14 & -1.06 \\ 2.14 & 3.37 & -1.02 \\ -1.06 & -1.02 & 2.45 \end{bmatrix}$$

Thus, the three phase levels are estimated respectively as $0.548(=0.919-0.371)$, 1.231 $(=0.919+0.312)$ and 0.978 $(=0.919+0.371-0.312)$. The corresponding standard errors are 0.10 (the square root of $10^{-3}(3.22+2\times 2.14+3.37)$), 0.06 and 0.07. A more formal assessment of differences between phases can be made by testing the hypothesis that $\beta_2=\beta_3=0$. We refer the Wald statistic (section 7.1.1) $W=\mathbf{c}^{\mathrm{T}}\mathbf{V}_{\mathrm{c}}^{-1}\mathbf{c}$, with $\mathbf{c}=[\tilde{\beta}_2\,\tilde{\beta}_3]^{\mathrm{T}}$ and $\mathbf{V}_c=\operatorname{var}(\mathbf{c})$, to χ_2^2; since $W=59.60$, it would be bad luck to be called upon to oppose the motion that the phase levels differ.

For comparison, the sample phase means are 0.547, 1.147 and 0.700, and a Hotelling's T^2 test for phase differences, which is based on the (unstructured) sample covariance matrix, gives $F(2,10)=31.7$. In general, similar conclusions from W and T^2 can be explained as follows. The statistics have the same basic quadratic form, the $\mathbf{c}$ and $\mathbf{V}_{\mathrm{c}}$ of W being replaced by a contrast vector of sample means and a sample covariance matrix in T^2. If the model fitted fits, the estimates used for $\mathbf{c}$ and $\mathbf{V}_{\mathrm{c}}$ will reproduce the sample versions well.

Although not of particular interest in this example, it is worth pointing out that variation between occasions within phases could also be appraised within the regression framework. The $\mathbf{X}_i$ matrix needs to be modified to accommodate seven occasions rather than three phases, so it becomes 7×7. Wald tests on the resulting parameter estimates can then be made as required.

The $\boldsymbol{\tau}$ estimates are $\tilde{\sigma}_e=0.252$, $\tilde{\rho}=6.1\times 10^{-5}$ (more precisely than given above) and $\tilde{\sigma}_I=0.187$, and the standard errors for these output by the program are 0.046, 0.008 and 0.057. However, as mentioned in section 7.1.4, it is not wise to put too much store by such standard errors. A log-likelihood ratio test for $\rho=0$ gives 3.025 which, under normality, would have an asymptotic χ_1^2 distribution leading to $0.05<p<0.10$. It seems that the rich structure of correlated autoregressive errors is not required to fit these data: a reduced covariance matrix, of compound symmetric type, would suffice, incidentally supporting the previous anova.

Example 7.2 Visual acuity

The data in Table A.9 first came to light in Crowder and Hand (1990) and we have already considered them in Example 2.2. The measurements are of electrical response time (milliseconds) at the back of the cortex when a light is flashed through a lens into the subject's eye. The within-subjects design is a 2×4 factorial, two eyes (factor E) by four lens strengths (factor L). As in Example 7.1, this is a single group of individuals with no between-subjects design. The results here may be compared with those given in Example 2.2 for these data.

The fitted $\boldsymbol{\mu}$ model is to contain an intercept term, an eye effect, a lens effect, and an eye by lens interaction. For this, the $\mathbf{X}_i$ matrix is set up as

$$\mathbf{X}_i = \begin{bmatrix} 1 & 1 & 1 & 0 & 0 & 1 & 0 & 0 \\ 1 & 1 & 0 & 1 & 0 & 0 & 1 & 0 \\ 1 & 1 & 0 & 0 & 1 & 0 & 0 & 1 \\ 1 & 1 & -1 & -1 & -1 & -1 & -1 & -1 \\ 1 & -1 & 1 & 0 & 0 & -1 & 0 & 0 \\ 1 & -1 & 0 & 1 & 0 & 0 & -1 & 0 \\ 1 & -1 & 0 & 0 & 1 & 0 & 0 & -1 \\ 1 & -1 & -1 & -1 & -1 & 1 & 1 & 1 \end{bmatrix}$$

For the covariance structure we begin by inspecting the sample covariance matrix for this random sample: the sample variances are

$$10^{-3}[1.25\ 1.26\ 7.75\ 1.29\ 6.13\ 5.79\ 7.28\ 8.31],$$

and the sample correlation matrix is

$$\mathbf{R} = \begin{bmatrix} 1.000 & 0.738 & -0.251 & 0.390 & 0.141 & -0.097 & -0.431 & -0.180 \\ 0.738 & 1.000 & 0.338 & 0.511 & 0.721 & 0.486 & 0.151 & 0.434 \\ -0.251 & 0.338 & 1.000 & 0.281 & 0.867 & 0.972 & 0.870 & 0.861 \\ 0.390 & 0.511 & 0.281 & 1.000 & 0.533 & 0.321 & 0.273 & 0.454 \\ 0.141 & 0.721 & 0.867 & 0.533 & 1.000 & 0.929 & 0.767 & 0.879 \\ -0.097 & 0.486 & 0.972 & 0.321 & 0.929 & 1.000 & 0.828 & 0.826 \\ -0.431 & 0.151 & 0.870 & 0.273 & 0.767 & 0.828 & 1.000 & 0.910 \\ -0.180 & 0.434 & 0.861 & 0.454 & 0.879 & 0.826 & 0.910 & 1.000 \end{bmatrix}$$

It is difficult to discern any familiar pattern here and one is tempted to proceed with a free, unstructured covariance model. However, there are eight measures, and an 8×8 cross-products matrix constructed from only seven individuals would necessarily be singular. Thus, to proceed with regression we need a structured covariance matrix. We will take $\mathbf{Z}_i$ to be the same as $\mathbf{X}_i$ but without the interaction, so column 1 of $\mathbf{Z}_i$ accounts for random individual levels, column 2 for E (eyes) and columns 3,4 and 5 for L (lenses). Correspondingly, $\mathbf{B} = \text{diag}[\sigma_I^2\ \sigma_E^2\ \mathbf{A}_L\mathbf{A}_L^T]$ for the random effects (section 6.3.2), where $\mathbf{A}_L$ is 3×3 lower triangular. The error covariance matrix will be taken as $\mathbf{E}_i = \sigma_e^2\mathbf{I}$ (section 6.2.1).

The parameter estimates (for the data scaled by 10^{-2}) turn out as

$$\tilde{\boldsymbol{\beta}} = 10^{-2}[113\ \ 0.909\ \ -0.770\ \ 0.0163\ \ -1.76\ \ 0.303\ \ 0.232\ \ -1.41]$$

$$\tilde{\sigma}_e = 0.0289, \quad \tilde{\sigma}_I = 0.0486, \quad \tilde{\sigma}_E = 0.0231$$

$$\tilde{\mathbf{A}}_L = 10^{-2}\begin{bmatrix} 2.30 & 0 & 0 \\ 1.03 & -0.472 & 0 \\ -3.29 & -0.438 & -0.004 \end{bmatrix}$$

The factorial effects can be tested as follows. The coefficient for the eye effect is 0.00909, with standard error 0.0096 and resulting normal deviate 0.95; this low level of significance ($p = 0.34$) gives little support to an eye level effect. On the other hand, a Wald test for differences between lenses, based on

$$W = \mathbf{c}^{\mathrm{T}}\mathbf{V}_{\mathbf{c}}^{-1}\mathbf{c} \quad \text{with} \quad \mathbf{c} = [\tilde{\beta}_3\ \tilde{\beta}_4\ \tilde{\beta}_5]^{\mathrm{T}}$$

gives $W = 13.0$ as χ_3^2 $(0.001 < p < 0.005)$ which does support the case for real differences. The corresponding estimates are 113.91 ($= \tilde{\beta}_1 + \tilde{\beta}_2$ for lens strength 1), 112.23 (lens 2), 113.02 (lens 3) and 112.84 (lens 4) with standard errors all about 0.02. The trend is not monotone in lens strength, but the post-mortem did not turn up any obvious mix-ups or a satisfying explanation. The test for interaction, based on

$$\mathbf{c} = [\tilde{\beta}_6\ \tilde{\beta}_7\ \tilde{\beta}_8]^{\mathrm{T}}$$

gives $W = 4.88$ as $\chi_3^2 (p = 0.18)$.

For comparison, Hotelling's T^2 tests (see, for example, Crowder and Hand, 1990, section 4.2) for the eye, lens and eye by lens effects gave $F(1,6) = 0.78$, $F(3,4) = 21.5$ (for which $0.01 < p < 0.025$) and $F(3,4) = 0.55$; see Example 2.2. The tests are based on the contrast vectors $\mathbf{H}_{\mathrm{E}}\bar{\mathbf{y}}$, $\mathbf{H}_{\mathrm{L}}\bar{\mathbf{y}}$, and $\mathbf{H}_{\mathrm{EL}}\bar{\mathbf{y}}$, respectively, where $\bar{\mathbf{y}}$ is the 8×1 sample mean vector

$$\mathbf{H}_{\mathrm{E}} = [1 \quad 1 \quad 1 \quad 1 \quad -1 \quad -1 \quad -1 \quad -1],$$

$$\mathbf{H}_{\mathrm{L}} = \begin{bmatrix} 1 & -1 & 0 & 0 & 1 & -1 & 0 & 0 \\ 1 & 0 & -1 & 0 & 1 & 0 & -1 & 0 \\ 1 & 0 & 0 & -1 & 1 & 1 & 0 & -1 \end{bmatrix}$$

$$\mathbf{H}_{\mathrm{EL}} = \begin{bmatrix} 1 & -1 & 0 & 0 & -1 & 1 & 0 & 0 \\ 1 & 0 & -1 & 0 & -1 & 0 & 1 & 0 \\ 1 & 0 & 0 & -1 & -1 & 0 & 0 & 1 \end{bmatrix}$$

Wald and Hotelling would evidently have agreed on this one.

Example 7.3 Rat growth

The data, reproduced from Crowder and Hand (1990), are given in Table A.1. They are body weights (grams) of rats measured over nine weeks with a change of diet regime during week 6.

The data plots in Fig. 7.1 show very clear differences between the groups, particularly in their overall levels; one must bear in mind that the groups had been on different diets for some time on the way in to the weigh-in. For illustration, we will focus upon the first six measures, days 1 to 36, and fit straight lines with a predetermined, though not unrealistic, covariance structure.

The $\boldsymbol{\mu}_i$ model is $\mu_{ij} = \alpha + \beta t_j$; here $t_j = \text{days}/10$, and α and β may differ between groups. The $\mathbf{X}_i$ matrices for the three groups are correspondingly set up as

$$\begin{bmatrix} 1 & 1 & 0 & 0.1 & 0.1 & 0 \\ 1 & 1 & 0 & 0.8 & 0.8 & 0 \\ 1 & 1 & 0 & 1.5 & 1.5 & 0 \\ 1 & 1 & 0 & 2.2 & 2.2 & 0 \\ 1 & 1 & 0 & 2.9 & 2.9 & 0 \\ 1 & 1 & 0 & 3.6 & 3.6 & 0 \end{bmatrix}, \begin{bmatrix} 1 & 0 & 1 & 0.1 & 0 & 0.1 \\ 1 & 0 & 1 & 0.8 & 0 & 0.8 \\ 1 & 0 & 1 & 1.5 & 0 & 1.5 \\ 1 & 0 & 1 & 2.2 & 0 & 2.2 \\ 1 & 0 & 1 & 2.9 & 0 & 2.9 \\ 1 & 0 & 1 & 3.6 & 0 & 3.6 \end{bmatrix}, \begin{bmatrix} 1 & -1 & -1 & 0.1 & -0.1 & -0.1 \\ 1 & -1 & -1 & 0.8 & -0.8 & -0.8 \\ 1 & -1 & -1 & 1.5 & -1.5 & -1.5 \\ 1 & -1 & -1 & 2.2 & -2.2 & -2.2 \\ 1 & -1 & -1 & 2.9 & -2.9 & -2.9 \\ 1 & -1 & -1 & 3.6 & -3.6 & -3.6 \end{bmatrix}$$

Inspection of the sample within-groups covariance matrices shows fairly uniform variance over the six measures and high correlations everywhere, the smallest being

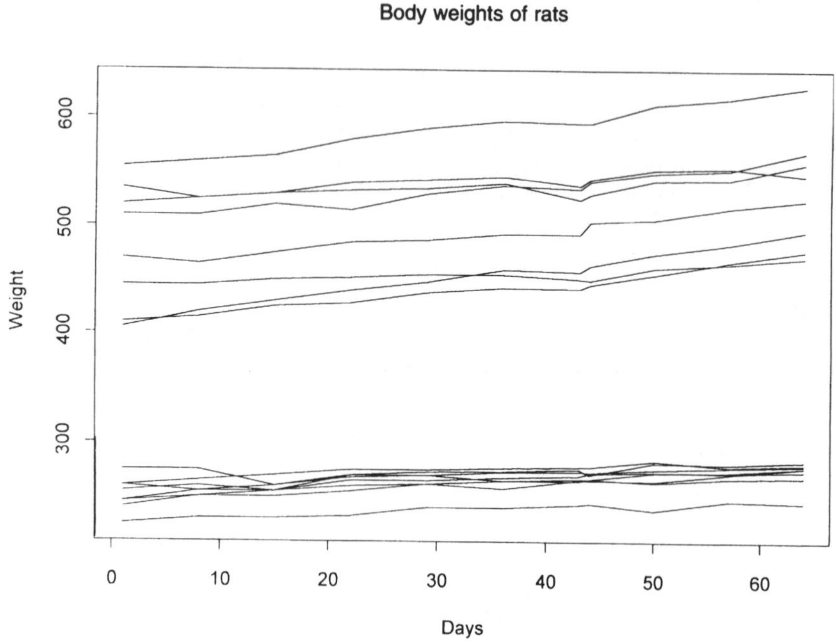

Figure 7.1. *Rat growth: body weights over nine weeks.*

0.74. The pooled sample covariance matrix, with the data scaled by 10^{-2}, is

$$\mathbf{S} = \begin{bmatrix} 0.143 & 0.137 & 0.129 & 0.136 & 0.136 & 0.136 \\ 0.137 & 0.134 & 0.127 & 0.135 & 0.135 & 0.135 \\ 0.129 & 0.127 & 0.123 & 0.130 & 0.130 & 0.130 \\ 0.136 & 0.135 & 0.130 & 0.139 & 0.139 & 0.139 \\ 0.136 & 0.135 & 0.130 & 0.139 & 0.140 & 0.141 \\ 0.136 & 0.135 & 0.130 & 0.139 & 0.141 & 0.143 \end{bmatrix}$$

The $\mathbf{Z}_i$ matrix will be taken to be $\mathbf{1}$, which yields the 'random individual levels' model (section 6.3.1), and the error covariance structure will be taken as stable autoregressive (section 6.2.3), in recognition of the settling-in period during which the error process has had time to stabilize.

The parameters are $\boldsymbol{\beta}$, of dimension 6, the set of linear regression coefficients appearing in $\boldsymbol{\mu}_i = \mathbf{X}_i\boldsymbol{\beta}$, and $\boldsymbol{\tau} = [\sigma_e \ \rho \ \sigma_1]$, the set of extra parameters governing the covariance structure $\boldsymbol{\Sigma}_i$. The final estimates are

$$\tilde{\boldsymbol{\beta}} = [4.04 \ -1.53 \ 0.491 \ 0.067 \ -0.026 \ 0.033], \quad \tilde{\boldsymbol{\tau}} = [0.050 \ 0.851 \ 0.325]$$

and the square roots of the diagonal elements of the asymptotic covariance matrix of $\tilde{\boldsymbol{\theta}} = [\tilde{\boldsymbol{\beta}} \ \tilde{\boldsymbol{\tau}}]$, $\mathbf{V}_\theta = \mathbf{D}^{-1}\mathbf{C}(\mathbf{D}^{-1})^{\mathrm{T}}$, yield the standard errors of the parameter estimates as

$$[0.11 \ 0.11 \ 0.21 \ 0.01 \ 0.01 \ 0.02 \ 0.00 \ 0.25 \ 0.10].$$

The group-1 intercept is estimated as $4.04 - 1.53 = 2.51$, with standard error 0.05. Likewise, the intercept for group 2 is $4.04 + 0.491 = 4.53$, with standard error 0.30, and that for group 3 is $4.04 + 1.53 - 0.491 = 5.08$, with standard error 0.12. Group 1 seems to stand apart from the other two, but any difference between groups 2 and 3 is not so clear. To compare these we need to contrast $\beta_1 + \beta_3$ (the group-2 intercept) with $\beta_1 - \beta_2 - \beta_3$ (the group-3 intercept). The difference, $\beta_2 + 2\beta_3$, is the linear parametric function $\mathbf{a}^T\boldsymbol{\beta}$ obtained by taking $\mathbf{a}^T = [0\ 1\ 2\ 0\ 0\ 0]$. This has estimate $\mathbf{a}^T\tilde{\boldsymbol{\beta}} = -0.548$ with standard error 0.327 (the square root of $\mathbf{a}^T\mathbf{V}_\beta\mathbf{a}$). The resulting normal deviate, $-0.548/0.327 = 1.68$, does not provide strong evidence for a real difference.

The fourth parameter, with estimate $\tilde{\beta}_4 = 0.067$ and standard error 0.01, is the regression coefficient for $x =$ time (days/10). It is the average slope over the three groups because the choice of dummy indicator variables in columns 4,5 and 6 of $\mathbf{X}_i$ makes the three group adjustments to β_4 add up to zero. The question is, then, whether the three slopes differ. This is equivalent to asking if $\beta_5 = \beta_6 = 0$, and to test this we can apply the Wald statistic (section 7.1.1): we refer $W = \mathbf{c}^T\mathbf{V}_c\mathbf{c}$ to χ_2^2, where $\mathbf{c} = [\tilde{\beta}_5\ \tilde{\beta}_6]^T$ and $\mathbf{V}_c$ is the corresponding 2×2 submatrix of $\mathbf{V}_\theta$. The result, $W = 6.61$, yields p value 0.036 and so suggests that the underlying slopes might well differ. The three slope estimates are actually 0.041, 0.10 and 0.059 with standard errors 0.007, 0.023 and 0.013, so group 2 looks like the odd one out.

The estimates for σ_e, ρ and σ_I are respectively 0.050, 0.851 and 0.325. The estimates for ρ and σ_I, together with their standard errors, suggest strong within-individual correlation between the repeated measures and strong between-individual variation in mean levels.

The sample covariance matrix $\mathbf{S}$ of the unstandardized residuals can be computed, and compared with the estimate $\tilde{\boldsymbol{\Sigma}}_i$ of the structured covariance matrix which, in this case, is the same for all i. To some extent, any similarity is self-fulfilling since they both arise from the fitted model. As noted in section 7.1.4, the sample covariance matrix of the standardized residuals should resemble a unit matrix. For the present example this is

$$\mathbf{S} = \begin{bmatrix} 1.56 & 0.16 & 0.27 & 0.16 & -0.15 & 0.07 \\ 0.16 & 1.27 & -0.47 & 0.20 & 0.19 & 0.26 \\ 0.27 & -0.47 & 1.03 & -0.38 & 0.04 & -0.37 \\ 0.16 & 0.20 & -0.38 & 0.60 & 0.21 & -0.02 \\ -0.15 & 0.19 & 0.04 & 0.21 & 0.41 & -0.01 \\ 0.07 & 0.26 & -0.37 & -0.02 & -0.01 & 1.08 \end{bmatrix}$$

The resemblance is not striking, but perhaps this is asking a bit much from a sample of size 16.

Residuals can be computed as described in section 7.1.4. The various residual plots which were done did not produce anything very exciting, i.e. no visually compelling evidence of lack of fit in the model. The normal plot of standardized residuals suggested an upper tail heavier than normal. Such non-normality has already been allowed for by using Gaussian estimation, but in a more extended study we might wish to search for an appropriate error distribution so that the full power of likelihood methods, frequentist or Bayesian, could be employed.

Example 7.4 Hip replacements

The data set given in Table A.14 come from Crowder and Hand (1990, Table 5.2). The figures comprise sex, age (years), and four repeated measurements of haematocrit, one recording before and three after the operation. The missing value code -9.0 is strongly represented, the reason for most being two different sources of data, unconnected with any outcome events. Also, the last figure in the measurements is always a 0 or a 5, indicating rounding. However, with four-figure measurements this should not be too catastrophic. The primary interest was in assessing age and sex differences. The data are plotted in Fig. 7.2.

Let us fit a linear model $\boldsymbol{\mu}_i = \mathbf{X}_i\boldsymbol{\beta}$ in which the rows of $\mathbf{X}_i$ contain terms for a constant (intercept), sex, occasion, sex by occasion interaction, age, and age by sex interaction. For example, the 4×10 $\mathbf{X}_i$ matrices thus generated for a male patient aged 66, and for a female one aged 57, are

$$\begin{bmatrix} 1 & 1 & 1 & 0 & 0 & 1 & 0 & 0 & 6.6 & 6.6 \\ 1 & 1 & 0 & 1 & 0 & 0 & 1 & 0 & 6.6 & 6.6 \\ 1 & 1 & 0 & 0 & 1 & 0 & 0 & 1 & 6.6 & 6.6 \\ 1 & 1 & -1 & -1 & -1 & -1 & -1 & -1 & 6.6 & 6.6 \end{bmatrix}$$

and

$$\begin{bmatrix} 1 & -1 & 1 & 0 & 0 & -1 & 0 & 0 & 5.7 & -5.7 \\ 1 & -1 & 0 & 1 & 0 & 0 & -1 & 0 & 5.7 & -5.7 \\ 1 & -1 & 0 & 0 & 1 & 0 & 0 & -1 & 5.7 & -5.7 \\ 1 & -1 & -1 & -1 & -1 & 1 & 1 & 1 & 5.7 & -5.7 \end{bmatrix}$$

The covariance structure will be taken initially as an equicorrelated model, resulting from random individual levels (section 6.3.1) plus conditional independence with homogeneity of variance (section 6.2.1), but parametrized by $[\sigma_e\, \rho]$ as in section 6.2.2. This assumed covariance model will be checked by inspecting residuals.

There are ten β's, since $\mathbf{X}_i$ is 4×10, and $\boldsymbol{\tau} = (\sigma_e, \rho)$. The model fitting gives the following estimates for the data, which have been scaled by 10^{-1}.

$$\tilde{\boldsymbol{\beta}} = [3.28\ \ 0.21\ \ 0.65\ \ -0.34\ \ -0.21\ \ 0.12\ \ -0.050\ \ -0.048\ \ 0.019\ \ -0.020]$$

$$\tilde{\boldsymbol{\tau}} = [0.40\ \ 0.23]$$

The estimated overall regression coefficient for gender is $\tilde{\beta}_2 = 0.21$, with standard error 0.43, and that for age is $\tilde{\beta}_9 = 0.019$, with standard error 0.06. Neither term, on its own, appears to be indispensable in the linear model. A Wald test for their joint presence can be made by referring the statistic $W = \mathbf{c}^{\mathrm{T}}\mathbf{V}_{\mathrm{c}}^{-1}\mathbf{c}$ to χ_2^2, where $\mathbf{c} = [\tilde{\beta}_2\, \tilde{\beta}_9]$. The result, $W = 0.29$, yields p value 0.86; apparently, the contributions of age and sex would not be sorely missed, which is exactly half true in life, I guess.

The presence of differences in mean level between the four occasions can be judged by testing the hypothesis that $\beta_3 = \beta_4 = \beta_5 = 0$. The Wald statistic turns out to be $W = 157.0$ which, as χ_3^2, seems pretty conclusive. The corresponding estimates are 3.93, 2.94, 3.07 and 3.18, with respective standard errors 0.44, 0.41, 0.44 and 0.45. The largest discrepancy is between occasions 1 and 2 which, not surprisingly, occur before and after the hip operation. In fact, a trend over occasions 2, 3 and 4 is

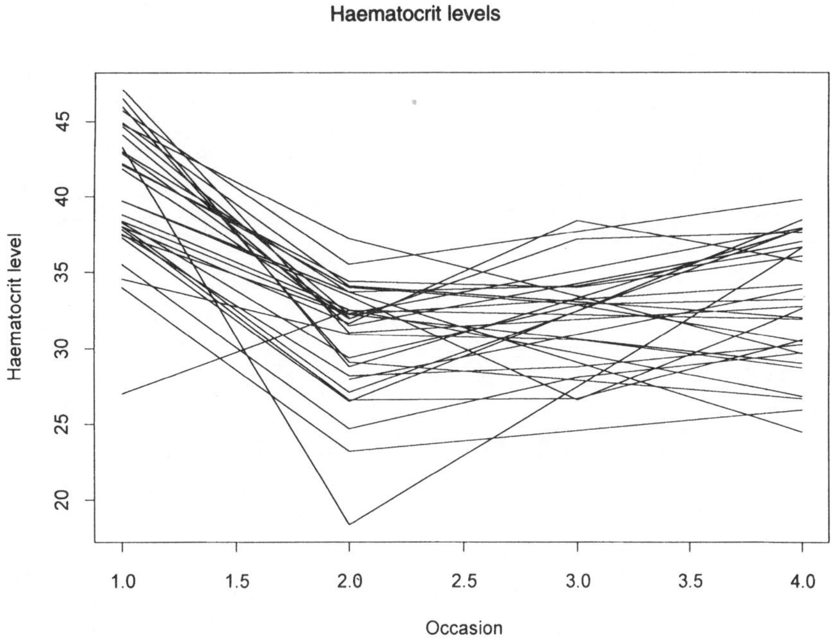

Figure 7.2. *Hip replacements: haematocrit levels before and after operation.*

apparent, climbing back up to the pre-operative value on occasion 1. Whether the differences are the same for men and women can be assessed by examining β_6, β_7 and β_8, the sex by occasions interaction coefficients. The estimates yield $W = 4.76$ as $\chi_3^2(p = 0.19)$, so an interaction is not established beyond any reasonable doubt. Another interaction effect to be scrutinized is β_{10}, the age by sex interaction, which measures the difference in age gradient between males and females. Thus, $\tilde{\beta}_{10} = -0.021$ with standard error 0.064, and the normal deviate is -0.33, implying agreement between the sexes (*sic*).

The estimated covariance matrix $\tilde{\boldsymbol{\Sigma}}_i$, the same for all i here, can be calculated from $\tilde{\tau}$ and, for comparison, the sample covariance matrix $\mathbf{S}$ of the unstandardized residuals can be computed:

$$\tilde{\boldsymbol{\Sigma}}_i = 10^{-2}\begin{bmatrix} 15.8 & 3.63 & 3.63 & 3.63 \\ 3.63 & 15.8 & 3.63 & 3.63 \\ 3.63 & 3.63 & 15.8 & 3.63 \\ 3.63 & 3.63 & 3.63 & 15.8 \end{bmatrix} \qquad \mathbf{S} = 10^{-2}\begin{bmatrix} 16.3 & 3.82 & 5.92 & 4.70 \\ 3.82 & 15.4 & 7.02 & 0.21 \\ 5.92 & 7.02 & 17.0 & 10.2 \\ 4.70 & 0.21 & 10.2 & 18.0 \end{bmatrix}$$

The resemblance between the two matrices is a bit dubious in places, but it is not easy to think of a natural, not over-elaborate, improvement on the covariance model which would achieve better matching.

In a normal plot of the 120 standardized residuals the lower tail looked a bit on the heavy side, but the Gaussian estimates of variability are designed to allow for non-normality so the various inferences should still be valid.

CHAPTER 8

Nonlinear models

In the examples presented in section 7.2, $\boldsymbol{\mu}_i$ was linear in the regression parameters, i.e. they were all linear models. The theory of section 7.1, however, is not so restricted, and we now extend the practice to nonlinear models.

8.1 Nonlinear regression

Let us begin by considering a simple prototype nonlinear regression model, the exponential curve $\beta_1 + \beta_2 e^{-\gamma x}$. The parameters β_1 and β_2 enter linearly and γ enters nonlinearly; for known γ this would just be a linear model. Suppose that observations y_{ij} $(j = 1,\ldots, p)$ have been recorded at settings x_{ij} for the ith individual, and that

$$y_{ij} = \beta_1 + \beta_2 \exp(-\gamma x_{ij}) + u_{ij} \tag{8.1}$$

u_{ij} being the usual zero-mean error term. In vector notation this is

$$\mathbf{y}_i = \mathbf{X}_i(\gamma)\boldsymbol{\beta} + \mathbf{u}_i \tag{8.2}$$

where $\mathbf{y}_i^{\mathrm{T}} = [y_{i1} \ldots y_{ip}]$, $\mathbf{u}^{\mathrm{T}} = [u_{i1} \ldots u_{ip}]$, $\boldsymbol{\beta}^{\mathrm{T}} = [\beta_1\, \beta_2]$ and $\mathbf{X}_i(\gamma)$ is $p \times 2$ with jth row $[1\; e^{-\gamma x_{ij}}]$. This contrived way of writing a nonlinear model, which goes back at least as far as Halperin (1963), has its advantages. It is now just a linear model with a small, but perfectly formed, modification–the design matrix $\mathbf{X}_i$ depends on an unknown parameter, γ, more generally a vector. If $\mathbf{u}_i$ has covariance matrix $\mathbf{U}_i(\boldsymbol{\beta},\gamma,\boldsymbol{\tau})$, depending on $\boldsymbol{\beta}$, γ and a further parameter $\boldsymbol{\tau}$, the Gaussian log-likelihood based on data vectors $\mathbf{y}_1,\ldots,\mathbf{y}_n$ is

$$Q(y; \boldsymbol{\beta}, \gamma, \boldsymbol{\tau}) = -\frac{1}{2}\sum_{i=1}^{n} \{\log\det(2\pi\mathbf{U}_i) + (y_i - \mathbf{X}_i\boldsymbol{\beta})^{\mathrm{T}}\mathbf{U}_i^{-1}(\mathbf{y}_i - \mathbf{X}_i\boldsymbol{\beta})\}$$

Where $\mathbf{X}_i$ is written for $\mathbf{X}_i(\gamma)$ and $\mathbf{U}_i$ for $\mathbf{U}_i(\boldsymbol{\beta}, \gamma, \boldsymbol{\tau})$. The Gaussian estimator for $\boldsymbol{\beta}$ can be obtained explicitly in terms of γ and $\boldsymbol{\tau}$ as

$$\tilde{\boldsymbol{\beta}}_{\gamma\tau} = (\Sigma\mathbf{X}_i^{\mathrm{T}}\mathbf{U}_i^{-1}\mathbf{X}_i)^{-1}\,\Sigma\mathbf{X}_i^{\mathrm{T}}\mathbf{U}_i^{-1}\mathbf{y}_i$$

This may be inserted into Q in place of $\boldsymbol{\beta}$, and the resulting profile log-likelihood maximized to calculate the estimates $\tilde{\gamma}$ and $\tilde{\boldsymbol{\tau}}$.

Not all nonlinear models present themselves in the standard form (8.2). For instance, the Bleasdale–Nelder form $(\alpha_1 + \alpha_2 x^{\phi})^{-\kappa}$ (Mead, 1970) has no linear parameters. However, it can be reformulated as $\beta(1 + \alpha x^{\phi})^{-\kappa}$, where $\beta = \alpha_1^{-\kappa}$ and $\alpha = \alpha_2/\alpha_1$, which has. Such a reparametrization would not be possible if we had started with $(1 + \alpha x^{\phi})^{-\kappa}$, say. In this case, an extended form, $\beta_0 + \beta_1(1 + \alpha x^{\phi})^{-\kappa}$, could be tried. One would then expect $\tilde{\beta}_0$ to be close to 0, and $\tilde{\beta}_1$ to 1; otherwise, the initial model would be called into question.

Example 8.1 Pill dissolution rates

The times taken for solid dose forms of medicine, i.e. pills, to dissolve are of importance in the pharmaceutical industry. The effects on dissolution times of various factors, such as formulation, shape, brand, manufacturing site, batch, etc., are much studied. In a typical experiment, the sequence of times taken until given fractions of a pill remain are observed. For instance, it might take 13 seconds until 90% of the pill remains, 16 seconds to get down to 70%, etc. Note that here the y variable is the time to dissolve, unlike the usual role of 'time' in longitudinal data; also, the x variable, fraction remaining, decreases over real 'time', though one could turn this round by replacing it by its complement, fraction dissolved, if preferred.

The data, listed in Table A.6, comprise four groups of pills. The profiles are plotted in Fig. 1.6. The times (in seconds) correspond to the fractions of pill remaining as given in the table.

The model to be fitted here is based on the classic cube root form of dissolution curve derived by Hixson and Crowell (1931): the time taken until only a fraction x of the pill remains is $t(1-x^{1/3})$, where $0<x<1$ and t is the total dissolution time. We will take a slightly generalized version of this model. First, the power 1/3 will be replaced by a parameter γ; this follows the discussion in Goyan (1965) of alternative γ values, 1/3, 1/2 and 2/3, which have been proposed in the literature. Secondly, $t(1-x^{\gamma})$ will be replaced by $\alpha + \beta(1-x^{\gamma})$; this reduces to the basic form if $x = 0$, but the relaxation of this constraint introduces an element of model robustness. The extra linear parameter, α, allows for a zero shift in the time recordings: ideally, it should be zero. In the current notation, for pill i the expected time y_{ij} until fraction x_{ij} remains is $\alpha + \beta(1-x_{ij}^{\gamma})$, with repeated observations for $j = 1,\dots,p$.

The parameters α and β will depend on various factors, as mentioned above. In the present example, there are four groups of pills which differ only in their storage conditions. Thus, the $\mathbf{X}_i(\gamma)$ matrix of the general outline of this section will have jth row $[\mathbf{s}_i \; \mathbf{s}_i(1-x_{ij}^{\gamma})]$, where $\mathbf{s}_i = [1\;1\;0\;0]$ if pill i is in group 1, $\mathbf{s}_i = [1\;0\;1\;0]$ for group 2, $\mathbf{s}_i = [1\;0\;0\;1]$ for group 3, and $\mathbf{s}_i = [1\;-1\;-1\;-1]$ for group 4. The corresponding vector of linear parameters can be designated as $[\alpha\;\alpha_1\;\alpha_2\;\alpha_3\;\beta\;\beta_1\;\beta_2\;\beta_3]$, giving the intercepts for the four groups as $\alpha+\alpha_1$, $\alpha+\alpha_2$, $\alpha+\alpha_3$ and $\alpha-\alpha_1-\alpha_2-\alpha_3$, and the slopes as $\beta+\beta_1$, $\beta+\beta_2$, $\beta+\beta_3$ and $\beta-\beta_1-\beta_2-\beta_3$.

For the covariance structure of the repeated time measures an initially plausible model would seem to be one of cumulative errors (section 6.2.3). If, for pill i, y_{ij} were unexpectedly large, say, then this random delay would be likely to be carried forward, i.e. the departures from expectation would accumulate. Again, to introduce some model robustness at the cost of one additional parameter, we will adopt the more general covariance form of section 6.2.3: $u_{ijk} = \sigma^2 \rho^{j-k}(1-\rho^{2k})/(1-\rho^2)$ for $j \geqslant k$.

For comparison, the within-groups covariance matrix shows increasing variance over time and generally high correlations.

The parameter vector is $\theta = [\alpha\ \alpha_1\ \alpha_2\ \alpha_3\ \beta_1\ \beta_2\ \beta_3\ \gamma\ \sigma\ \rho]$ and the estimates, for the data scaled by 10^{-1}, come out as

$$\tilde{\theta} = [1.06\ 0.22\ -\ 0.10\ -0.05\ 2.71\ -0.54\ -0.12\ 0.41\ 0.69\ 0.087\ 1.11]$$

The $\tilde{\gamma}$ value, 0.69, with standard error 0.04, is credibly close to 2/3, the value suggested by one of the protagonists in the literature. On the other hand, the $\tilde{\rho}$ value, 1.11, with standard error 0.01, is not credibly close to 1, the cumulative error value. In fact, the suggestion is that the errors are being magnified as time goes on, which, in turn, suggests fanning out of the individual profiles as in growth curves. This is borne out by an informal inspection of the unstandardized residuals. The question will be taken up in more detail in the next section.

It is differences between groups that are of most interest to the pharmaceutical scientist. First, a test for differences between the intercepts can be based on $W = \mathbf{c}^T \mathbf{V}_c^{-1} \mathbf{c}$, with $c = [\tilde{\alpha}_1\ \tilde{\alpha}_2\ \tilde{\alpha}_3]$; this yields $W = 12.90$ as χ_3^2 ($p < 0.005$). The evidence points strongly towards different intercepts, and the corresponding estimates for the four groups are 1.28, 0.96, 1.01 and 0.99 with standard errors 0.08, 0.05, 0.03 and 0.07. Second, the same process for the slope coefficients yields $W = 45.4$ as χ_3^2 ($p < 0.0005$). This argues strongly for differences in the total dissolution time between groups, and the corresponding estimates are 2.17, 2.58, 3.12 and 2.97 with standard errors 0.19, 0.08, 0.09 and 0.24.

Finally, the possibility of zero intercepts looks pretty unlikely from the estimated values: the Wald test for this hypothesis yields 1.5×10^3 as χ_4^2. Interestingly, the fact that α is about 1, with α_1, α_2 and α_3 much smaller, suggests the possibility that the measurements are all about 10 seconds too large, if the basic model is to be believed (although, for this hypothesis, $W = 13.3$ as χ_4^2, $p < 0.01$). Why this should be so is, unfortunately, lost in the mists of time–the data were acquired many years ago and now one can only speculate.

Example 8.2 Timber slip under loading

In civil engineering construction wood is often anchored to concrete and it is of some importance to know how strong such bonds are. Some experiments have been conducted at Surrey University in which a timber plank is sandwiched between two concrete surfaces, with anchoring provided by four screws. An increasing load is applied to the timber and the slippage (in millimetres) is recorded at intervals of 2 seconds. This results in 200–300 pairs of measurements, of slip and load, for a given experimental run. Apparently, it is routine to prune the data by selecting those pairs in which the slip takes predetermined values. Thus, in statistical terms, the variables swop roles: the observed slip becomes x and the applied load becomes y. The data given in Table A.15 are from eight experimental runs and are plotted in Fig. 8.1.

There is a well-established empirical law relating the two: $y = \alpha(1 - e^{-\gamma x})$. In the same spirit of model-robustness as applied in Example 8.1, we will use the slightly extended version $\mu_{ij} = \alpha + \beta e^{-\gamma x_{ij}}$ and then look at the possibility that $\beta = -\alpha$. For the nonlinear model form $\mathbf{X}_i(\gamma)$ is set up with jth row $[1\quad e^{-\gamma x_{ij}}]$. The covariance structure will be specified as the unstable autoregressive scheme of section 6.2.3. This is really in the nature of a first stab because, after some discussion with the engineer concerned, there did not seem to be much to go on for this aspect.

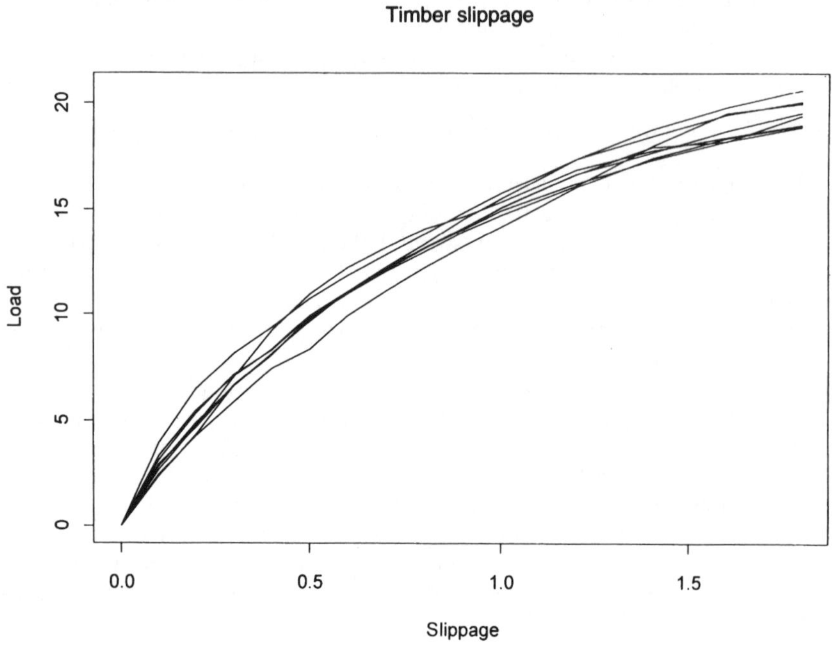

Figure 8.1. *Timber slip under loading: load versus slippage.*

The estimates obtained from the data, scaled by 10^{-1}, are

$$\tilde{\theta}=[\tilde{\alpha}\ \tilde{\beta}\ \tilde{\gamma}\ \tilde{\sigma}\ \tilde{\rho}]=[2.21\quad -2.21\quad 1.20\quad 0.0288\quad 0.909]$$

Straight away, $\tilde{\beta}=-\tilde{\alpha}$ hits you in the eye, at least to three-figure accuracy. A nervous check on the program reveals no obvious blunder that would artificially induce this constraint, so it is almost looking like an embarrassing success here–agreement between statistics and engineering. However, a formal Wald test for this linear hypothesis yields $W=40.8$ as χ_1^2 $(p<0.0005)$; we can breathe again–order is restored in the universe. On the other hand, the engineer maintains that $\beta=-\alpha$ has been confirmed by the data–this is a nice illustration of the distinction between statistical significance and practical significance. The reason for the improbably large value of W is that the standard error of $\tilde{\alpha}+\tilde{\beta}$ comes out very small, 1.2×10^{-4}, i.e. the intercept seems to be very well determined by the data plus model. This in turn is connected with pooling eight data sets, giving 120 observations, and a small residual error, $\tilde{\sigma}=0.0288$.

Coming on to the covariance structure now, the autoregressive parameter $\tilde{\rho}=0.909$ is quite high but, with a standard error 0.03, is not credibly close to the value 1 which reflects cumulative errors (section 6.2.3). It could represent a systematic lack of fit of a single curve to all eight runs. A growth curve approach for these data will be taken up in section 8.3.

As a prelude to the growth curve analysis, let us refit the basic model but allowing the linear parameters, α and β, to differ between runs. This requires $\mathbf{X}_i(\gamma)$ to be set up as $[\mathbf{A}_i\ \mathbf{A}_{i\gamma}]$, where $\mathbf{A}_i$ has 15 identical rows, each being the indicator vector for experimental run i out of 8, for example [1 1 0 0 0 0 0 0] for run 1; $\mathbf{A}_{i\gamma}$ starts out with the same rows but then has the j th multiplied by $e^{-\gamma x_{ij}}$. The resulting estimates for the linear parameters are

$$2.21,\quad 0.01,\quad -0.001,\quad -0.02,\quad -0.11,\quad 0.08,\quad 0.09,\quad -0.004,$$

$$-2.21,\quad -0.02,\quad 0.008,\quad 0.02,\quad 0.10,\quad -0.09,\quad -0.08,\quad 0.005$$

the constraint $\beta = -\alpha$ looks plausible for most of the runs, but a Wald test yields $W = 2.27 \times 10^7$ as χ_8^2, which does not inspire much confidence in the hypothesis. For information, the hypothesis tested is $\mathbf{H}\boldsymbol{\theta}_1 = \mathbf{0}$, where $\boldsymbol{\theta}_1$comprises the first 16 components of $\boldsymbol{\theta}$, i.e. the linear parameters, and

$$\mathbf{H} = \begin{bmatrix} 1 & 1 & 0 & 0 & 0 & 0 & 0 & 0 & 1 & 1 & 0 & 0 & 0 & 0 & 0 & 0 \\ 1 & 0 & 1 & 0 & 0 & 0 & 0 & 0 & 1 & 0 & 1 & 0 & 0 & 0 & 0 & 0 \\ 1 & 0 & 0 & 1 & 0 & 0 & 0 & 0 & 1 & 0 & 0 & 1 & 0 & 0 & 0 & 0 \\ 1 & 0 & 0 & 0 & 1 & 0 & 0 & 0 & 1 & 0 & 0 & 0 & 1 & 0 & 0 & 0 \\ 1 & 0 & 0 & 0 & 0 & 1 & 0 & 0 & 1 & 0 & 0 & 0 & 0 & 1 & 0 & 0 \\ 1 & 0 & 0 & 0 & 0 & 0 & 1 & 0 & 1 & 0 & 0 & 0 & 0 & 0 & 1 & 0 \\ 1 & 0 & 0 & 0 & 0 & 0 & 0 & 1 & 1 & 0 & 0 & 0 & 0 & 0 & 0 & 1 \\ 1 & -1 & -1 & -1 & -1 & -1 & -1 & -1 & 1 & -1 & -1 & -1 & -1 & -1 & -1 & -1 \end{bmatrix}$$

The remaining parameter estimates are $\tilde{\gamma} = 1.19$, $\tilde{\sigma} = 0.028$, and $\tilde{\rho} = 0.803$.

More important, from the statistical point of view, is how we deal with α's and β's differing between what are intended to be replicate experimental runs. There appears to be little doubt that there are real differences if one is to go by Wald tests: these give $W = 3.3 \times 10^8$ as χ_7^2 for the α's and $W = 3.2 \times 10^8$ as χ_7^2 for the β's. The huge size of these W-statistics gives one pause. To compute them, a large covariance matrix $\mathbf{V}_\theta = \mathbf{D}^{-1}\mathbf{C}(\mathbf{D}^{-1})^{\mathrm{T}}$ has to be estimated from a limited amount of data, and then 7×7 chunks of it inverted. It is possible that this all leads to rather large numerical fluctuations. If $Q(\mathbf{y}\,;\boldsymbol{\beta},\gamma,\tau)$ were the true likelihood function, then we could perform a likelihood ratio test of the homogeneity hypothesis, which gives $\chi_{14}^2 = 9.45$, a very different result. However, this latter test also falls under suspicion from the point of view of fitting many parameters to a limited sample, and maybe also a serious lack of test power compounded by non-normality. When the tests are repeated under the model constraint $\rho = 0$, i.e. taking $\mathbf{U}_i = \sigma^2\mathbf{I}$, we obtain similarly huge W-statistics and $\chi_{14}^2 = 118.88$ for the likelihood ratio. Evidently, the conflict between the Wald and likelihood ratio tests disappears when the serial correlation parameter is suppressed. But, judging by $\tilde{\rho}/\mathrm{se}\,(\tilde{\rho}) = 33.99$ (from the fit with homogeneous α and β) and $\tilde{\rho}/\mathrm{se}\,(\tilde{\rho}) = 17.05$ (from the fit with separate α's and β's), and the corresponding likelihood ratios, $\chi_1^2 = 166.29$ and $\chi_1^2 = 56.85$, ρ is not zero. It is all a little unsatisfactory.

Although such questions might be fascinating statistically, and indicate a need for theoretical work in this area, the engineer will view it all as statisticians arguing about the number of hypothesis tests that one can balance on a pin head. The fitted curves, even from the restricted model with homogeneous parameters, follow the data very closely, and inspection of the residuals does not show up any glaring deficiencies.

Whatever the true reasons for the conflicting evidence above, we still have a problem in principle of non-homogeneity of parameters between different experimental runs. Such variation is not put down to any changes of treatment or conditions, and so is to be regarded as unassigned, or random, variation. This leads us logically into the next section.

8.2 Nonlinear regression with random coefficients

The basic nonlinear regression model of section 8.1 is now extended to cover the case of random regression coefficients. This is more or less essential for biological data, in view of the large natural variation among living things, and the framework has mostly been developed in that context – hence the name 'growth curves'. It is for this reason that no examples involving biological data appeared in section 8.1 – the nonlinear model with parameters common to all individuals is just not sensible for such data. However, such random variation also occurs in engineering and physical sciences, as was seen in Example 8.2.

8.2.1 Nonlinear parameter fixed

Suppose that the $p_i \times 1$ vector $\mathbf{y}_i$ of measurements is recorded for individual i $(i = 1, \ldots, n)$ and that a within-individuals nonlinear regression model like (8.2) applies, but extended in the manner of the Laird-Ware two-stage model (section 5.1):

$$\mathbf{y}_i = \mathbf{X}_i(\gamma)\boldsymbol{\beta} + \mathbf{Z}_i(\gamma)\mathbf{b}_i + \mathbf{e}_i \tag{8.3}$$

where the $\mathbf{e}_i$ are independent with means $\mathbf{0}$ and covariance matrices $\mathbf{E}_i(\boldsymbol{\tau})$. Now suppose, as in section 5.1, that the linear parameters $\mathbf{b}_i$ are independently distributed over the population of individuals with mean $\mathbf{0}$ and covariance matrix $\mathbf{B}(\boldsymbol{\tau})$, and are independent of the $\mathbf{e}_i$. In consequence, the $\mathbf{y}_i$ are independent $p_i \times 1$ vectors with means and covariance matrices given by

$$\boldsymbol{\mu}_i = E(\mathbf{y}_i) = \mathbf{X}_i(\gamma)\boldsymbol{\beta}, \quad \boldsymbol{\Sigma}_i = \text{cov}(\mathbf{y}_i) = \mathbf{Z}_i(\gamma)\mathbf{B}(\boldsymbol{\tau})\mathbf{Z}_i(\gamma)^{\mathrm{T}} + \mathbf{E}_i(\boldsymbol{\tau}) \tag{8.4}$$

The Gaussian log-likelihood based on the data $(\mathbf{y}_1, \ldots, \mathbf{y}_n)$ is then

$$Q(\mathbf{y};\boldsymbol{\beta}, \gamma, \boldsymbol{\tau}) = -\frac{1}{2}\sum_{i=1}^{n} \{\log\det(2\pi\boldsymbol{\Sigma}_i) + (\mathbf{y}_i - \boldsymbol{\mu}_i)^{\mathrm{T}}\boldsymbol{\Sigma}_i^{-1}(\mathbf{y}_i - \boldsymbol{\mu}_i)\} \tag{8.5}$$

In (8.4) the nonlinear regression parameter γ is taken to be homogeneous over individuals, like $\boldsymbol{\tau}$ in the covariance structure. The more complicated situation, of randomly varying γ's, will be considered in section 8.2.2. Homogeneity is tenable when γ and $\boldsymbol{\tau}$ are somehow more fundamental than the $\mathbf{b}_i$'s, i.e. more like constants of nature than the individually varying $\mathbf{b}_i$

characteristics. For instance, in the exponential curve (8.1) $\beta_1$and β_2 are respectively location and scale parameters for y_{ij}, whereas γ determines the intrinsic shape of the regression model. It is always possible to specify models for nonlinear parameters, like γ and τ, involving explanatory variables and more basic parameters common to all individuals. However, unless the sample size is huge, such modelling runs the risk of being over-elaborate.

Gaussian estimation will be used, rather than maximum likelihood, since the error distribution is often clearly non-normal. For a simple example, take the exponential curve (8.1) with $\beta_1 = 0$, $\beta_2 > 0$ and $\gamma > 0$. Then, as x_{ij} increases $\mu_{ij} = E(y_{ij})$ tends to zero and, if the y measurements are necessarily positive, negative values of e_{ij} become squeezed: e_{ij} cannot be less than $-\mu_{ij}$. On the other hand, in a situation in which there is real confidence in an assumption of normality, the Gaussian estimation formulae can easily be simplified to obtain those for a normal likelihood analysis.

Example 8.1 (cont'd) Pill dissolution rates

It is possible that there will be random variation between nominally identical pills in respect of their dissolution curves. Various factors could be implicated: the manufacturing process and the precise conditions of the experiment are prime suspects to be held for questioning. The general outline above is framed to cover such situations, at least for random variation in the linear parameters. Thus, in the basic model for individual i,

$$y_{ij} = (\beta_{1i} + b_{1i}) + (\beta_{2i} + b_{2i})(1 - x_{ij}^{\gamma}) + e_{ij}$$

we take $[\beta_{1i}\ \beta_{2i}]$ to be $[\alpha + \alpha_g \quad \beta + \beta_g]$ when individual i is in group g; the required set of linear regression coefficients is then $\boldsymbol{\beta} = [\alpha\ \alpha_1\ \alpha_2\ \alpha_3\ \beta\ \beta_1\ \beta_2\ \beta_3]^{\mathrm{T}}$. Also, over the pill population, $[b_{1i}\, b_{2i}]$ has mean $\mathbf{0}$ and 2×2 covariance matrix $\mathbf{B}$, which we will assume not to differ between the pill groups. To express this within the general framework (8.3) we take $\mathbf{X}_i(\gamma)$ as previously, and $\mathbf{Z}_i(\gamma)$ as of similar form but without the grouping indicators, i.e. $\mathbf{Z}_i(\gamma)$ is $p \times 2$ with jth row $[1, 1 - x_{ij}^{\gamma}]$.

The estimates obtained are

$$\tilde{\boldsymbol{\beta}} = [1.06\ \ 0.22\ \ -0.094\ \ -0.046\ \ 2.73\ \ -0.55\ \ -0.098\ \ 0.43]$$

for the linear regression coefficients, $\tilde{\gamma} = 0.69$, $\tilde{\sigma}_e = 0.057$, $\tilde{\rho} = 0.79$, and

$$\tilde{\mathbf{B}} = 10^{-2}\begin{bmatrix} 1.44 & 3.05 \\ 3.05 & 6.47 \end{bmatrix}$$

The estimates of the 'old' parameters, i.e. all but $\mathbf{B}$, are similar to those obtained previously where, in effect, $\mathbf{B}$ was constrained to be $\mathbf{0}$. However, looking tentatively at the standard errors associated with $\tilde{\mathbf{B}}$, it seems that $\mathbf{B}$ is probably not equal to $\mathbf{0}$.

The parameter γ, estimated as 0.69 with standard error 0.04, is close to the postulated value 2/3, as previously. On the other hand, ρ has estimate 0.79 with standard error 0.32, so this is within credible distance of the value 1, unlike previously. The reason for the larger standard error here could be the sharing of fitted serial correlation between $\mathbf{E}_i$ and $\mathbf{Z}_i\ \mathbf{B}\mathbf{Z}_i^{\mathrm{T}}$: there could be an aliasing effect, meaning that the parameters involved are estimated with less precision.

Differences between groups can be tested as before. The Wald statistic for the α's gives $W = 12.8$ as χ_3^2 and that for the β's gives $W = 55.9$ as χ_3^2. These results are very similar to the previous ones.

Example 8.2 (cont'd) Timber slip under loading

Random variation between timber samples is not at all beyond the bounds of reasonable belief. After all, they were once 'living things' (see above). To apply the growth curve model outlined above to the present data, we have to write the basic model as

$$y_{ij} = (\beta_1 + b_{1i}) + (\beta_2 + b_{2i})e^{-\gamma x_{ij}} + e_{ij}$$

where the linear regression coefficients (b_{1i}, b_{2i}) vary randomly over the population of planks with mean $\mathbf{0}$ and covariance matrix $\mathbf{B}$. Thus, $\mathbf{Z}_i(\gamma)$ is set up as $\mathbf{X}_i(\gamma)$, there being no grouping here.

The estimates turn out as $\tilde{\boldsymbol{\beta}} = [2.21 \quad -2.21]$, $\tilde{\gamma} = 1.20$, $\tilde{\sigma}_e = 0.0288$, $\tilde{\rho} = 0.909$, and

$$\tilde{\mathbf{B}} = 10^{-10}\begin{bmatrix} 1.82 & -2.04 \\ -2.04 & 2.34 \end{bmatrix}$$

Immediately noticeable is the minute variation in $[b_{1i} \quad b_{2i}]$ given by $\mathbf{B}$: the planks might once have been subject to biological variation, but they show precious little evidence of it in death. The point $\tilde{\boldsymbol{\theta}}$ found in the search for the maximum of $Q(\mathbf{y}; \boldsymbol{\beta}, \gamma, \boldsymbol{\tau})$ is essentially the same as that in the previous treatment of these data in section 8.1 where $\mathbf{B}$ was effectively constrained to be $\mathbf{0}$. As noted in the continuation of Example 8.1, on pill dissolution rates, there could be an aliasing problem here, with serial correlation being shared between $\mathbf{B}$ and the parameter ρ in $\mathbf{E}_i$. Let us refit the data with ρ set to zero. This delivers $\tilde{\boldsymbol{\beta}} = [2.25 \quad -2.21]$, $\tilde{\gamma} = 1.11$, $\tilde{\sigma}_e = 0.0377$ and

$$\tilde{\mathbf{B}} = 10^{-3}\begin{bmatrix} 2.80 & -2.05 \\ -2.05 & 3.60 \end{bmatrix}$$

Now, there's nothing much to choose between the fitted curves, they are all very close to the data. However, the Gaussian log-likelihood is dramatically reduced, from 255.4553 (ρ unconstrained) to 206.4537 ($\rho = 0$). If we were relying on likelihood theory this would be pretty conclusive, with $\chi_1^2 = 98.00$: we would have to accept the first fit, with ρ unconstrained. This one refuses to lie down and behave like a textbook example – perhaps we should therefore have honoured tradition and left it out.

Example 8.3 Four physical tasks

Crowder (1978) gave data from seven female student volunteers. Heart rate (y beats per minute) and energy expenditure (w calories per minute) were each measured during the performance of four standard physical tasks (lying, sitting, walking, skipping). Some biological evidence was quoted in the paper to support the assumption that y is linearly related to $x = w/(\text{weight})^{0.75}$, using the subject's weight to scale the energy expenditure, and the conjecture that the individual lines might be concurrent at some point. Having set out these profound thoughts, it had to be admitted that this was not the most successful of the applications given in the paper. In fact, a plot of the data revealed that the regression line for subject 1 was clearly anomalous, compared with the others, and that the hypothesis of concurrence was not overwhelmingly supported, the remaining lines being near-parallel, if anything. Neverthless, it will provide a useful illustration here of nonlinear growth curves and, maybe more importantly, of things not going strictly according to plan.

The data are reproduced in Table A.16: the rows of the data matrix contain the subject number, weight (in kilograms), energy expenditures (four tasks), and corresponding heart rates (four tasks). A plot of y against x is given in Fig. 8.2.

The general equation for straight lines of slopes b_i, all passing through the point $[\xi \quad \eta]$, is $y = \eta + b_i(x - \xi)$. Thus, the model to be fitted here is of the form

$$y_{ij} = (\beta_1 + b_{1i}) + (\beta_2 + b_{2i})(x_{ij} - \gamma_2) + e_{ij}$$

where $x_{ij} = w/(\text{weight})^{\gamma_1}$. The power constant 0.75, resulting from earlier estimation exercises recorded in the biosciences literature, has been replaced by a parameter γ_1. The random linear coefficients, b_{1i} and b_{2i}, vary over the population of subjects with mean $\mathbf{0}$ and 2×2 covariance matrix $\mathbf{B}$. For strict concurrence of the individual lines, at the point $[\gamma_2 \; \beta_1]$, b_{1i} would have to be zero for all individuals, i.e. $B_{11} = 0$. The tasks were not too demanding, and were separated by adequate rest periods, so we will entertain the 'conditional-independence' model (section 6.2.1), $\mathbf{E}_i = \sigma_e^2\,\mathbf{I}$.

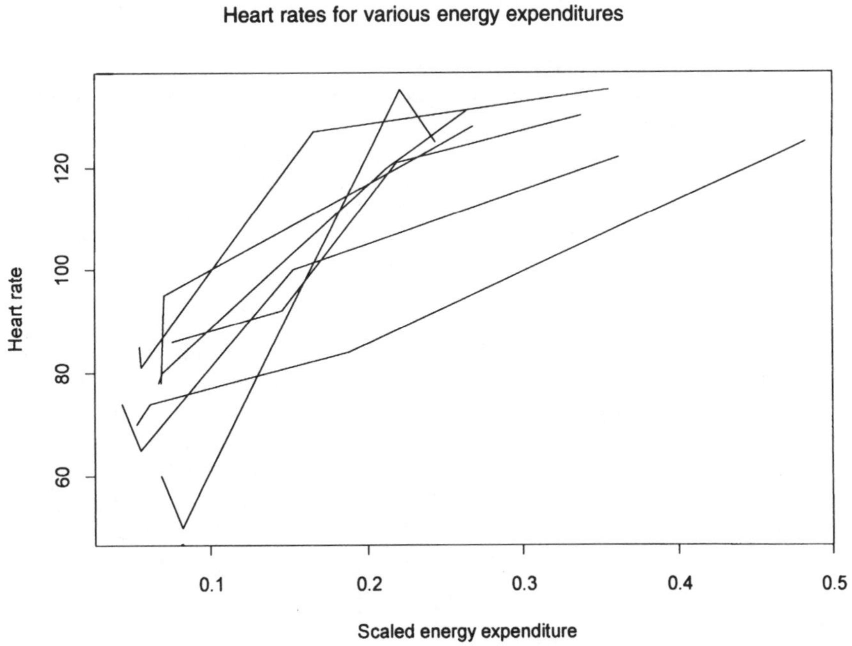

Figure 8.2. *Four physical tasks: heart rates for various energy expenditures.*

The matrices $\mathbf{X}_i$ and $\mathbf{Z}_i$ are set up as $p \times 2$ with jth row $[1 \quad x_{ij} - \gamma_2]$. The parameter set is

$$\boldsymbol{\theta} = [\beta_1\ \beta_2\ \gamma_1\ \gamma_2\ \sigma_e\ A_{11}\ A_{21}\ A_{22}]$$

where $\mathbf{A}\mathbf{A}^{\mathrm{T}} = \mathbf{B}$.

The result of fitting the model to the data (scaled by 10^{-1}) from subjects 2–7, subject 1 having been sent off for anomalous behaviour, was

$$\tilde{\boldsymbol{\theta}} = [-7.60 \quad 2.45 \quad 1.92 \quad -3.39 \quad 0.08 \quad 1.07 \quad -0.33 \quad 0.00]$$

Now, although the NAG optimization routine *e04jbf* reported a successful search for $\tilde{\boldsymbol{\theta}}$, the NAG matrix inversion routine *f01abf* reported that $\mathbf{D} = -\partial^2 Q/\partial\boldsymbol{\theta}^2$ was not positive definite at $\tilde{\boldsymbol{\theta}}$. At a genuine minimum of $-Q$, $\mathbf{D}$ should be positive definite, so something is wrong somewhere. Since $\mathbf{D}$ cannot be inverted, we cannot evaluate $V_\theta = \mathbf{D}^{-1}\mathbf{C}(\mathbf{D}^{-1})^{\mathrm{T}}$, the estimated covariance matrix of $\tilde{\boldsymbol{\theta}}$, and so cannot give standard errors or perform parametric tests.

A likely explanation for the difficulty is as follows. The individually fitted straight lines for subjects 2–7 form a thin pencil pointing south–west (see Crowder, 1978). It is not clear where the point of concurrence should be, even if one exists: it could be a long way down, even infinitely far so that the lines are in fact parallel. In terms of Q this means that the profile Gaussian log-likelihood $Q(\gamma_2, \beta_1)$, obtained by maximizing Q over all the other parameters, has a long ridge pointing south–west. The ridge could have a shallow peak at some point, or just continue to rise forever as one travels down it. Evidently, the precision resulting from the combined effects of the many numerical operations involved, including finite differencing for the derivatives, is not sufficient to give a definitive fix on a possible maximum point.

Example 8.4 Blood glucose levels over five hours

Crowder and Tredger (1981) and Crowder (1983) fitted an exponentially damped polynomial model, of the form $\beta_1 + \beta_2 x^3 \mathrm{e}^{-\gamma x}$, to blood glucose levels from seven volunteers (including one of the present authors). The subject took alcohol at time 0 and, at 14 times over a period of 5 hours, gave a blood sample. The whole session was repeated at a later date with the same subjects but with a dietary additive. The data are given in Table A.17 and plotted in Fig. 8.3.

The model to be fitted has the form

$$y_{ij} = (\beta_{1j} + b_{1i}) + (\beta_{2j} + b_{2i}) x_{ij}^3 \mathrm{e}^{-\gamma_j x_{ij}} + e_{ij}$$

for $i = 1, \ldots, 7$ and $j = 1, \ldots, 28$, and where $[\beta_{1j}\ \beta_{2j}\ \gamma_j]$ takes one value for $j = 1, \ldots, 14$ (the first session) and another for $j = 15, \ldots, 28$ (the second session); also, the random coefficients $[b_{1i}\ b_{2i}]$ will be taken to vary over subjects with mean $\mathbf{0}$ and 2×2 covariance matrix $\mathbf{B}$. Thus $\mathbf{X}_i(\boldsymbol{\gamma})$ is to be set up as a 28×4 matrix with jth row $[1, 1, x_{ij}^3 \mathrm{e}^{-\gamma_1 x_{ij}}, 0]$ for $j = 1, \ldots, 14$ and $[1, -1, 0, x_{ij}^3 \mathrm{e}^{-\gamma_2 x_{ij}}]$ for $j = 15, \ldots, 28$: this will make $\mu_{ij} = E(y_{ij})$ equal to $\beta_1 + \beta_2 + \beta_3 x_{ij}^3 e^{-\gamma_1 x_{ij}}$ for session 1, and to $\beta_1 - \beta_2 + \beta_4 x_{ij}^3 e^{-\gamma_2 x_{ij}}$ for session 2. Similarly, $\mathbf{Z}_i(\boldsymbol{\gamma})$ will be 28×2 with rows 1 to 14 set up as $[1, x_{ij}^3 e^{-\gamma_1 x_{ij}}]$ and rows 15 to 28 as $[1, x_{ij}^3 e^{-\gamma_2 x_{ij}}]$. For $\mathbf{E}_i$ we will adopt the form

$$\begin{bmatrix} \mathbf{E}_{i1} & \mathbf{0} \\ \mathbf{0} & \mathbf{E}_{i1} \end{bmatrix}$$

where $\mathbf{E}_{i1}$ follows the stable Markov model (section 6.2.4): $\mathbf{E}_{i1_{jk}} = \sigma_e^2 \rho^{|t_j - t_k|}/(1 - \rho^2)$.

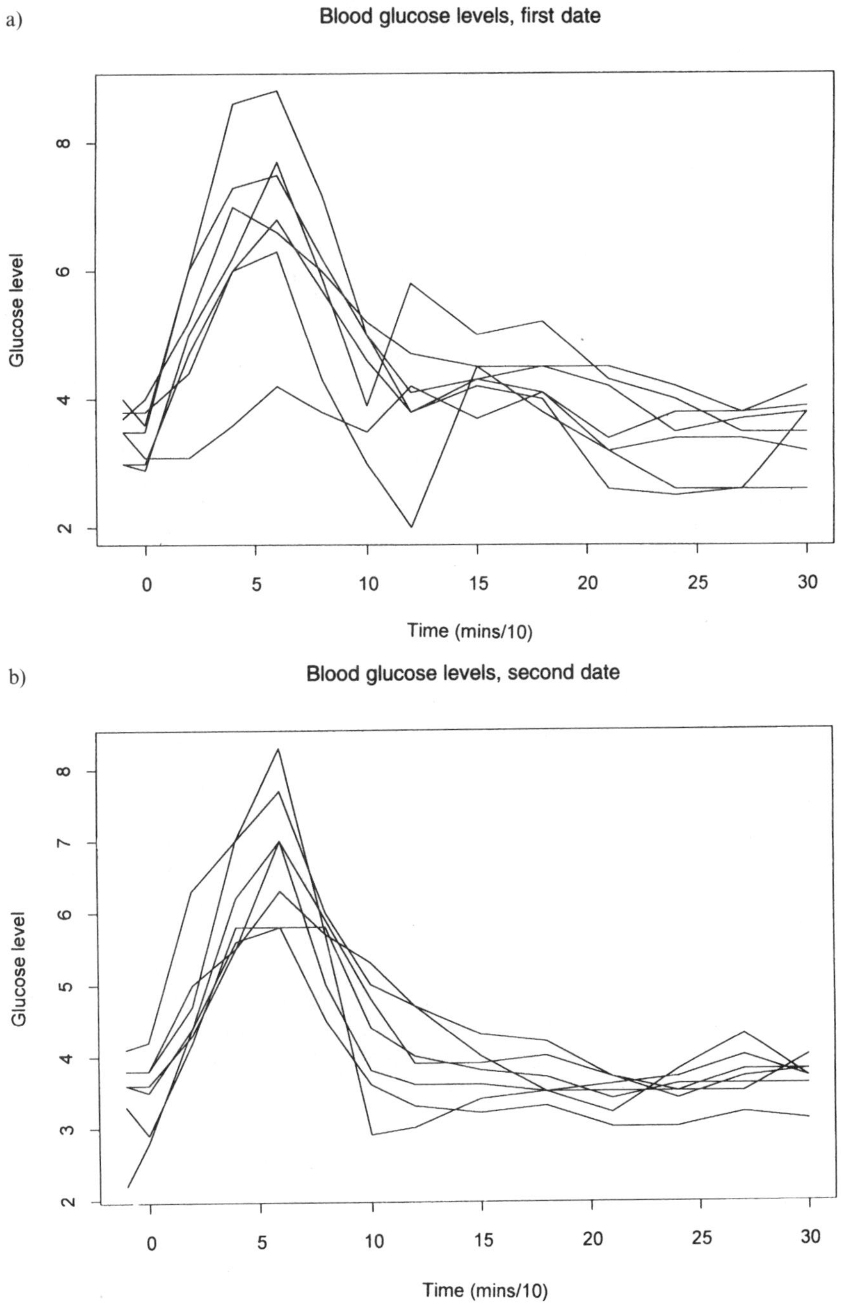

Figure 8.3. *Blood glucose levels over five hours: (a) first date; (b) second date.*

The parameter vector $\boldsymbol{\theta}$ then comprises $[\beta_1 \ldots \beta_4]$, $[\gamma_1\ \gamma_2]$, $[\sigma_e\ \rho]$ and three A components for $\mathbf{AA}^{\mathrm{T}} = \mathbf{B}$.

The numerical results from fitting the model are

$$\tilde{\boldsymbol{\beta}} = [4.86\quad 0.025\quad -0.86\quad -1.78]$$

$$\tilde{\boldsymbol{\gamma}} = [0.93\quad 1.22], \quad [\tilde{\sigma}_e\ \tilde{\rho}] = [1.07\quad 0.095]$$

$$\tilde{\mathbf{B}} = 10^{-10} \begin{bmatrix} 11.75 & -7.97 \\ -7.97 & 37.38 \end{bmatrix}$$

Uppermost on the agenda is the establishment of a difference between the sessions, i.e. between β_2 and 0, between β_3 and β_4, and between γ_1 and γ_2. The ratio $\tilde{\beta}_2/\mathrm{se}(\tilde{\beta}_2)$ comes out as 0.81, $(\tilde{\beta}_3 - \tilde{\beta}_4)/\mathrm{se}(\tilde{\beta}_3 - \tilde{\beta}_4)$ as 0.17, and $(\tilde{\gamma}_1 - \tilde{\gamma}_2)/\mathrm{se}(\tilde{\gamma}_1 - \tilde{\gamma}_2)$ as 0.04. These results are disappointing for the Few who selflessly sacrificed their bodies to alcohol abuse in the noble cause of medical research (and we never got paid!). The minute size of $\tilde{\mathbf{B}}$, is reminiscent of Example 8.2 earlier in this section. As there, refitting the model under the constraint $\rho = 0$ produces a more substantial $\tilde{\mathbf{B}}$, a much lower Gaussian log-likelihood, and little change in anything else.

In a normal plot of the standardized residuals there was a suggestion of normality, possibly reflecting the high calibre of the volunteer subjects involved.

8.2.2 Nonlinear parameter random

The model of section 8.2.1 can be extended to accommodate between-individuals variation of the nonlinear parameter γ. Harking back to the exponential curve, we are now admitting that individuals can have different shapes as well as sizes. In fact, recognition of 'all shapes and sizes' seems perfectly natural, even unavoidable, in many contexts. Unfortunately, this introduces a higher level of difficulty. However, the preliminary fitting of individual curves (which is one of the range of exploratory techniques which should be employed prior to any global analysis) might indicate that such variability of γ needs to be accommodated. By this is meant just that an informal inspection of the fitted individual γ-parameters shows variation in excess of that suggested by the standard errors of the estimates.

The within-individuals model, corresponding to (8.3), is

$$\mathbf{y}_i = \mathbf{X}_i(\mathbf{g}_i)\boldsymbol{\beta} + \mathbf{Z}_i(\mathbf{g}_i)\mathbf{b}_i + \mathbf{e}_i \tag{8.6}$$

Note that $\boldsymbol{\gamma}$ has been replaced by $\mathbf{g}_i$, the nonlinear parameter of individual i.

Let

$$E(\mathbf{b}_i \mid \mathbf{g}_i) = \bar{\mathbf{b}}_i(\mathbf{g}_i), \quad \mathrm{var}\,(\mathbf{b}_i \mid \mathbf{g}_i) = \mathbf{B}(\mathbf{g}_i) \tag{8.7}$$

be the indicated conditional mean and covariance matrix of $\mathbf{b}_i$ given $\mathbf{g}_i$. Then, from (8.6), $\mathbf{y}_i$ has conditional mean and covariance

$$E(\mathbf{y}_i \mid \mathbf{g}_i) = \mathbf{X}_i(\mathbf{g}_i)\boldsymbol{\beta} + \mathbf{Z}_i(\mathbf{g}_i)\bar{\mathbf{b}}_i(\mathbf{g}_i)$$

$$\text{var}\,(\mathbf{y}_i \mid \mathbf{g}_i) = \mathbf{Z}_i(\mathbf{g}_i)\mathbf{B}(\mathbf{g}_i)\mathbf{Z}_i(\mathbf{g}_i)^{\mathrm{T}} + \mathbf{E}_i(\boldsymbol{\tau}) \tag{8.8}$$

assuming that the $\mathbf{e}_i$ and $(\mathbf{b}_i, \mathbf{g}_i)$ are uncorrelated.

From (8.8) the unconditional mean and covariance matrix of $\mathbf{y}_i$ are

$$\boldsymbol{\mu}_i = E\{\mathbf{X}_i(\mathbf{g}_i)\boldsymbol{\beta} + \mathbf{Z}_i(\mathbf{g}_i)\bar{\mathbf{b}}_i(\mathbf{g}_i)\}$$

$$\boldsymbol{\Sigma}_i = E\{\mathbf{Z}_i(\mathbf{g}_i)\mathbf{B}(\mathbf{g}_i)\mathbf{Z}_i(\mathbf{g}_i)^{\mathrm{T}}\} + \mathbf{E}_i(\boldsymbol{\tau}) + \text{var}\{\mathbf{X}_i(\mathbf{g}_i)\boldsymbol{\beta} + \mathbf{Z}_i(\mathbf{g}_i)\bar{\mathbf{b}}_i(\mathbf{g}_i)\} \tag{8.9}$$

where the operations $E\{\cdot\}$ and $\text{var}\{\cdot\}$ are taken over the $\mathbf{g}_i$ distribution. To evaluate these, some distributional specifications will have to be made for $\mathbf{g}_i$. This will depend on the context, i.e. on what kind of parameter $\boldsymbol{\gamma}$ is; for instance, if γ were a rate parameter in an exponential decay model such as (8.1), a distribution on $(0, \infty)$ would be appropriate. In most practical cases the integrals implicit in (8.9) will have to be computed numerically. Once $\boldsymbol{\mu}_i$ and $\boldsymbol{\Sigma}_i$ are made available, via a customized subroutine, say, Gaussian estimation can be applied.

Normal likelihood theory

For the between-individuals model, which describes the distribution of parameters over the population of individuals, suppose that $(\mathbf{b}_i, \mathbf{g}_i)$ has a joint probability density $f_{bg}(\mathbf{b}_i, \mathbf{g}_i)$. Then, denoting the conditional density of $\mathbf{y}_i$ given the individual parameters by $f(\mathbf{y}_i \mid \mathbf{b}_i, \mathbf{g}_i)$, the log-likelihood is $\log \boldsymbol{\Pi} f(\mathbf{y}_i)$, where $f(\mathbf{y}_i)$ is the unconditional density of $\mathbf{y}_i$ obtained by integration over $(\mathbf{b}_i, \mathbf{g}_i)$:

$$f(\mathbf{y}_i) = \int f(\mathbf{y}_i \mid \mathbf{b}_i, \mathbf{g}_i) f_{bg}(\mathbf{b}_i, \mathbf{g}_i)\, \mathrm{d}\mathbf{b}_i\, \mathrm{d}\mathbf{g}_i \tag{8.10}$$

The integral is usually intractable for realistic models.

A certain amount of simplification is possible in some cases. If the conditional distribution of $\mathbf{b}_i \mid \mathbf{g}_i$ is taken as normal, then so is that of $\mathbf{y}_i \mid \mathbf{g}_i$ from (8.6). Then (8.10) can be reduced by integrating out $\mathbf{b}_i$:

$$f(\mathbf{y}_i) = \int f(\mathbf{y}_i \mid \mathbf{b}_i, \mathbf{g}_i) f_{b|g}(\mathbf{b}_i \mid \mathbf{g}_i) f_g(\mathbf{g}_i) \mathrm{d}\mathbf{b}_i\, \mathrm{d}\mathbf{g}_i$$

$$= \int f(\mathbf{y}_i \mid \mathbf{g}_i) f_g(\mathbf{g}_i) \mathrm{d}\mathbf{g}_i \tag{8.11}$$

here $f(\mathbf{y}_i|\mathbf{g}_i)$ is the multivariate normal density with mean and covariance matrix given by (8.8), and $f_g(\mathbf{g}_i)$ is the marginal density of $\mathbf{g}_i$ in the $(\mathbf{b}_i,\mathbf{g}_i)$ distribution. The integral in (8.11), over $\mathbf{g}_i$ only, is of smaller dimension than that in (8.10). If there were no between-individuals variation in $\mathbf{g}_i$, $f_g(\mathbf{g}_i)$ would be concentrated at a single point, say γ, and then the integral (8.11) would reduce to $f(\mathbf{y}_i|\gamma)$, the density of $N(\boldsymbol{\mu}_i,\boldsymbol{\Sigma}_i)$ with $\boldsymbol{\mu}_i$ and $\boldsymbol{\Sigma}_i$ given in (8.4). The assumption of conditional normality of $\mathbf{b}_i$ given $\mathbf{g}_i$ is just a natural extension of the usual assumption of unconditional normality of $\mathbf{b}_i$ given a fixed γ.

In general, $\mathbf{y}_i$ will not have a normal distribution, even though $\mathbf{y}_i|\mathbf{g}_i$ is conditionally normal. Once $f_g(\mathbf{g}_i)$ has been specified, $f(\mathbf{y}_i)$ can be computed (numerically) from (8.11), and a likelihood function thus obtained for inference.

8.2.3 Mean curves

We end this section with an aspect peculiar to nonlinear growth curves, where the $\mathbf{g}_i$ have a non-degenerate distribution over individuals.

■ *Illustration 8.1* (Crowder, 1983) Consider again a simple exponential within-individuals model:

$$E(y_{ij}|g_i, \mathbf{b}_i) = b_{i1} + b_{i2}\exp(-g_i x_{ij})$$

Suppose that the conditional mean of $\mathbf{b}_i = (b_{i1}, b_{i2})^{\mathrm{T}}$ given g_i, $\bar{\mathbf{b}}_i(g_i)$ of (8.8), has components linear in g_i:

$$E(b_{i1}|g_i) = c_1 + c_2 g_i, \quad E(b_{i2}|g_i) = d_1 + d_2 g_i$$

Then

$$E(y_{ij}|g_i) = (c_1 + c_2 g_i) + \{d_1 + d_2 g_i \exp(-g_i x_{ij})\}$$

Suppose further that g_i has an exponential distribution across individuals with density $f_g(g_i) = \xi e^{-\xi g_i}$. It follows from $E(g_i) = 1/\xi$ and

$$E\{g_i \exp(-g_i x_{ij})\} = \xi/(\xi + x_{ij})^2$$

that, unconditionally,

$$E(y_{ij}) = (c_1 + c_2/\xi) + \{d_1 + d_2\xi/(\xi + x_{ij})^2\}$$

The point is that $E(y_{ij})$, unlike $E(y_{ij}|g_i, \mathbf{b}_i)$, is not an exponential function of x_{ij}. The form of the curve, $E(y_{ij})$ versus x_{ij}, is determined by the form of $E\{g_i \exp(-g_i x_{ij})\}$. ■

In general, the conditional mean $E(y_{ij}|\mathbf{g}_i, \mathbf{b}_i)$, regarded as a function of x, will define the form of the individual curves. Let us use the phrase **mean curve** to describe the curve with the same form but with the mean parameters $\{E(\mathbf{g}), E(\mathbf{b})\}$ inserted. Thus, in Illustration 8.1,

$$E(g_i) = \eta/\xi \quad \text{and} \quad E(b_{ir}) = a_r + b_r\eta/\xi$$

so the mean curve there has the form

$$y = (a_1 + b_1\eta/\xi) + (a_2 + b_2\eta/\xi)\exp(-\eta x/\xi)$$

On the other hand, the population mean value $E(y_{ij})$, as a function of x, has a different form, which we call the **curve of means**. It defines average y

responses at given x values, whereas the mean curve exhibits the characteristics of individual trajectories such as turning points and asymptotes.

Having identified two different types of mean curve, the question arises as to which one should be used. The reply, evasive as ever, is that it depends on what you are trying to describe, typical individual curves (mean curve) or population averages at given x values (curve of means). This is a simple example of what has been called 'deconstructing statistical questions' (Hand, 1994).

In linear models, and more generally in cases where $\mathbf{g}_i$ is homogeneous over individuals as in section 8.2.1, the mean curve and the curve of means coincide. This can be seen from the expression for $\boldsymbol{\mu}_i$ (the curve of means) in (8.9); if $\mathbf{g}_i = \boldsymbol{\gamma}$ for all i, the reduces to $\mathbf{X}_i(\boldsymbol{\gamma})\boldsymbol{\beta} + \mathbf{Z}_i(\boldsymbol{\gamma})\bar{\mathbf{b}}_i(\boldsymbol{\gamma})$, the mean curve.

8.3 Transformations with Gaussian estimation

Many measures are necessarily positive and it is common practice to log-transform the observations as a prelude to applying normal linear models. Box and Cox (1964) investigated a more general idea where the transformation belongs to a parametric class. In particular, they considered transforming y to $(y^\lambda - 1)/\lambda$, which is basically a transformation to the λth power, but fixed up to tend continuously to $\log y$ as λ tends to 0. A normal linear model can now be fitted to the data, estimating the parameter λ in the process. For $\lambda \neq 0$ the analysis is necessarily approximate because, for $y > 0$, $(y^\lambda - 1)/\lambda$ can only take values in $(-1/\lambda, \infty)$ when $\lambda > 0$, and in $(-\infty, 1/\lambda)$ when $\lambda < 0$, not the whole range $(-\infty, \infty)$ of a normal distribution. Nevertheless, the technique has been widely used with success.

The effect of a nonlinear transformation should not be underestimated. It distorts the scale of y, so linear-model parameters are not comparable in different scales – they have different meanings. The main purpose of transformation, as outlined by Box and Cox (1964), is to simplify models–in particular, to linearize (the mean function), homogenize (the variance) and normalize (the error distribution). For instance, it should not take the reader long to figure out a power transformation which will remove the interaction effect from the following 2×2 table of observations:

$$\begin{bmatrix} 1 & 4 \\ 4 & 9 \end{bmatrix}$$

The serious point is that the linear effects in the two models, the ones describing the raw and transformed data, are on non-commensurate scales.

8.3.1 Univariate observations

We consider first the univariate case before going on to longitudinal data. Thus, we have independent observations y_i $(i = 1, \ldots, n)$ and the transformed version $y_{i\lambda}=(y_i^\lambda-1)/\lambda$ is approximately $N(\mu_i,\sigma^2)$, where $(\mu_i=\mu_i(\mathbf{x}_i;\boldsymbol{\beta})$ is a function of the vectors $\mathbf{x}_i$ of explanatory variables and $\boldsymbol{\beta}$ of regression coefficients, but not necessarily of linear form $\mathbf{x}_i^{\mathrm{T}}\boldsymbol{\beta}$; this is the basic set-up in which σ and λ are homogeneous over individuals, not varying with $\mathbf{x}_i$. Gaussian estimation can be applied to allow for the non-exact normality. The Gaussian log-likelihood function becomes, under the transformation to $y_{i\lambda}$,

$$Q(\mathbf{y};\boldsymbol{\theta})=\log\prod_{i=1}^{n}[(2\pi\sigma^2)^{-1/2}\,y_i^{\lambda-1}\exp\{-(y_{i\lambda}-\mu_i)^2/2\sigma^2\}]$$

$$=\sum_{i=1}^{n}\left\{-\frac{1}{2}\log(2\pi\sigma^2)+(\lambda-1)\log y_i-(y_{i\lambda}-\mu_i)^2/2\sigma^2\right\} \tag{8.12}$$

Differentiation of Q with respect to $\boldsymbol{\theta}=[\boldsymbol{\beta}\ \sigma\ \lambda]$ produces the Gaussian score function components

$$\mathbf{q}_{\boldsymbol{\beta}}(\mathbf{y};\boldsymbol{\theta})=\sigma^{-2}\sum_{i=1}^{n}(y_{i\lambda}-\mu_i)\boldsymbol{\mu}_i',\quad q_\sigma(\mathbf{y};\boldsymbol{\theta})=\sigma^{-3}\left\{\sum_{i=1}^{n}(y_{i\lambda}-\mu_i)^2-n\sigma^2\right\}$$

$$q_\lambda(\mathbf{y};\boldsymbol{\theta})=\sum_{i=1}^{n}\{\log y_i-(y_{i\lambda}-\mu_i)(y_i^\lambda\log y_i-y_{i\lambda})/\lambda\sigma^2\} \tag{8.13}$$

Equating $\mathbf{q}_{\boldsymbol{\beta}}(\mathbf{y};\boldsymbol{\theta})$ to $\mathbf{0}$, and $q_\sigma(y;\boldsymbol{\theta})$ to 0, defines the estimators $\tilde{\boldsymbol{\beta}}_\lambda$ and $\tilde{\sigma}_\lambda^2$ in terms of λ, the second explicitly as

$$\tilde{\sigma}_\lambda^2=n^{-1}\sum(y_{i\lambda}-\tilde{\mu}_i)^2$$

In the case of a linear model, $\mu_i=\mathbf{x}_i^{\mathrm{T}}\boldsymbol{\beta}$, $\tilde{\boldsymbol{\beta}}_\lambda$ can also be expressed explicitly as

$$\tilde{\boldsymbol{\beta}}_\lambda=\left\{\sum\mathbf{x}_i\mathbf{x}_i^{\mathrm{T}}\right\}^{-1}\sum y_{i\lambda}\mathbf{x}_i$$

in this case, to compute $\tilde{\lambda}$ we can insert these forms into $Q(\mathbf{y};\boldsymbol{\theta})$ and maximize the resulting profile log-likelihood with respect to the single variable λ.

Box and Cox (1964; 1982) and Hinkley and Runger (1984) took a slightly offbeat attitude towards λ: inferences are made conditional on $\tilde{\lambda}$. Justification of this 'invalid' approach is given in the papers cited. Effectively, λ ceases to be treated formally as a parameter and becomes part of the model structure, i.e. the functional forms assumed for the mean and variance of y_i. Thus, a λ value is to be found for which the model fits the data acceptably. The easiest route, if available, is to use a normal-models package which allows a user-defined transformation of y: different values of λ are tried until Q is maximized, i.e. the residual sum of squares is minimized, to an acceptable tolerance. The λ value so located is then taken as 'known', not estimated, and inference proceeds based on (8.12) with $\boldsymbol{\theta}=[\boldsymbol{\beta}\ \ \sigma]$. The asymptotic covariance matrix $\mathbf{V}_\theta$ of $\tilde{\boldsymbol{\theta}}$ is found from

$$\mathbf{D}=E\{-\partial^2Q/\partial(\boldsymbol{\beta},\sigma)^2\}=\sigma^{-2}\begin{bmatrix}\mathbf{M} & \mathbf{0}\\ \mathbf{0} & 2n\end{bmatrix}$$

$$\mathbf{C} = \operatorname{cov}\begin{bmatrix} \mathbf{q}_\beta(\mathbf{y};\boldsymbol{\theta}) \\ -q_\sigma(\mathbf{y}:\boldsymbol{\theta}) \end{bmatrix} = \sigma^{-2}\begin{bmatrix} \mathbf{M} & \sum\gamma_{1i}\boldsymbol{\mu}_i' \\ \sum\gamma_{1i}\boldsymbol{\mu}_i'^{\mathrm{T}} & \sum(\gamma_{2i}+2) \end{bmatrix}$$

and

$$\mathbf{V}_\theta = \mathbf{D}^{-1}\mathbf{C}(\mathbf{D}^{-1})^{\mathrm{T}} = \sigma^2\begin{bmatrix} \mathbf{M}^{-1} & \mathbf{M}^{-1}\boldsymbol{\Sigma}\gamma_{1i}\boldsymbol{\mu}_i'/(2n) \\ \sum\gamma_{1i}\boldsymbol{\mu}_i'^{\mathrm{T}}\mathbf{M}^{-1}/(2n) & \sum(\gamma_{2i}+2)/(2n)^2 \end{bmatrix} \tag{8.14}$$

where $\mathbf{M} = \sum\boldsymbol{\mu}_i'\boldsymbol{\mu}_i'^{\mathrm{T}}$. The asymptotic covariance matrix of $\tilde{\boldsymbol{\beta}}_\lambda$ is thus $\sigma^2\mathbf{M}^{-1}$ in which $\mathbf{M}$ can be estimated as $\sum\boldsymbol{\mu}_i'(\tilde{\boldsymbol{\beta}}_\lambda)\boldsymbol{\mu}_i'(\tilde{\boldsymbol{\beta}}_\lambda)^{\mathrm{T}}$. In the particular case of a linear model, $\mu_i = \mathbf{x}_i^{\mathrm{T}}\boldsymbol{\beta}, \mathbf{M} = \sum\mathbf{x}_i\mathbf{x}_i^{\mathrm{T}}$. For the other entries in $\mathbf{V}_\theta$, γ_{1i} and γ_{2i} are needed. The normal values, $\gamma_{1i} = \gamma_{2i} = 0$, can be used as approximations, or estimates based on sample moments can be substituted. The latter option gives an estimate which is more robust to non-normality, but not very reliable unless the sample size is large.

A useful variant which has emerged in recent years is the 'transform both sides' approach (Carroll and Ruppert, 1984). Suppose that we have a nonlinear form $h(x;\boldsymbol{\phi})$ with respectable theoretical justification for describing the response y. On the other hand, suppose that our knowledge of the nature of random variation in the observations of y is weak. Then we take for our model $y_{i\lambda} = h_{i\lambda} + e_i$, where $h_{i\lambda}$ is the same transformation of $h(x;\boldsymbol{\phi})$ as $y_{i\lambda}$ is of y_i. The situation, as far as computation is concerned, is now not very different from that described above though we might well have to introduce an extra linear parameter, using the device mentioned in section 8.1.

8.3.2 Multivariate observations

There are various ways, of varying complexity, in which transformations can be applied to positive repeated measures. Let us consider a fairly basic situation in which the $\mathbf{y}_i$'s have similar structure, for instance y_{ij} is recorded at the same time x_j for every i. Then each $\mathbf{y}_i$ will have p components, though possibly with some missing values assumed to be missing at random, i.e. the fact that they are absent has no implications for the inferences to be made. Consider first transforming the jth component y_{ij} of $\mathbf{y}_i$ to $(y_{ij}^{\lambda_j} - 1)/\lambda_j$, thus allowing the λ parameters to vary between components. This might be deemed necessary because the marginal distributions of the components are thought to differ systematically, perhaps even with some smooth trend over time which could be modelled into the λ_j sequence. On the other hand, the transformed components would then be on different measurement scales so any comparative assessments, for example of corresponding regression coefficients, would be effectively ruled out. We will use a common value of λ for all components. This will certainly suffice when they have similar univariate distributions and, in any case, will give results capable of straightforward interpretation.

It is appropriate to apply Gaussian estimation to the transformed vector $\mathbf{y}_{i\lambda}$ because, as noted above, we cannot achieve exact normality of the components, and probably still less of the dependence structure. The Gaussian log-likelihood is

$$Q(\mathbf{y};\boldsymbol{\theta}) = \log \prod_{i=1}^{n} \left[\det(2\pi\boldsymbol{\Sigma}_i)^{-1/2} (\Pi_j\, y_{ij}^{\lambda-1}) \exp\left\{ -\frac{1}{2}(\mathbf{y}_{i\lambda} - \boldsymbol{\mu}_i)^{\mathrm{T}} \boldsymbol{\Sigma}_i^{-1} (\mathbf{y}_{i\lambda} - \boldsymbol{\mu}_i) \right\} \right]$$

$$= \sum_{i=1}^{n} \left\{ -\frac{1}{2} \log\det(2\pi\boldsymbol{\Sigma}_i) + (\lambda - 1)\Sigma_j \log y_{ij} - \frac{1}{2}(\mathbf{y}_{i\lambda} - \boldsymbol{\mu}_i)^{\mathrm{T}} \boldsymbol{\Sigma}_i^{-1} (\mathbf{y}_{i\lambda} - \boldsymbol{\mu}_i) \right\} \qquad (8.15)$$

where Π_j and $\sum_j$ denote the product and sum over the non-missing components of $\mathbf{y}_i$, and $\boldsymbol{\mu}_i$ and $\boldsymbol{\Sigma}_i$ have entries deleted in accordance with missing components.

Assume that $\boldsymbol{\mu}_i = \boldsymbol{\mu}_i(\mathbf{x}_i;\boldsymbol{\beta})$ and $\boldsymbol{\Sigma}_i = \boldsymbol{\Sigma}_i(\mathbf{x}_i;\boldsymbol{\beta},\boldsymbol{\tau})$, and treat λ as described in section 8.3.1, i.e. estimate it and then take the value as known. Then $\boldsymbol{\theta} = [\boldsymbol{\beta} \quad \boldsymbol{\tau}]$ and (8.15) has the same form as (7.7) except that $\mathbf{y}_i$ there is replaced by $\mathbf{y}_{i\lambda}$ here, and (8.15) contains the additional term $(\lambda - 1)\sum_j \log y_{ij}$; however, this term plays no part in the inference because it does not involve $\boldsymbol{\theta}$. So, what this all boils down to is that we proceed just as in section 7.1.3 but with transformed y measurements.

Example 8.1 (cont'd) Pill dissolution rates

In addition to the straightforward use of transformations, as described above, they can be used to assess model adequacy. Thus, if a proposed model turns out to fit the transformed data better than it does the untransformed data, something is wrong somewhere.

We consider the pill dissolution times in this way. We have already found that, after subtracting about 10 seconds from the recorded times, the model fits best with γ close to 2/3, one of the theoretically predicted values. The result of refitting the model with a Box – Cox transformation is as follows. With $\mathbf{Z}_i = \mathbf{0}$, allowing no between-individual variation in the regression coefficients as in section 8.1, the maximized Gaussian log-likelihood is 104.1182 with λ constrained to be 1 (i.e. for the untransformed data), and 104.4710 with λ estimated as $\tilde{\lambda} = 1.12$. With $\mathbf{Z}_i$ having jth row $(1, 1 - x_{ij}^{\gamma})$, allowing for random regression coefficients as in section 8.2, the Gaussian log-likelihood is 123.2313 for $\lambda = 1$ and 123.8082 for $\tilde{\lambda} = 0.81$. In either case, the departure of λ from the value 1, and the associated increase in Gaussian log-likelihood, are unconvincing so no serious lack of model fit is evident in this particular direction. Also, in either case, none of the other parameters is much changed when λ moves from 1 to $\tilde{\lambda}$ except for α, the average intercept. For the first fit, $\tilde{\alpha}$ goes from 1.06 for untransformed data down to 0.056 at $\lambda = 1.12$; for the second, $\tilde{\alpha}$ goes from 1.06 for untransformed data down to 0.068 at $\lambda = 0.81$. It seems that the 10-second penalty more or less disappears under the transformation, though why this should be so was a mystery to the author. (Strictly between ourselves, it occurred to him later that the Box–Cox transformation incorporates the subtraction of 1 from y_{ij}^{λ}.)

8.4 Further reading

For more discussion, applications and worked examples of nonlinear growth curves the following papers are representative: Fisk (1967), Crowder and Tredger (1981), Berkey (1982), Crowder (1983), Racine-Poon (1985), Racine–Poon *et al.* (1986), Berkey and Laird (1986), Rudemo, Ruppert and Streiberg (1989), Lindstrom and Bates (1990), Palmer, Phillips and Smith (1991), Vonesh and Carter (1992), Davidian and Giltinan (1993) and McCabe and Leybourne (1993). Sandland and McGilchrist (1979) discussed growth models based on stochastic differential equations. Davidian and Gallant (1993) presented a semiparametric analysis in which the individual regression curve has parametrically specified form, but the individual regression coefficients, linear and nonlinear, depend on random effects whose between-individuals distribution is only assumed to satisfy a certain smoothness condition which enables it to be estimated via a certain series expansion.

The aliasing problem between parameters involved in the serial correlation (Examples 8.1 and 8.2 in section 8.2) has been discussed by Jones (1990): he gave an example together with some recommendations for model choice based on the Akaike information criterion.

Transformations have long been applied to statistical data, but the idea of using a parametrized class seems to have originated with Box and Cox (1964). Bickel and Doksum (1981) showed that blind application of standard asymptotic theory can give unsatisfactory results: in particular, allowing λ to vary can induce extremely large variances for the $\hat{\beta}$'s. Box and Cox (1982) and Hinkley and Runger (1984) pointed out that β's under different λ values are not commensurate and so it does not make sense to allow λ to vary when comparing β values. The set of papers makes entertaining reading, and the discussion of the last one brings out some interesting points for debate.

CHAPTER 9

Generalized linear models and maximum quasi-likelihood estimation

This chapter begins with a basic outline of generalized linear models (GLMs) and maximum quasi-likelihood estimation (MQLE) for univariate, independent observations. It then goes on to the multivariate case, as required for repeated measures or longitudinal data, and more general models and methods. The main focus for modelling and fitting is, as previously, on means, variances and covariances among the measures.

9.1 GLMs and MQLE for univariate data

Generalized linear models have had a major impact on the teaching of theoretical Statistics over the past twenty years. In association with this, the GLIM computer package provides the means for turning theory into practice with real data. The source papers are Nelder and Wedderburn (1972) and Wedderburn (1974), and the standard text is McCullagh and Nelder (1989).

9.1.1 Basic structure

We assume that data are available comprising independent, univariate observations y_i $(i=1, \ldots, n)$, where y_i has mean $\mu_i = E(y_i)$, variance $\sigma_i^2 = \mathrm{var}(y_i)$, and is escorted by a $q \times 1$ vector $\mathbf{x}_i$ of explanatory variables. The GLM/MQLE structure has the following components: (a) model specifications for μ_i and σ_i^2, $g(\mu_i) = \mathbf{x}_i^{\mathrm{T}}\boldsymbol{\beta}$ and $\sigma_i^2 = \phi v(\mu_i)$, where g is the **link function**, $\boldsymbol{\beta}$ is the $q \times 1$ vector of regression coefficients, $v(\cdot)$ is the **variance function**, and ϕ is a scale factor, either a known constant or an unknown parameter; and (b) a **quasi-score function** for $\boldsymbol{\beta}$,

$$\mathbf{q}(\mathbf{y}; \boldsymbol{\beta}) = \sum v(\mu_i)^{-1}(y_i - \mu_i)\boldsymbol{\mu}_i' \tag{9.1}$$

where $\mathbf{y} = (y_1, \ldots, y_n)^{\mathrm{T}}$ and $\boldsymbol{\mu}_i'$ is the $q \times 1$ vector with jth element $\partial\mu_i/\partial\beta_j$.

The term 'GLM' relates specifically to the μ_i specification in (a), the generalization being the introduction of a link function g. The phrase 'maximum quasi-likelihood estimation' refers to the estimating equation, $\mathbf{q}(\mathbf{y};\boldsymbol{\beta}) = \mathbf{0}$, used to obtain the MQL estimator $\hat{\boldsymbol{\beta}}$ of $\boldsymbol{\beta}$. If y_i has a linear exponential family distribution with specifications (a), then $\mathbf{q}(\mathbf{y};\boldsymbol{\beta})$ is proportional to the true score function for $\boldsymbol{\beta}$, i.e. $\mathbf{q}(\mathbf{y};\boldsymbol{\beta}) \propto \partial \log L/\partial\boldsymbol{\beta}$, where L is the likelihood function based on the data $\mathbf{y}$. In that case, $\hat{\boldsymbol{\beta}}$ is a genuine maximum likelihood estimator (MLE). Otherwise, the quasi-score function retains key properties of a true score function which give the MQLE $\hat{\boldsymbol{\beta}}$ asymptotic (large-sample) properties similar to those of a true MLE.

In one simple case MQLE coincides with weighted least squares. Let

$$Q(\mathbf{y};\boldsymbol{\beta}) = \sum (y_i - \mu_i)^2/\sigma_i^2$$

Provided that the σ_i are not functions of $\boldsymbol{\beta}$, $\mathbf{q}(\mathbf{y};\boldsymbol{\beta})$ is proportional to $\partial Q/\partial\boldsymbol{\beta}$, so the MQLE is obtained by minimizing Q. On the other hand, if the σ_i are dependent on $\boldsymbol{\beta}$, for example as in (a), the differentiation would introduce extra terms involving $\partial\sigma_i/\partial\boldsymbol{\beta}$. In any case, the equation $\mathbf{q}(\mathbf{y};\boldsymbol{\beta}) = \mathbf{0}$ is a weighted method of moments, equating $\sum \mathbf{a}_i y_i$ to its expected value, $\sum \mathbf{a}_i \mu_i$, with vector weights $\mathbf{a}_i = v(\mu_i)^{-1}\boldsymbol{\mu}_i'$.

■ *Illustration 9.1 Some exponential family models*

1. **Binomial logit model:** y_i is the number of successes in m trials each with probability π_i of success, and $\log\{\pi_i/(1-\pi_i)\} = \mathbf{x}_i^{\mathrm{T}}\boldsymbol{\beta}$. In GLM terms, $\mu_i = m\pi_i$, $g(\mu) = \log\{\mu/(m-\mu)\}$, and $\sigma_i^2 = \phi v(\mu_i)$ with $v(\mu) = \mu(1-\mu/m)$ and $\phi = 1$.
2. **Poisson loglinear model:** y_i has Poisson distribution with mean μ_i governed by $\log \mu_i = \mathbf{x}_i^{\mathrm{T}}\boldsymbol{\beta}$. In GLM terms, $g(\mu) = \log \mu$, and $\sigma_i^2 = \phi v(\mu_i)$ with $v(\mu) = \mu$ and $\phi = 1$.
3. **Exponential loglinear model:** y_i has exponential distribution with mean μ_i governed by $\log \mu_i = \mathbf{x}_i^{\mathrm{T}}\boldsymbol{\beta}$. In GLM terms, $g(\mu) = \log \mu$ and $\sigma_i^2 = \phi v(\mu_i)$ with $v(\mu)\,\mu^{-2}$ and $\phi = 1$.
4. **Normal linear homoscedastic model:** y_i is $N(\mu_i, \sigma^2)$ with $\mu_i = \mathbf{x}_i^{\mathrm{T}}\boldsymbol{\beta}$. In GLM terms, $g(\mu) = \mu$ and $\sigma_i^2 = \phi v(\mu_i)$ with $v(\mu) = 1$. ■

It can be verified that the score function for each of these examples has the form of $\mathbf{q}(\mathbf{y};\boldsymbol{\beta})$ in (b) and that, therefore, the MQLE $\hat{\boldsymbol{\beta}}$ is a true MLE.

9.1.2 Inference based on MQLE

Provided that μ_i is correctly specified in (a), the quasi-score equation $\mathbf{q}(\mathbf{y};\boldsymbol{\beta}) = \mathbf{0}$ will generally produce a consistent, asymptotically normal estimator for $\boldsymbol{\beta}$ (section 7.1.1). If σ_i^2 is also correctly specified the estimating equation $\mathbf{q}(\mathbf{y};\boldsymbol{\beta}) = \mathbf{0}$ is asymptotically optimal in the sense that, among all such estimating equations which are linear in the observations y_i, it yields the estimator with the smallest asymptotic variance (McCullagh, 1983, section 4). In this case, the MQLE $\hat{\boldsymbol{\beta}}$ has large-sample distribution approximately $N(\boldsymbol{\beta}, \mathbf{V}_{\beta})$, with $\mathbf{V}_{\beta}=\phi(\mathbf{X}^{\mathrm{T}}\mathbf{W}\mathbf{X})^{-1}$; here $\mathbf{X}$ is the $n\times q$ matrix with ith row $\mathbf{x}_i^{\mathrm{T}}$, $\mathbf{W}=\mathrm{diag}\ [w_i]$ and $w_i^{-1}=v(\mu_i)g'(\mu_i)^2$. Tests and confidence regions for $\boldsymbol{\beta}$ follow from this in the usual way (section 7.1.1). To compute an estimate of $\mathbf{V}_{\beta}$ we substitute $\hat{\mu}_i=\mu_i(\hat{\boldsymbol{\beta}})$ for μ_i in w_i, and use an estimator for ϕ. If $\mathbf{q}(\mathbf{y};\boldsymbol{\beta})$ is a true score function the MLE for ϕ is a natural choice. Otherwise, the one usually recommended is

$$\hat{\phi}=(n-q)^{-1}\sum(y_i-\hat{\mu}_i)^2/v(\hat{\mu}_i) \tag{9.2}$$

the divisor incorporating a degrees-of-freedom correction for estimating $\boldsymbol{\beta}$ in $\mu_i(\boldsymbol{\beta})$.

Assume now that y_i does have an exponential family distribution with mean μ_i and variance σ_i^2 correctly specified in (a). The likelihood function L, for the observations $\mathbf{y}=[y_1\ldots y_n]^{\mathrm{T}}$, depends on the parameter $\boldsymbol{\beta}$ through $\boldsymbol{\mu}=[\mu_1\ldots\mu_n]$ and, for the moment, we will write it as $L(\mathbf{y};\boldsymbol{\mu},\phi)$ rather than in the 'correct' form $L(\mathbf{y};\boldsymbol{\beta},\phi)$. In a **saturated model** the completely unconstrained estimate y_i is used for μ_i, producing likelihood $L(\mathbf{y};\mathbf{y},\phi)$. Under the model specification (a) the maximized likelihood is $L(\mathbf{y};\hat{\boldsymbol{\mu}},\phi)$, where $\hat{\boldsymbol{\mu}}=\boldsymbol{\mu}(\hat{\boldsymbol{\beta}})$. In GLM terminology the corresponding scaled log-likelihood ratio,

$$D(\mathbf{y};\hat{\boldsymbol{\mu}})=2\phi\{\log L(\mathbf{y};\mathbf{y},\phi)-\log L(\mathbf{y};\hat{\boldsymbol{\mu}},\phi)\} \tag{9.3}$$

is called the **deviance**. Suppose that a hypothesis H places a constraint on $\boldsymbol{\beta}$, effectively reducing the parameter dimension from q to q_1, say. Denote by $\hat{\boldsymbol{\beta}}_H$ the MLE of $\boldsymbol{\beta}$ under H, and let $\hat{\boldsymbol{\mu}}_H=\boldsymbol{\mu}(\hat{\boldsymbol{\beta}}_H)$. The log-likelihood ratio statistic for testing H is

$$2\{\log L(\mathbf{y};\hat{\boldsymbol{\mu}},\phi)-\log L(\mathbf{y};\hat{\boldsymbol{\mu}}_H,\phi)\}=\phi^{-1}\{D(\mathbf{y};\hat{\boldsymbol{\mu}}_H)-D(\mathbf{y};\hat{\boldsymbol{\mu}})\}$$

By standard asymptotic likelihood theory, this is approximately distributed as χ^2 with degrees of freedom $q-q_1$ in large samples. If ϕ is unknown an estimate needs to be inserted, such as that described above, and then one can refer $\hat{\phi}^{-1}\{D(\mathbf{y};\hat{\boldsymbol{\mu}}_H)-D(\mathbf{y};\hat{\boldsymbol{\mu}})\}$ to the F distribution with $q-q_1$ (numerator) and $n-q$ (denominator) degrees of freedom (Jorgensen, 1987, section 4.3).

9.1.3 Goodness of fit

Comprehensive accounts of general methods for assessing how well regression models fit data are given in Cook and Weisberg (1982), Atkinson (1985) and Belsley, Kuh and Welsch (1980). Some summary notes relevant to our context only will be given here.

Residuals

The traditional standardized residual corresponding to y_i is $(y_i - \mu_i)/\sigma_i$. In the quasi-likelihood case this will have unknown distribution but can be used for plotting. With estimates inserted, we obtain $(y_i - \hat{\mu}_i)/\hat{\sigma}_i$, where $\hat{\mu}_i = \mu_i(\hat{\boldsymbol{\beta}})$ and $\hat{\sigma}_i = \hat{\phi} v(\hat{\mu}_i)$. An improved version, which has variance closer to 1, is

$$r_i = (y_i - \hat{\mu}_i)/\{\hat{\sigma}_i(1 - \hat{h}_{ii})^{1/2}\} \tag{9.4}$$

where h_{ii} is the (i, i)th element of the **hat matrix**

$$\mathbf{H} = \mathbf{W}^{1/2}\mathbf{X}(\mathbf{X}^{\mathrm{T}}\mathbf{W}\mathbf{X})^{-1}\mathbf{X}^{\mathrm{T}}\mathbf{W}^{1/2}$$

r_i is called a **standardized Pearson residual.**

Uses of residuals include the identification of outliers or unexpected patterns, or general misspecification of relationships. Useful plots include those of residuals versus (a) fitted values, (b) x's already in the model, (c) x's not yet in the model, and (d) preceding residuals (to detect serial dependence for time-ordered observations). Residuals can also be used to construct more formal goodness-of-fit tests, such as the Pearson chi-square.

In the true likelihood case residuals can also be defined from the contributions to the deviance. Thus, $D(\mathbf{y}; \hat{\boldsymbol{\mu}})$ in (9.3) is the sum of n terms, say $D_i (i = 1, \ldots, n)$, and the ith **deviance residual** is $s_i D_i^{1/2}$, where $s_i = \operatorname{sgn}(y_i - \hat{\mu}_i)$; the variance-corrected version is $s_i D_i^{1/2}(1 - \hat{h}_{ii})^{-1/2}$.

The precise definition used for residuals is probably less important than using some definition and then inspecting the residuals carefully. When the observations have a very restricted set of values, such as with binary data, residual plots are less helpful.

Useful references in the present context are Williams (1984), Pierce and Schafer (1986), Davison and Snell (1991, section 4.4) and Weiss and Lazaro (1992). Chapter 5 of Collett's (1991) excellent book gives a very readable account of goodness of fit for binomial and binary data.

Extended models

A general approach for validating the current parametric model is to embed it in a more general one of which it is a submodel. Then a parametric test for degeneracy to the special case can be applied. Examples include the trial introduction of additional explanatory variables, Box–Cox transformation of the response variable or of covariates, Lehmann alternatives (where the fitted distribution function $F(y)$ is enhanced to $F(y)^{\phi}$), and Neyman 'smooth tests' (where the embedding of the fitted density $f(y)$ is done in another way, adding terms to its expression in exponential form).

In the GLM context we might try extended versions of the link function or the variance function. An example of the former is Aranda-Ordaz's (1981) family:

$$g_\lambda(\mu) = \log[\lambda^{-1}\{(1 - \mu)^{-\lambda} - 1\}]$$

This is useful for binary data since it reduces to the logit transform for $\lambda = 1$, and to the complementary log-log transform as $\lambda \to 0$. The extended model can be fitted repeatedly to find the best-fitting λ value. Then, assuming that we have a true likelihood, likelihood ratio tests can be applied to decide whether λ equal to 1 or 0 will suffice. Extension of the variance function can be performed likewise. Of course, if an extended model is deemed to be necessary this might take us outside the GLM framework.

Model improvement

As a result of examining residuals and testing for extended models, and perhaps using other diagnostics, it might become apparent that the model is in need of improvement. There might be further explanatory variables to be brought in, or modification of certain functional forms defining the model. The process of fitting, checking and improving is iterative, but can be overdone–statistical judgement plays an important part here.

9.1.4 Comments

The GLM/MQLE framework is a useful vehicle for introducing simple exponential family models and the theory that goes with them. The GLIM package enlivens the learning process by facilitating practical work with simple data sets. However, the framework is limited, and one soon comes up against the boundaries when one has to deal with real-life data. A major limitation is that GLM/MQLE focuses strongly on the mean μ_i and the associated parameter vector $\boldsymbol{\beta}$, relegating the variance structure merely to a known function of the mean. In most practical situations, however, $\boldsymbol{\beta}$ forms only one part of the full parameter vector, and the mean forms only one part of the data structure, even though it might be of primary interest.

■ *Illustration 9.2 GLMs with additional parameters* The following models are commonly used extensions of those in Illustration 9.1.

1. **Beta-binomial logit model:** $\mu_i = m\pi_i$ with $\log\{\pi_i/(1-\pi_i)\} = \mathbf{x}_i^{\mathrm{T}}\boldsymbol{\beta}$ as before, but now $\sigma_i^2 = m\pi_i(1-\pi_i)\{1+(m-1)\rho\}$, where ρ is the intraclass correlation coefficient.
2. **Negative binomial loglinear model:** $\log\mu_i = \mathbf{x}_i^{\mathrm{T}}\boldsymbol{\beta}$ as for the Poisson loglinear model, but now $\sigma_i^2 = \mu_i + \nu\mu_i^2$.
3. **Weibull loglinear model for survival data:** y_i^{ϕ} has an exponential distribution with mean $\exp(\mathbf{x}_i^{\mathrm{T}}\boldsymbol{\beta})$.
4. **Box–Cox transformed normal linear homoscedastic model:** $(y_i^{\lambda}-1)/\lambda$ has distribution $N(\mathbf{x}_i^{\mathrm{T}}\boldsymbol{\beta}, \sigma^2)$. ■

Suppose that $\mathbf{q}(\mathbf{y};\boldsymbol{\beta})$ in (9.1) is based on a true likelihood function, $L(\mathbf{y};\boldsymbol{\theta})$, with $\boldsymbol{\theta} = [\boldsymbol{\beta}, \boldsymbol{\tau}]$ and $\mathbf{q}(\mathbf{y};\boldsymbol{\beta}) = \partial \log L/\partial\boldsymbol{\beta}$. Then, to obtain the complete MLE for $\boldsymbol{\theta}$, $\mathbf{q}(\mathbf{y};\boldsymbol{\beta})$ can be supplemented with the rest of the score vector, $\partial \log L/\partial\boldsymbol{\tau}$. On the other hand, if MQLE is being used without references to a specific likelihood, then ad hoc estimates will be needed for $\boldsymbol{\tau}$, like the one in (9.2) for ϕ.

Much time and ingenuity has been spent figuring out how to use the GLIM computer package for fitting models outside the GLM/MQLE framework. This often involves some cunning algebraic manipulation to make the desired application resemble a GLM in some way. However, although a valid estimate of $\boldsymbol{\beta}$ might be teased out of the package, other quantities, such as standard errors, are likely to be returned wrongly and then yet more investigation and remedial action will be required. The analysis of censored survival data is a case in point (McCullagh and Nelder, 1989, Chapter 13). On the whole, it is our opinion that a good system like GLIM should be used for what it does well, and not for what it does not.

9.2. GLMs and MQLE for repeated measures

9.2.1 The problem of covariance structures

Suppose now that the data comprise independent vector observations $\mathbf{y}_i$ $(i = 1, \ldots, n)$, where $\mathbf{y}_i$ is $p_i \times 1$ with mean $\boldsymbol{\mu}_i = E(\mathbf{y}_i)$ and $p_i \times p_i$ covariance matrix $\boldsymbol{\Sigma}_i = \text{cov}(\mathbf{y}_i)$. In principle, though not usually in practice, the components of $\mathbf{y}_i$ could be of different types–some continuous, some binary, and others categorical converted to binary (section 10.3.2). The GLM/MQLE set-up can be extended to cover the multivariate case as follows (McCullagh, 1983): (a) the model specifications for $\boldsymbol{\mu}_i$ and $\boldsymbol{\Sigma}_i$ are $g_j(\mu_{ij}) = \mathbf{x}_{ij}^{\mathrm{T}}\boldsymbol{\beta}$ and $\boldsymbol{\Sigma}_i = \phi\mathbf{V}(\boldsymbol{\mu}_i)$, where the g_j are link functions, the $\mathbf{x}_{ij}$ are $q \times 1$ covariate vectors, and $\mathbf{V}(\cdot)$ is a matrix of variance and covariance functions; (b) estimation of $\boldsymbol{\beta}$ is based on the quasi-score function

$$\mathbf{q}(\mathbf{y}; \boldsymbol{\beta}) = \sum \boldsymbol{\mu}_i'^{\mathrm{T}} \mathbf{V}(\boldsymbol{\mu}_i)^{-1}(\mathbf{y}_i - \boldsymbol{\mu}_i) \tag{9.5}$$

where $\boldsymbol{\mu}_i'$ is $p_i \times q$ with (j, k)th element $\partial\mu_{ij}/\partial\beta_k$.

The properties of MQLE go through much as in the univariate case. For instance, the MQLE $\hat{\boldsymbol{\beta}}$ is consistent and asymptotically $N(\boldsymbol{\beta}, \mathbf{V}_{\boldsymbol{\beta}})$, with $\mathbf{V}_{\boldsymbol{\beta}} = \phi\{\sum\boldsymbol{\mu}_i'^{\mathrm{T}}\mathbf{V}(\boldsymbol{\mu}_i)^{-1}\boldsymbol{\mu}_i'\}^{-1}$, and ϕ can be estimated by

$$\hat{\phi} = (p_+ - q)^{-1} \sum (\mathbf{y}_i - \hat{\boldsymbol{\mu}}_i)^{\mathrm{T}} \mathbf{V}(\hat{\boldsymbol{\mu}}_i)^{-1}(\mathbf{y}_i - \hat{\boldsymbol{\mu}}_i)$$

where $p_+ = p_1 + \cdots + p_n$, and use of the divisor $p_+ - q$, instead of just p_+, reflects the degrees of freedom adjustment which gives unbiased variance estimators in linear models.

Unfortunately, for situations of realistic practical interest $\boldsymbol{\Sigma}_i$ usually needs to incorporate additional parameters. The examples tackled in previous chapters illustrate this. Then, in order to make progress without a true likelihood function, we have to go outside the strict GLM/MQLE framework. This view was taken by Jorgensen (1987) who described some multivariate models based on GLMs and linear exponential families, but concluded that they were limited by the inflexibility of the correlation structure's being determined by the mean.

In general, to overcome the main limitations of the GLM/MQLE framework we need a strategy for dealing with additional parameters which occur in the model, particularly in the covariance structure of longitudinal data. We assume that a true likelihood function is not available, otherwise we could employ it in the usual way. Various approaches are possible. One such is Gaussian estimation: as noted in section 7.1, the MQL score function is part of the $\boldsymbol{\beta}$ component of the Gaussian score function, and the additional parameters can be estimated by bringing the rest of the Gaussian score function into play. (In fact, under the univariate GLM specification $\sigma_i^2 = \phi v(\mu_i)$, where ϕ has just one component, this yields the estimator (9.2) for ϕ, albeit with divisor n instead of $n - q$.) In the present chapter, however, we are concerned with extensions of GLM/MQLE to cope with longitudinal data.

Suppose that $\mathbf{y}_i$ has mean $\boldsymbol{\mu}_i(\boldsymbol{\beta})$ and covariance matrix $\boldsymbol{\Sigma}_i(\boldsymbol{\theta})$, with $\boldsymbol{\theta} = (\boldsymbol{\beta}, \boldsymbol{\tau})$, so that $\boldsymbol{\tau}$ is the vector of additional parameters in the covariance structure. The quasi-score function for $\boldsymbol{\beta}$ is

$$\mathbf{q}_{\boldsymbol{\beta}}(\mathbf{y};\boldsymbol{\theta}) = \sum_{i=1}^{n} \boldsymbol{\mu}_i'^{\mathrm{T}} \boldsymbol{\Sigma}_i^{-1}(\mathbf{y}_i - \boldsymbol{\mu}_i)$$

and let $\mathbf{q}_{\boldsymbol{\tau}}(\mathbf{y};\boldsymbol{\theta})$ be the additional vector estimating function proposed for estimation of $\boldsymbol{\tau}$. Then, under regularity conditions (section 7.1.1.), the estimator $[\hat{\boldsymbol{\beta}}\ \hat{\boldsymbol{\tau}}]$ is consistent and asymptotically normal with mean $(\boldsymbol{\beta}, \boldsymbol{\tau})$ and covariance matrix $\mathbf{D}^{-1}\mathbf{C}(\mathbf{D}^{-1})^{\mathrm{T}}$, where

$$\mathbf{C} = \mathrm{var}\begin{bmatrix} \mathbf{q}_{\boldsymbol{\beta}}(\mathbf{y};\boldsymbol{\theta}) \\ \mathbf{q}_{\boldsymbol{\tau}}(\mathbf{y};\boldsymbol{\theta}) \end{bmatrix} = \begin{bmatrix} \mathbf{C}_{\boldsymbol{\beta\beta}} & \mathbf{C}_{\boldsymbol{\beta\tau}} \\ \mathbf{C}_{\boldsymbol{\tau\beta}} & \mathbf{C}_{\boldsymbol{\tau\tau}} \end{bmatrix}$$

$$\mathbf{D} = E\left\{ -\frac{\partial}{\partial\boldsymbol{\theta}} \begin{bmatrix} \mathbf{q}_{\boldsymbol{\beta}}(\mathbf{y};\boldsymbol{\theta}) \\ \mathbf{q}_{\boldsymbol{\tau}}(\mathbf{y};\boldsymbol{\theta}) \end{bmatrix} \right\} = \begin{bmatrix} \mathbf{D}_{\boldsymbol{\beta\beta}} & \mathbf{D}_{\boldsymbol{\beta\tau}} \\ \mathbf{D}_{\boldsymbol{\tau\beta}} & \mathbf{D}_{\boldsymbol{\tau\tau}} \end{bmatrix}$$

with

$$\mathbf{D}_{\boldsymbol{\beta\beta}} = \sum \boldsymbol{\mu}_i'^{\mathrm{T}} \boldsymbol{\Sigma}_i^{-1} \boldsymbol{\mu}_i', \quad \mathbf{D}_{\boldsymbol{\beta\tau}} = \mathbf{0}, \quad \mathbf{D}_{\boldsymbol{\tau\beta}} = E\{-\partial \mathbf{q}_{\boldsymbol{\tau}}(\mathbf{y};\boldsymbol{\theta})/\partial\boldsymbol{\beta}\}$$

$$\mathbf{D}_{\boldsymbol{\tau\tau}} = E\{-\partial \mathbf{q}_{\boldsymbol{\tau}}(\mathbf{y};\boldsymbol{\theta})/\partial\boldsymbol{\tau}\}$$

$$\mathbf{C}_{\boldsymbol{\beta\beta}} = \sum \boldsymbol{\mu}_i'^{\mathrm{T}} \boldsymbol{\Sigma}_i^{-1} \boldsymbol{\mu}_i', \quad \mathbf{C}_{\boldsymbol{\beta\tau}} = \mathbf{C}_{\boldsymbol{\tau\beta}}^{\mathrm{T}} = \mathrm{cov}\{\mathbf{q}_{\boldsymbol{\beta}}(\mathbf{y};\boldsymbol{\theta})\}\, \mathbf{q}_{\boldsymbol{\tau}}(\mathbf{y};\boldsymbol{\theta})\}$$

$$\mathbf{C}_{\boldsymbol{\tau\tau}} = \mathrm{var}\{\mathbf{q}_{\boldsymbol{\tau}}(\mathbf{y};\boldsymbol{\theta})\}$$

Because $\mathbf{D}_{\boldsymbol{\beta\tau}} = \mathbf{0}$, the asympototic covariance matrix $\mathbf{V}_{\boldsymbol{\beta}}$ of $\hat{\boldsymbol{\beta}}$ reduces to $\mathbf{D}_{\boldsymbol{\beta\beta}}^{-1} \mathbf{C}_{\boldsymbol{\beta\beta}} (\mathbf{D}_{\boldsymbol{\beta\beta}}^{-1})^{\mathrm{T}} = \mathbf{D}_{\boldsymbol{\beta\beta}}^{-1}$, the same as if $\boldsymbol{\tau}$ were known rather than estimated.

9.2.2 *MQLE with a 'working' correlation matrix*

Suppose that the jth component of $\mathbf{y}_i$, y_{ij}, is subject to a GLM, i.e. $g_j(\mu_{ij}) = \mathbf{x}_{ij}^{\mathrm{T}}\boldsymbol{\beta}$ and $\sigma_{ij}^2 = \phi v_j(\mu_{ij})$, where $\mu_{ij} = E(y_{ij})$ and $\sigma_{ij}^2 = \mathrm{var}(y_{ij})$, and the g_j and v_j are specified link and variance functions, respectively. Further, suppose that knowledge about the system falls short of the covariances, or correlations, between the components. Now, $\boldsymbol{\Sigma}_i = \mathrm{cov}(\mathbf{y}_i)$ can be expressed as $\phi\mathbf{S}_i\mathbf{R}_i\mathbf{S}_i$, where $\mathbf{R}_i$ is the correlation matrix of $\mathbf{y}_i$, and $\mathbf{S}_i$ is diagonal, containing the standard deviations of the y_{ij} scaled by $\phi^{-1/2}$, i.e. $S_{ijj} = \sqrt{v_j(\mu_{ij})}$. The quasi-score function (9.5) becomes

$$\mathbf{q}(\mathbf{y};\boldsymbol{\beta}) = \sum \boldsymbol{\mu}_i'^{\mathrm{T}}(\mathbf{S}_i\mathbf{R}_i\mathbf{S}_i)^{-1}(\mathbf{y}_i - \boldsymbol{\mu}_i) \tag{9.6}$$

However, since $\mathbf{R}_i$ is not available, this is not usable as it stands. Liang and Zeger's (1986) suggestion was to replace $\mathbf{R}_i$ in $\mathbf{q}(\mathbf{y};\boldsymbol{\beta})$ by a 'working' correlation matrix, $\bar{\mathbf{R}}_i$, not assumed to be precisely correct. Before going on to apply the method to data we will look at this suggestion in stages: (a) $\bar{\mathbf{R}}_i$ completely specified; (b) $\bar{\mathbf{R}}_i$ completely unspecified; (c) $\bar{\mathbf{R}}_i$ specified parametrically, correctly; (d) $\bar{\mathbf{R}}_i$ specified parametrically, incorrectly.

(a) Suppose that $\bar{\mathbf{R}}_i$ is taken to be some completely known matrix, containing no unknown parameters. For instance, the choice $\bar{\mathbf{R}}_i = \mathbf{I}_{p_i}$, the $p_i \times p_i$ unit matrix, would be correct if the components of $\mathbf{y}_i$ were actually independent. Again, previous experience of similar data could sensibly suggest a more appropriate choice such as a structured form (Chapter 6) with the parameters fixed at some previous estimate. Then (9.6) could be solved numerically to yield the estimator $\hat{\boldsymbol{\beta}}$. Provided that the mean specification of $\boldsymbol{\mu}_i$ is correct, the estimating equation $\mathbf{q}(\mathbf{y};\boldsymbol{\beta}) = \mathbf{0}$ will be unbiased, and then $\hat{\boldsymbol{\beta}}$ will be asymptotically normal with mean $\boldsymbol{\beta}$ and covariance matrix $\mathbf{V}_{\boldsymbol{\beta}} = \mathbf{D}^{-1}\mathbf{C}(\mathbf{D}^{-1})^{\mathrm{T}}$, where (section 7.1.1)

$$\mathbf{C} = \mathrm{var}\{\mathbf{q}(\mathbf{y};\boldsymbol{\beta})\} = \sum_{i=1}^{n} \boldsymbol{\mu}_i'^{\mathrm{T}}(\mathbf{S}_i\bar{\mathbf{R}}_i\mathbf{S}_i)^{-1}\boldsymbol{\Sigma}_i(\mathbf{S}_i\bar{\mathbf{R}}_i\mathbf{S}_i)^{-1}\boldsymbol{\mu}_i'$$
$$\mathbf{D} = E\{-\mathbf{q}'(\mathbf{y};\boldsymbol{\beta})\} = \sum_{i=1}^{n} \boldsymbol{\mu}_i'^{\mathrm{T}}(\mathbf{S}_i\bar{\mathbf{R}}_i\mathbf{S}_i)^{-1}\boldsymbol{\mu}_i' \tag{9.7}$$

To reiterate, it is not assumed that $\mathbf{R}_i$ is correctly specified here. If it were, then would have $\boldsymbol{\Sigma}_i = \phi\mathbf{S}_i\bar{\mathbf{R}}_i\mathbf{S}_i$ and, in consequence $\mathbf{C} = \phi\mathbf{D}$ as for MQLE. Nevertheless, we can still obtain a valid estimate for the asymptotic covariance of $\hat{\boldsymbol{\beta}}$ by inserting $\hat{\boldsymbol{\beta}}$ for $\boldsymbol{\beta}$ everywhere, and $(y_{ij} - \hat{\mu}_{ij})(y_{ik} - \hat{\mu}_{ik})$ for $\boldsymbol{\Sigma}_{ijk}$ in $\mathbf{C}$ and $\mathbf{D}$. Similar use of model-based and sample-based estimates was made in Chapter 7.

(b) Suppose next that $\mathbf{R}_i$ is allowed to be completely unknown, the $p_i(p_i - 1)/2$ distinct entries being constrained only by the requirement to form a valid correlation matrix. This is the opposite extreme to the situation in (a). Now, to allow complete freedom of the $\bar{\mathbf{R}}_i$ would involve $p_{++} = \sum p_i(p_i - 1)/2$ parameters, which is unmanageable. However, it is often reasonable to assume homogeneity, i.e. that $\mathbf{R}_i$ is a submatrix of a common $p \times p$ correlation matrix $\mathbf{R}$ which has entries $R_{jk} = \rho_{jk}$; if $p_i < p$ it is only because $\mathbf{y}_i$ contains missing components. The situation is then as described at the beginning of this section: we have additional parameters in the correlation structure, and so we need to supplement the estimating equations. A natural way of estimating the ρ_{jk} is to focus upon the cross-products of the components of $\mathbf{y}_i$. Let

$$\bar{y}_{ij} = (y_{ij} - \mu_{ij})/S_{ijj}$$

denote the standardized version of y_{ij}, so

$$E(\bar{y}_{ij}) = 0, \quad E(\bar{y}_{ij}^2) = \phi \quad \text{and} \quad E(\bar{y}_{ij}\bar{y}_{ik}) = \phi\rho_{jk}$$

Simple moment estimators can now be used for ϕ and the ρ_{jk}, based on the $\bar{y}_{ij}$ with $\hat{\boldsymbol{\beta}}$ inserted for $\boldsymbol{\beta}$. For ϕ we have the estimator

$$\hat{\phi} = (p_+ - q)^{-1} \sum_{i=1}^{n} \sum_{j=1}^{p_i} \bar{y}_{ij}^2$$

For $\hat{\rho}_{jk}$ the natural moment estimator is r_{jk}, the sample average correlation, given by

$$r_{jk} = (n-q)^{-1} \hat{\phi}^{-1} \sum_{i=1}^{n} \bar{y}_{ij}\bar{y}_{ik}$$

the summation here is understood to ignore missing components $\bar{y}_{ij}$, with the divisor $n - q$ reduced accordingly.

(c) Liang and Zeger (1986) went a step further: they proposed the use of a parametrized form for $\bar{\mathbf{R}}_i$ designed to capture the anticipated form of dependence between the components of $\mathbf{y}_i$. Typical choices for $\bar{\mathbf{R}}_i$, depending on a parameter τ, say, include the following: (i) $\bar{R}_{ijk} = \tau$, an equicorrelated structure, as occurs in 'split plots' or 'clustered data' (section 6.2.2); (ii) $\bar{R}_{ijk} = \tau^{|t_{ij} - t_{ik}|}$, representing a geometrically decreasing correlation between observations made at times t_{ij} and t_{ik}, as occurs in autoregressive and Markov models (section 6.2.4). Liang and Zeger suggested simple *ad hoc* estimators for these forms. For (i) r_{jk}, as defined in (b), can be averaged over j and k to yield $\hat{\tau}$. For (ii), assuming that $|t_{ij} - t_{ik}|$ is the same for all i, $\tau^{|t_{ij} - t_{ik}|}$ can be matched to r_{jk} by fitting a straight line, nominally of slope log τ, to a plot of log r_{jk} versus $|t_{ij} - t_{ik}|$; such a plot also serves as a rough check on the assumed form for $\mathbf{R}_i$.

Suppose that the form taken for $\bar{\mathbf{R}}_i$ is actually correct. Then, under regularity conditions (section 7.1.1), $[\hat{\boldsymbol{\beta}} \ \hat{\tau}]$ is consistent and asymptotically normal, and the asymptotic covariance matrix of $\hat{\boldsymbol{\beta}}$, $\boldsymbol{\beta}$ being the parameter of main interest, is $\mathbf{V}_{\boldsymbol{\beta}} = \mathbf{D}^{-1}\mathbf{C}(\mathbf{D}^{-1})^{\mathrm{T}}$ with $\mathbf{C}$ and $\mathbf{D}$ as given in (9.7) (Liang and Zeger, 1986, Theorem 2).

(d) Finally, let us turn to what is probably the case of most practical importance, i.e. when $\bar{\mathbf{R}}_i$ is not correctly specified. Thus, detailed knowledge is not assumed and $\bar{\mathbf{R}}_i$ is only a 'working' correlation structure regarded as a reasonable guess but unlikely to be precisely correct. The idea is that the closer $\bar{\mathbf{R}}_i$ is to the true underlying form, the more efficient the estimator $\hat{\boldsymbol{\beta}}$. Liang and Zeger (1986) suggested that we obtain $\hat{\tau}$ and $\hat{\boldsymbol{\beta}}$ for this case as described in (c).

Of the four strategies for choosing $\bar{\mathbf{R}}_i$, (a) is the safe bet. The contributions $(\mathbf{y}_i - \boldsymbol{\mu}_i)$ to $\mathbf{q}(\mathbf{y}; \boldsymbol{\beta})$ in (9.6) will not be summed with the optimal weighting matrices, but there are no complications such as having additional parameters to estimate. In (b) we are trying to obtain the correct matrices $\mathbf{R}_i$ without making any assumptions about their form. The price is that the strategy is restricted to the homogeneous case, and there are $p(p-1)/2$ correlation parameters to estimate, with the resulting weakening of the inferences (section 6.1). In case (c) we are lucky enough to know the correct covariance structure governed by the parameter τ, and then standard asymptotic properties for the estimators follow. Case (d), of particular interest in practice, is not so easily dealt with. This is because the status of $\hat{\tau}$ is not now clear-cut: it is not the estimate of some 'true' parameter because the $\bar{\mathbf{R}}_i(\tau)$ are not the true correlation matrices. Some care is needed in the interpretation of the procedure and more work is needed in this area. Broadly, if the supplementary estimating equations ensure that $\hat{\tau}$ exists and converges to some value τ as $n \to \infty$, and if $\bar{\mathbf{R}}_i(\tau)$ is not too dissimilar to $\mathbf{R}_i$, then the aim will have been achieved.

9.2.3 *General covariance structures*

Liang and Zeger (1986) were first in the field with the use of working correlation matrices for longitudinal data, and the idea, once suggested, is easy to appreciate. They have used the description 'generalized estimating equations' for these MQL-based estimating equations.

Naturally, more sophisticated versions have followed. The assumption that the form of $\boldsymbol{\Sigma}_i$ is governed by $\boldsymbol{\mu}_i$, via specified variance functions, is unnecessarily restrictive, even if it only applies to the diagonal elements as in section 9.2.2. We will now follow Prentice and Zhao (1991) in allowing completely arbitrary covariance structures. The quasi-score function can then be written as

$$\mathbf{q}(\mathbf{y}; \boldsymbol{\beta}) = \sum \boldsymbol{\mu}_i'^{\mathrm{T}} \boldsymbol{\Sigma}_i^{-1} (\mathbf{y}_i - \boldsymbol{\mu}_i) \tag{9.8}$$

where it is recognized that we are unsure about the precise details of the covariance structure. Suppose that the $\boldsymbol{\Sigma}_i$ are replaced by working covariance matrices $\bar{\boldsymbol{\Sigma}}_i(\boldsymbol{\beta}, \boldsymbol{\tau})$ depending on $\boldsymbol{\beta}$ and an additional parameter vector $\boldsymbol{\tau}$. The resulting working score function for $\boldsymbol{\beta}$ is then, in the notation of section 7.1,

$$\mathbf{q}_{\boldsymbol{\beta}}(\mathbf{y}; \boldsymbol{\beta}, \boldsymbol{\tau}) = \sum \boldsymbol{\mu}_i'^{\mathrm{T}} \bar{\boldsymbol{\Sigma}}_i^{-1} (\mathbf{y}_i - \boldsymbol{\mu}_i) \tag{9.9}$$

We now need some additional estimating equations for $\boldsymbol{\tau}$, say $\mathbf{q}_{\boldsymbol{\tau}}(\mathbf{y}; \boldsymbol{\beta}, \boldsymbol{\tau}) = \mathbf{0}$. As suggested in section 7.1, we could use the corresponding components of the Gaussian score function. An alternative suggestion is to use a working quasi-score function for $\boldsymbol{\tau}$ based on the cross-products of the components y_{ij}.

Let $\mathbf{u}_i$ be the vector containing the $p_i(p_i-1)/2$ distinct products $y_{ij}y_{ik}$, and let $\boldsymbol{\upsilon}_i$ have components $\mu_{ij}\mu_{ik}+\bar{\Sigma}_{ijk}$, i.e. $\boldsymbol{\upsilon}_i=E(\mathbf{u}_i)$ under the working model. Then a corresponding working quasi-score function for $\mathbf{u}_i$ is

$$\mathbf{q}_\tau(\mathbf{y};\boldsymbol{\beta},\boldsymbol{\tau})=\sum \boldsymbol{\upsilon}_i'^{\mathrm{T}}\mathbf{W}_i^{-1}(\mathbf{u}_i-\boldsymbol{\upsilon}_i) \tag{9.10}$$

where $\boldsymbol{\upsilon}_i'=\partial\boldsymbol{\upsilon}_i/\partial\boldsymbol{\tau}$ and $\mathbf{W}_i$ is a working covariance matrix for $\mathbf{u}_i$; although $\boldsymbol{\upsilon}_i$ depends on $\boldsymbol{\beta}$ as well as on $\boldsymbol{\tau}$, $\partial\boldsymbol{\upsilon}_i/\partial\boldsymbol{\beta}$ is not used in (9.10) – we already have the estimating function $\mathbf{q}_\beta(\mathbf{y};\boldsymbol{\beta},\boldsymbol{\tau})$ for $\boldsymbol{\beta}$. In fact, the same quasi-score function $\mathbf{q}_\tau(\mathbf{y};\boldsymbol{\beta},\boldsymbol{\tau})$ would be obtained had we started with a vector $\mathbf{u}_i$ of mean-centred components $(y_{ij}-\mu_{ij})(y_{ik}-\mu_{ik})$ (Zhao and Prentice, 1990, Appendix 1). Prentice and Zhao (1991) made some suggestions for choice of the working covariance matrices $\bar{\boldsymbol{\Sigma}}_i$ and $\mathbf{W}_i$.

Under regularity conditions (section 7.1.1), and if $\bar{\boldsymbol{\Sigma}}_i(\boldsymbol{\beta},\boldsymbol{\tau})$ is correctly specified, $[\hat{\boldsymbol{\beta}}\ \hat{\boldsymbol{\tau}}]$ will be jointly asymptotically normal with mean $(\boldsymbol{\beta},\boldsymbol{\tau})$ and covariance matrix $\mathbf{V}_\theta=\mathbf{D}^{-1}\mathbf{C}(\mathbf{D}^{-1})^{\mathrm{T}}$, where, with $\mathbf{C}$ and $\mathbf{D}$ partitioned as in (7.5),

$$\mathbf{D}_{\beta\beta}=\sum \boldsymbol{\mu}_i'^{\mathrm{T}}\bar{\boldsymbol{\Sigma}}_i^{-1}\boldsymbol{\mu}_i',\quad \mathbf{D}_{\tau\tau}=\sum \boldsymbol{\upsilon}_i'^{\mathrm{T}}\mathbf{W}_i^{-1}\boldsymbol{\upsilon}_i',\quad \mathbf{D}_{\beta\tau}=\mathbf{0},\quad \mathbf{D}_{\tau\beta}=\sum \boldsymbol{\upsilon}_i'^{\mathrm{T}}\mathbf{W}_i^{-1}(\partial\boldsymbol{\upsilon}_i/\partial\boldsymbol{\beta})$$

$$\mathbf{C}_{\beta\beta}=\sum \boldsymbol{\mu}_i'^{\mathrm{T}}\bar{\boldsymbol{\Sigma}}_i^{-1}\boldsymbol{\Sigma}_i\bar{\boldsymbol{\Sigma}}_i^{-1}\boldsymbol{\mu}_i',\quad \mathbf{C}_{\tau\tau}=\sum \boldsymbol{\upsilon}_i'^{\mathrm{T}}\mathbf{W}_i^{-1}\mathbf{U}_{1i}\mathbf{W}_i^{-1}\boldsymbol{\upsilon}_i'$$

$$\mathbf{C}_{\beta\tau}=\sum \boldsymbol{\mu}_i'^{\mathrm{T}}\bar{\boldsymbol{\Sigma}}_i^{-1}\mathbf{U}_{2i}\mathbf{W}_i^{-1}\boldsymbol{\upsilon}_i'$$

where $\mathbf{U}_{1i}=\mathrm{var}(\mathbf{u}_i)$ and $\mathbf{U}_{2i}=\mathrm{cov}(\mathbf{y}_i,\mathbf{u}_i)$. The asymptotic covariance matrix of $\hat{\boldsymbol{\beta}}$, for the parameter of main interest, is $\mathbf{V}_\beta=\mathbf{D}_{\beta\beta}^{-1}\mathbf{C}_{\beta\beta}(\mathbf{D}_{\beta\beta}^{-1})^{\mathrm{T}}$, and this can itself be estimated by inserting the estimates $\hat{\boldsymbol{\beta}}$ and $\hat{\boldsymbol{\tau}}$.

9.3 Some examples

The four examples of section 7.2 will now be refitted by MQLE based on (9.9) and (9.10); in the latter, $\mathbf{W}_i$ will be taken as $\mathbf{I}$, a unit matrix, throughout. The parameter estimates and associated tests produced by MQLE will be compared with those formerly produced by Gaussian estimation (GE). It turns out that, for these examples, the estimates of the linear regression coefficients, i.e. $\boldsymbol{\beta}$ in $\boldsymbol{\theta}=[\boldsymbol{\beta}\ \boldsymbol{\tau}]$, do not differ much between GE and MQLE. This can be explained to some extent by noting that in the decoupled case (section 7.1.3), which obtains in all four examples, the estimating function $\mathbf{q}_\beta(\mathbf{y};\boldsymbol{\theta})$ has the same form for GE and MQLE. However, differences do appear in the estimates of the covariance parameters $\boldsymbol{\tau}$, reflecting the different forms of estimating function $\mathbf{q}_\tau(\mathbf{y};\boldsymbol{\theta})$ used by GE and MQLE.

Example 9.1 Reactions of 12 patients to treatment

The data were subjected to Gaussian estimation in Example 7.1. The numerical results from fitting the model by GE and MQLE are, for comparison,

$$\tilde{\boldsymbol{\beta}}(\mathrm{GE})=[0.919\ \ -0.371\ \ 0.312],\quad \tilde{\boldsymbol{\tau}}(\mathrm{GE})=[0.252\ \ 0.00\ \ 0.187]$$

$$\tilde{\boldsymbol{\beta}}(\mathrm{MQLE})=[0.795\ \ -0.252\ \ 0.344],\quad \tilde{\boldsymbol{\tau}}(\mathrm{MQLE})=[0.152\ \ 0.002\ \ 0.00]$$

and

$$\mathbf{V}_\beta(\text{GE}) = 10^{-3}\begin{bmatrix} 3.84 & 2.05 & -0.56 \\ 2.05 & 2.87 & -0.97 \\ -0.56 & -0.97 & 1.74 \end{bmatrix}$$

$$\mathbf{V}_\beta(\text{MQLE}) = 10^{-3}\begin{bmatrix} 3.98 & 2.04 & -0.84 \\ 2.04 & 2.68 & -0.77 \\ -0.84 & -0.77 & 1.53 \end{bmatrix}$$

The estimates from the two methods are broadly similar, particularly those for $\boldsymbol{\beta}$ and $\mathbf{V}_\beta$ which relate to the aspects of main interest here. The Wald test for $H: \beta_2 = \beta_3 = 0$ based on the MQL estimates gives $W = 80.10$ as χ^2_2, compared with $W = 59.60$ previously produced by GE.

The $\boldsymbol{\tau}$ estimates here, $\tilde{\sigma}_e = 0.152$, $\tilde{\rho} = 0.002$, and $\tilde{\sigma}_\mathrm{I} = 0.00$, are different from the Gaussian estimates in that the correlations in the data are now accounted for by the parameter ρ rather than σ_I as previously. This seems to reflect a degree of aliasing between these covariance parameters as noted in Examples 8.1 and 8.2 in section 8.2.1. However, this fortunately does not seem to affect the inferences about $\boldsymbol{\beta}$ very much.

Example 9.2 Visual acuity

The parameter estimates for these data, previously appearing in Example 7.2, are

$$\tilde{\boldsymbol{\beta}}(\text{GE}) = 10^{-2}[113 \;\; 0.909 \;\; -0.770 \;\; 0.0163 \;\; -1.76 \;\; 0.303 \;\; 0.232 \;\; -1.41]$$

$$\tilde{\boldsymbol{\beta}}(\text{MQLE}) = 10^{-2}[113 \;\; 0.911 \;\; -0.768 \;\; 0.0179 \;\; -1.77 \;\; 0.304 \;\; 0.232 \;\; -1.41]$$

$$\tilde{\sigma}_e(\text{GE}) = 0.0289, \quad \tilde{\sigma}_\mathrm{I}(\text{GE}) = 0.0486, \quad \tilde{\sigma}_\mathrm{E}(\text{GE}) = 0.0231$$

$$\tilde{\mathbf{A}}_\mathrm{L}(\text{GE}) = 10^{-2}\begin{bmatrix} 2.30 & 0 & 0 \\ 1.03 & -0.472 & 0 \\ -3.29 & -0.438 & -0.004 \end{bmatrix}$$

$$\tilde{\sigma}_e(\text{MQLE}) = 0.0292, \quad \tilde{\sigma}_\mathrm{I}(\text{MQLE}) = 0.00, \quad \tilde{\sigma}_\mathrm{E}(\text{MQLE}) = 0.00$$

$$\tilde{\mathbf{A}}_\mathrm{L}(\text{MQLE}) = 10^{-2}\begin{bmatrix} 1.79 & 0 & 0 \\ 2.28 & 0.00 & 0 \\ 0.00 & 0.00 & 0.00 \end{bmatrix}$$

As in Example 9.1, the $\boldsymbol{\beta}$ estimates are quite similar for GE and MQLE, but differences do show up in the $\boldsymbol{\tau}$ estimates. However, again, the various inferences about the $\boldsymbol{\beta}$ parameters are not much affected by such differences: for the eyes effect, residing in parameter β_2, the normal deviate is now 0.95, the same as by GE; the Wald test statistic for the lens effect is $W = 113.0$ as χ^2_3, compared with $W = 13.0$ by GE; for the eyes by lenses interaction, $W = 2.90$ as χ^2_3, compared with $W = 4.88$ by GE.

Example 9.3 Rat growth

The parameters for these data, formerly of Example 7.3, are estimated as

$$\tilde{\boldsymbol{\beta}}(\text{GE}) = [4.04 \;\; -1.53 \;\; 0.491 \;\; 0.067 \;\; -0.026 \;\; 0.033]$$

$$\tilde{\boldsymbol{\tau}}(\text{GE}) = (0.050\ \ 0.851\ \ 0.325)$$

$$\tilde{\boldsymbol{\beta}}(\text{MQLE}) = [4.04\ \ -1.53\ \ 0.492\ \ 0.067\ \ -0.026\ \ 0.033]$$

$$\tilde{\boldsymbol{\tau}}(\text{MQLE}) = (0.007\ \ 0.863\ \ 0.00)$$

As in the previous examples, the differences between GE and MQLE show up mainly in the covariance parameters. The previous inference of strong between-individual variation in mean levels ($\tilde{\sigma}_{\text{I}}(\text{GE}) = 0.325$) seems to be somewhat watered down now ($\tilde{\sigma}_{\text{I}}(\text{MQLE}) = 0.00$).

The difference between groups 2 and 3, previously assessed by the normal deviate 1.68 under GE, gives 1.69 under MQLE. The question of differences between groups of the regression coefficient for x = time gave Wald statistic $W = 6.61$ as χ_2^2 under GE, and now gives $W = 6.64$ under MQLE.

Example 9.4 Hip replacements
The data, analysed previously in Example 7.4, yield estimates

$$\tilde{\boldsymbol{\beta}}(\text{GE}) = [3.28\ 0.21\ 0.65\ -0.34\ -0.21\ 0.12\ -0.050\ -0.048\ 0.019\ -0.020]$$

$$\tilde{\boldsymbol{\beta}}(\text{MQLE}) = [3.23\ \ 0.22\ \ 0.65\ \ -0.34\ \ -0.19\ \ 0.11\ \ -0.061\ \ -0.035\ \ 0.026\ \ -0.020]$$

$$\tilde{\boldsymbol{\tau}}(\text{GE}) = [0.40\ 0.23],\quad \tilde{\boldsymbol{\tau}}(\text{MQLE}) = [0.41, 0.86]$$

The major difference between the two sets of estimates is, again, in $\tilde{\boldsymbol{\tau}}$ with $\tilde{\rho}(\text{GE}) = 0.23$ and $\tilde{\rho}(\text{MQLE}) = 0.86$; the MQL estimate implies much stronger constant correlation within individuals.

The Wald test for the joint effect of age and sex gave $W = 0.29$ as χ_2^2 by GE and now gives $W = 0.55$ by MQLE. The hypothesis that $\beta_3 = \beta_4 = \beta_5 = 0$, concerning differences in mean level between the four occasions, gave $W = 157.0$ as χ_3^2 under GE, and now gives $W = 159.0$ under MQLE. Likewise, for the sex by occasions interaction, $W(\text{GE}) = 4.76$ as χ_3^2 and $W(\text{MQLE}) = 4.74$ and, for the age by sex interaction, the normal deviate is -0.33 by both GE and MQLE.

9.4 GLMs with random effects

All the GLMs so far covered in this chapter have been marginal models which specify the unconditional means $E(y_{ij})$ and variances $\text{var}(y_{ij})$ directly. We now describe briefly some models for means and variances conditioned on certain individual-specific random effects.

9.4.1 *GLMs with random regression coefficients*

Suppose that a GLM holds for y_{ij}, the observation from individual i on occasion j, of the following form:

$$g_j(m_{ij}) = \mathbf{x}_{ij}^{\text{T}}\boldsymbol{\beta} + \mathbf{z}_{ij}^{\text{T}}\mathbf{b}_i \quad \text{and} \quad s_{ij}^2 = \phi v_j(m_{ij}) \tag{9.11}$$

Here, $m_{ij} = E(y_{ij}|\mathbf{b}_i)$ and $s_{ij}^2 = \text{var}(y_{ij}|\mathbf{b}_i)$ are the indicated conditional mean and variance, modelled via the link function g_j and variance function v_j; the $\mathbf{b}_i$ are random regression coefficients with some distribution over the population of individuals; $\mathbf{x}_{ij}$ and $\mathbf{z}_{ij}$ are covariate vectors which affect y_{ij}. To complete the picture, i.e. to supplement the above specification of conditional means and variances, it is assumed that the conditional covariance between any two components, y_{ij} and y_{ik}, is zero: $s_{ijk} = \text{cov}(y_{ij}, y_{ik}|\mathbf{b}_i) = 0$. Then, the only source of correlation between components is the shared random effect $\mathbf{b}_i$.

The model described here was proposed by Zeger, Liang and Albert (1988). The analogy with the two-stage model of section 5.1 is close. A regression model is specified within individuals, containing a fixed coefficient $\boldsymbol{\beta}$ and a random one $\mathbf{b}_i$, as above, and then a between-individuals distribution of the random effects is assumed, as below. When the two stages are put together the overall model, averaged across individuals, can be derived.

The unconditional mean μ_{ij} and variance σ_{ij}^2 of y_{ij}, and the unconditional covariance σ_{ijk} between y_{ij} and y_{ik}, can be obtained by integrating over the distribution of the random effects $\mathbf{b}_i$. Let $f(\mathbf{b}_i)$ be the probability density function of $\mathbf{b}_i$ which defines its distribution over the population of individuals. A popular choice for this distribution is $N(\mathbf{0}, \mathbf{B})$. Then, for the unconditional mean,

$$\mu_{ij} = E(m_{ij}) = \int m_{ij} f(\mathbf{b}_i) \mathrm{d}\mathbf{b}_i \tag{9.12}$$

where m_{ij} is evaluated, via the inverse link function g_j^{-1}, as

$$m_{ij} = g_j^{-1}(\mathbf{x}_{ij}^{\mathrm{T}}\boldsymbol{\beta} + \mathbf{z}_{ij}^{\mathrm{T}}\mathbf{b}_i)$$

For the unconditional variance,

$$\begin{aligned} \sigma_i^2 &= E(s_i^2) + \text{var}(m_{ij}) \\ &= \int \phi v_j(m_{ij}) f(\mathbf{b}_i)\, \mathrm{d}\mathbf{b}_i + \int (m_{ij} - \mu_{ij})^2 f(\mathbf{b}_i) \mathrm{d}\mathbf{b}_i \end{aligned} \tag{9.13}$$

For the unconditional covariance

$$\sigma_{ijk} = E(s_{ijk}) + \text{cov}(m_{ij}, m_{ik}) = 0 + \int (m_{ij} - \mu_{ij})(m_{ik} - \mu_{ik}) f(\mathbf{b}_i) \mathrm{d}\mathbf{b}_i \tag{9.14}$$

Specification of $f(\mathbf{b}_i)$ thus gives the forms of $\boldsymbol{\mu}_i = E(\mathbf{y}_i)$, with components μ_{ij}, and $\boldsymbol{\Sigma}_i(p_i \times p_i) = \text{var}(\mathbf{y}_i)$, with elements σ_{ij}^2 and σ_{ijk}. Then, MQLE can be performed based on the multivariate quasi-score function (9.5). However, in practice the assumed form for $f(\mathbf{b}_i)$ will contain unknown parameters to be estimated, in particular $\mathbf{B} = \text{cov}(\mathbf{b}_i)$, which describes the variation of $\mathbf{b}_i$ across individuals; it is usual to take $E(\mathbf{b}_i) = \mathbf{0}$ since any non-zero mean can be absorbed into $\mathbf{x}_{ij}^{\mathrm{T}}\boldsymbol{\beta}$ in (9.11). Now, (9.13) and (9.14) give $\boldsymbol{\Sigma}_i$ in terms of $f(\mathbf{b}_i)$, so the parameters in $f(\mathbf{b}_i)$ should be estimable by a method of moments which matches $\boldsymbol{\Sigma}_i$ to $(\mathbf{y}_i - \boldsymbol{\mu}_i)(\mathbf{y}_i - \boldsymbol{\mu}_i)^{\mathrm{T}}$ and then averages over i. An estimating function of type (9.10) would serve, in which $\boldsymbol{\upsilon}_i$ has components $\boldsymbol{\Sigma}_{ijk}$ and $\boldsymbol{\upsilon}_i'$ is the Jacobian matrix of their derivatives with respect to the parameters in $f(\mathbf{b}_i)$.

Zeger, Liang and Albert (1988) derived some approximations to yield $\mathbf{B}$ explicitly.

9.4.2 Two-stage likelihood models

If the model for $\mathbf{y}_i$ is associated with an identified conditional distribution, say with joint density $f(\mathbf{y}_i|\mathbf{b}_i)$, then a log-likelihood function can be constructed as

$$Q(\mathbf{y}; \boldsymbol{\beta}, \phi, \mathbf{B}) = \log \prod_{i=1}^{n} \int f(\mathbf{y}_i|\mathbf{b}_i) f(\mathbf{b}_i) d\mathbf{b}_i \qquad (9.15)$$

Provided that the integral can be evaluated numerically without too much trouble, likelihood-based inference can be conducted. An even more attractive option would be to find, if possible, a realistic combination of $f(\mathbf{y}_i|\mathbf{b}_i)$ and $f(\mathbf{b}_i)$ for which the integral can be evaluated analytically.

A commonly applied specification has the y_{ij} conditionally independent, given $\mathbf{b}_i$, so

$$f(\mathbf{y}_i|\mathbf{b}_i) = \prod_{j=1}^{p_i} f_j(y_{ij}|\mathbf{b}_i)$$

Suppose that, after some standardization of the y_{ij} not involving $\mathbf{b}_i$, say to $\bar{y}_{ij}$, the forms of the f_j are all the same, say f_0. Then

$$f(\bar{\mathbf{y}}_i|\mathbf{b}_i) = \prod_{j=1}^{p_i} f_0(\bar{y}_{ij}|\mathbf{b}_i)$$

where $\bar{\mathbf{y}}_i$ has components $\bar{y}_{ij}$, and the $\bar{y}_{ij}$ are exchangeable in the resulting likelihood. For instance, in split-plots anova the f_j are normal densities and the usual linear standardization, $\bar{y}_{ij} = (y_{ij} - m_{ij})/s_{ij}$, operates. Exchangeability then reduces to the familiar equicorrelated structure. Again, for conditionally independent Weibull variates, as used in multivariate survival analysis, standardization involves power-transformation of the y_{ij}. Conversely, the two-stage normal covariance form $\boldsymbol{\Sigma} = \mathbf{ZBZ}^{\mathrm{T}} + \mathbf{E}$ (section 5.1) can be held up to see that, even if the y_{ij} are conditionally equicorrelated, through $\mathbf{E}$, this would only carry over unconditionally if the rows of $\mathbf{Z}$ were all the same. Such a case is $\mathbf{Z} = \mathbf{1}$, the form for random individual levels (section 6.3.1). Otherwise, a standardization independent of $\mathbf{b}_i$ cannot be found.

9.5 Further reading

For a comprehensive, general treatment of GLMs and MQLE the book by McCullagh and Nelder (1989) probably comes first on most reference lists. Several other books have appeared over the past few years, some of these being concerned more with the use of the GLIM computer package for processing data. These include Gilchrist, Francis and Whittaker (1985), Aitkin *et al.*

(1989) and Dobson (1990). Firth (1991) gives a useful review of recent developments of the theory.

Wei and Stram (1988) and Moulton and Zeger (1989) suggested a way of avoiding the problem of covariance structures in longitudinal data (section 9.2.1) as follows. They used standard MQLE, based on the univariate version of (9.5), to estimate $\boldsymbol{\beta}$ from the data at each separate time point, thus obtaining $\hat{\boldsymbol{\beta}}_1, \ldots, \hat{\boldsymbol{\beta}}_p$. This enables examination of possible trends in $\boldsymbol{\beta}$ over time; in the absence of such trend an overall estimate may be computed by averaging the $\hat{\boldsymbol{\beta}}_j$'s in some way. The distributional modelling, upon which the estimation is based, is thus marginal, i.e. for separate components of $\mathbf{y}_i$ as in Liang and Zeger (1986) and Zeger and Liang (1986), rather than multivariate.

Zeger and Liang (1986) gave a more discursive version of Liang and Zeger (1986) and applied the methodology to some data connecting children's morbidity to mothers' stress.

Zeger and Karim (1991) applied Gibbs sampling for a Bayesian analysis of the two-stage likelihood model described in section 9.4.2. They took the components y_{ij} as conditionally independent, given $\mathbf{b}_i$, with exponential family distributions governed by m_{ij} and s_{ij}^2 as given in section 9.4.1. The $\mathbf{b}_i$ distribution was taken as $N(\mathbf{0}, \mathbf{B})$, and a non-informative prior for $[\boldsymbol{\beta}, \phi, \mathbf{B}]$ was used. They showed how to simulate the conditional distributions required for the Gibbs sampling.

Aspects of computation for the kinds of analysis described in this chapter are tackled by Gay and Welsch (1988), and Breslow and Clayton (1993).

CHAPTER 10

Binary and categorical measures

10.1 Regression models for binary data

10.1.1 Univariate measures

Suppose that data $y_1, \ldots, y_n$ are available, the y_i's being independent, binary observations with $\mathrm{pr}(y_i = 1) = \pi_i$ and $\mathrm{pr}(y_i = 0) = 1 - \pi_i$. Suppose also that the parameter π_i is related to the explanatory vector $\mathbf{x}_i$ via a regression model $\pi_i = h(\mathbf{x}_i, \boldsymbol{\beta})$. Commonly, h has the form of a distribution function and $\mathbf{x}_i$ and $\boldsymbol{\beta}$ appear in the linear combination $\mathbf{x}_i^T\boldsymbol{\beta}$. In that case the model $\pi_i = h(\mathbf{x}_i^T\boldsymbol{\beta})$ is a generalized linear model (section 9.1) with mean $\mu_i = \pi_i$, inverse link function h, and variance function $v(\mu) = \mu(1-\mu)$. Typical forms for h include the following:

■ *Illustration 10.1*

1. $h = \Phi$, the standard normal distribution function, leading to **probit analysis** (Finney, 1952).
2. $h(z) = (1 + \mathrm{e}^{-z})^{-1}$, the standard logistic distribution function, leading to **logit analysis** (Cox, 1970); equivalently, $\log\{\pi_i/(1-\pi_i)\} = \mathbf{x}_i^T\boldsymbol{\beta}$, a linear model for the log-odds ratio.
3. $h(z) = 1 - \exp(-\mathrm{e}^z)$, one of the extreme-value distribution functions; equivalently, $\log\{-\log(1-\pi_i)\} = \mathbf{x}_i^T\boldsymbol{\beta}$, a linear model for the **complementary log-log** transform of π_i, leading to what might be called **evit analysis**, but has not, as far as I know.
4. $h(z) = (1 + \sin z)/2$ for $-\pi/2 < z < \pi/2$; equivalently, $\sin^{-1}(2\pi_i - 1) = \mathbf{x}_i^T\boldsymbol{\beta}$, a linear model for the arcsine or angular transform of π_i. ■

The forms 2 and 3 in Illustration 10.1 are probably the ones most commonly applied nowadays: they are explicitly computable functions of z, unlike Φ in form 1, and are not range-restricted like form 4.

The log-likelihood function for this set-up is

$$Q(\mathbf{y};\boldsymbol{\theta}) = \log \prod_{i=1}^{n} \{\pi_i^{y_i}(1-\pi_i)^{1-y_i}\} \tag{10.1}$$

where $\boldsymbol{\theta}$ contains $\boldsymbol{\beta}$ together with any additional parameters occurring in h. If there are none such, the model is a pure GLM and GLIM need not be abused.

10.1.2 *Longitudinal data: GLMs with a 'working' correlation matrix*

The data now comprise vector observations $\mathbf{y}_i (i = 1,\ldots, n)$ in which the components y_{ij} are binary; $\mathbf{y}_i$ is of length p_i, not necessarily the same for all individuals. Let $\pi_{ij} = \text{pr}(y_{ij} = 1)$ and $\boldsymbol{\pi}_i = [\pi_{i1} \ldots \pi_{ip_i}]^{\mathrm{T}}$, and suppose that $\pi_{ij} = h_j(\mathbf{x}_{ij}, \boldsymbol{\beta})$, where h_j is some specified inverse link function, $\mathbf{x}_{ij}$ is a vector of covariates, and $\boldsymbol{\beta}$ is a $q \times 1$ vector of regression coefficients. For h_j one of the forms suggested in Illustration 10.1 can be used, probably the most common choice in practice being the linear logit form:

$$h_j(\mathbf{x}_{ij}, \boldsymbol{\beta}) = \{1 + \exp(-\mathbf{x}_{ij}^{\mathrm{T}}\boldsymbol{\beta})\}^{-1} \quad \text{or} \quad \log\{\pi_{ij}/(1-\pi_{ij})\} = \mathbf{x}_{ij}^{\mathrm{T}}\boldsymbol{\beta}$$

For estimation of $\boldsymbol{\beta}$ the situation is covered by the general outline of section 9.2.2. The observation $\mathbf{y}_i$ has $p_i \times 1$ mean vector $\boldsymbol{\mu}_i = \boldsymbol{\pi}_i$ and $p_i \times p_i$ covariance matrix $\boldsymbol{\Sigma}_i = \phi \mathbf{S}_i \mathbf{R}_i \mathbf{S}_i$, where $\mathbf{S}_i$ is the $p_i \times p_i$ diagonal matrix with elements $S_{ijj} = \sqrt{\pi_{ij}(1-\pi_{ij})}$, $\mathbf{R}_i$ is the $p_i \times p_i$ correlation matrix of $\mathbf{y}_i$, and ϕ is included to allow for over- or under-dispersion. The working score function for $\boldsymbol{\beta}$ is

$$\mathbf{q}_{\boldsymbol{\beta}}(\mathbf{y}; \boldsymbol{\beta}, \boldsymbol{\tau}) = \sum_{i=1}^{n} \boldsymbol{\pi}_i'^{\mathrm{T}} (\mathbf{S}_i \bar{\mathbf{R}}_i \mathbf{S}_i)^{-1} (\mathbf{y}_i - \boldsymbol{\pi}_i) \tag{10.2}$$

as in (9.6) or (9.9), where $\boldsymbol{\pi}_i'$ is $p_i \times q$ with (j, k)th element $\partial\pi_{ij}/\partial\beta_k$. A working score function for $\boldsymbol{\tau}$, the parameter governing $\bar{\mathbf{R}}_i$, can be taken as in (9.9):

$$\mathbf{q}_{\boldsymbol{\tau}}(\mathbf{y}; \boldsymbol{\beta}, \boldsymbol{\tau}) = \sum_{i=1}^{n} \boldsymbol{\upsilon}_i'^{\mathrm{T}} \mathbf{W}_i^{-1} (\mathbf{u}_i - \boldsymbol{\upsilon}_i) \tag{10.3}$$

where $\mathbf{u}_i$ is the vector of length $p_i(p_i - 1)/2$ whose elements are the cross-products $y_{ij}y_{ik}$, $\boldsymbol{\upsilon}_i = E(\mathbf{u}_i)$, $\boldsymbol{\upsilon}_i' = \partial\boldsymbol{\upsilon}_i/\partial\boldsymbol{\tau}$, and $\mathbf{W}_i$ is a working covariance matrix for $\mathbf{u}_i$.

Example 10.1 Pill palatability

The British are well known as a nation of animal lovers and in many a household the pets rule the roost. Occasionally, these reigning cats and dogs might feel a little off-colour and medication is called for. Of some concern, then, particularly in veterinary circles, is the palatability of circular oral treatments – pills, for short. (One of our previous dogs was a past mistress in the art of deception: however much the pill was wrapped up in tasty morsels, it would be found some time later down behind the sofa or in someone's shoe.)

The data in Table A.18 result from trying out four different formulations of a drug on each of 27 cats. The rows of the data matrix contain five case variables: cat number, sex, breed, weight and age. These are followed by eight pairs (f, y) recorded respectively on days 1, 2, 3, 4, 8, 9, 10 and 11: f is the formulation and y is the observed palatability scored as 1 = readily accepted, 2 = accepted, 3 = reluctantly accepted, 4 = accepted in food, and 5 = refused.

A quick scan of the data shows the frequencies of scores 1–5 as 72, 13, 6, 68 and 57. The scores 2 and 3 are a bit thin on the ground, maybe because it is not easy to decide between 'accepted' and 'reluctantly accepted'. Also, scores 4 and 5 both mean that the pill itself was not acceptable. Let us make a first analysis of the data converted to binary form, grouping scores 1, 2 and 3 as 'accepted', and 4 and 5 as 'not accepted'. We will use the logit model with $\mathbf{x}_{ij} = [1 \text{ gender } \mathbf{f}_j]^{\mathrm{T}}$, where $\mathbf{f}_j$ is an indicator vector representing formulation: $\mathbf{f}_j = [1\ 0\ 0], [0\ 1\ 0], [0\ 0\ 1]$ and $[-1\ -1 -1]$ respectively for formulations 1, 2, 3 and 4. Hence, $\mathbf{y}_i$ (8×1), $\boldsymbol{\pi}_i$ (8×1) and $\mathbf{S}_i$ (8×8) are defined for the quasi-score function (10.2), as are $\mathbf{u}_i$ (28×1) and $\mathbf{v}_i$ (28×1) for (10.3). There remain $\bar{\mathbf{R}}_i$ (8×8) and $\mathbf{W}_i$ (28×28). For $\bar{\mathbf{R}}_i$ we will take the 'safe option' (a), referred to in section 9.2.2, i.e. the 'independence' model with $\bar{\mathbf{R}}_i = \mathbf{I}$. Likewise, we will take $\mathbf{W}_i = \mathbf{I}$, the most basic choice, as in section 9.3.

In order to fit the model to the data some sweeping assumptions have been made. The first was that only a small proportion of the information would be lost by reducing the scores to binary, though this was supported by the original experimenters. Further, no allowance has been made for weight, age and breed of cat. The model also carries no 'day' effect, so a learning curve is not envisaged, nor a carry-over effect in addition to the serial correlation; the latter omission is reasonable if, for example, refusal or acceptance of the pill has no consequences for the normal daily routine of feeding, etc. A certain amount of tentative fitting of submodels to subsets of the data can be carried out to check these assumptions but, as always, it is easy to over-interrogate a limited data set.

The estimate produced is

$$\tilde{\boldsymbol{\beta}} = [-0.39\ \ 0.33\ \ 0.33\ \ -0.22\ \ 0.022]$$

The main interest is in the effect of formulation: for the null hypothesis $H\colon \beta_3 = \beta_4 = \beta_5 = 0$ the Wald test gives $W = 4.19$ as $\chi_3^2(p > 0.10)$ so the effect of formulation is not convincingly demonstrated. For gender, the ratio of $\tilde{\beta}_1$ to its standard error yields normal deviate -1.22.

The corresponding estimate from applying Gaussian estimation is

$$\tilde{\boldsymbol{\beta}} = [-0.41\ \ 0.35\ \ 0.36\ \ -0.23\ \ 0.037]$$

The normal deviate for gender is -1.08 and the Wald test for formulations gives $W = 4.64$ as χ_3^2. These figures are not very different to those obtained via MQLE.

10.1.3 Longitudinal data: random effects models

The data are as described in section 10.1.2, vector observations $\mathbf{y}_i(p_i \times 1)$ of binary components y_{ij}. However, we now consider a random effects model of the type covered in section 9.4.1. Thus, applying (9.11) to the binary case, $m_{ij} = \mathrm{pr}(y_{ij} = 1 \mid \mathbf{b}_i)$ and a popular choice for the link function g_j is the logit (Illustration 10.1(2)). A common choice for the variance function is $s_{ij}^2 = \phi m_{ij}(1 - m_{ij})$. This is a reasonable but not obligatory choice: the implicit assumption is that ϕ is homogeneous over i and j, but not necessarily equal to 1 as in the strict binomial case. The conditional covariances s_{ijk} are taken to be zero, and then (9.12), (9.13) and (9.14) yield the unconditional mean and covariance structure of $\mathbf{y}_i$ once $f(\mathbf{b}_i)$ is specified, $N(\mathbf{0}, \mathbf{B})$ being the usual choice. Estimation by MQL can now proceed as described in section 9.4.2.

If a complete specification is made for $f(\mathbf{y}_i|\mathbf{b}_i)$, rather than just the conditional mean and covariance matrix, then the procedure of section 9.4.2 applies. Probably the most natural full specification is that, conditional on $\mathbf{b}_i$, the $y_{ij}(j = 1,\ldots,p_i)$ are independent binomial variates, so

$$f(\mathbf{y}_i|\mathbf{b}_i) = \prod_{j=1}^{p_i} m_{ij}^{y_{ij}}(1 - m_{ij})^{1-y_{ij}} \tag{10.4}$$

Then, with

$$g_j(m_{ij}) = \mathbf{x}_{ij}^{\mathrm{T}}\boldsymbol{\beta} + \mathbf{z}_{ij}^{\mathrm{T}}\mathbf{b}_i$$

for some chosen g_j, and with $f(\mathbf{b}_i)$ chosen as $N(\mathbf{0}, \mathbf{B})$, for instance, the full likelihood function (9.15) can be used for inference in the usual way.

10.2 Binary data from groups of individuals

Suppose that the individuals are grouped into S subpopulations but otherwise undifferentiated with respect to covariate information. For example, human subjects might be classified by gender (male, female) and blood group (A, B, AB, O). Then there would be $S = 8$ subpopulations. If age were also to be included the individuals would then be differentiated more finely than the grouping and what follows would not apply: we would have to go back to the regression models of section 10.1.

10.2.1 Univariate measures

The log-likelihood (10.1) for the grouped case reduces to

$$Q(\mathbf{y}; \boldsymbol{\theta}) = \log \prod_{s=1}^{S} \pi_s^{r_s}(1 - \pi_s)^{n_s - r_s} \tag{10.5}$$

where $\pi_s = h(\mathbf{x}_s^{\mathrm{T}}\boldsymbol{\beta})$, n_s is the sample size in group s, and r_s is the number in group s with y value 1. This is the log-product of S binomial likelihoods. If the π_s are unconstrained, their MLEs are $\hat{\pi}_s = p_s$, p_s being the proportion r_s/n_s observed in group s. Otherwise, Q is to be maximized over $\boldsymbol{\theta}$, which usually requires an iterative computation.

■ *Illustration 10.2* Consider a two-way classification, for example male/female by treatments 1 and 2. There are then $S = 4$ groups, and the set of $\mathbf{x}$ vectors can be taken as [1 0 0 0], [1 1 0 0], [1 0 1 0] and [1 1 1 1]. In a saturated model $\boldsymbol{\beta} = [\alpha \;\; \beta_1 \;\; \beta_2 \;\; \gamma]$, where α represents the regression intercept, β_1 and β_2 the treatment effects, and γ an interaction term. ■

An alternative, non-iterative method of estimation, due to Berkson (1953), is based on the so-called **empirical logit transform** of p_s:

$$g_s = \log\{p_s/(1 - p_s)\} = \log\{r_s/(n_s - r_s)\}$$

The method is least-squares based on the weighted sum of squares

$$Q(\mathbf{y};\boldsymbol{\beta}) = \sum_{s=1}^{S} w_s^{-1}(g_s - \mathbf{x}_s^{\mathrm{T}}\boldsymbol{\beta})^2$$

where the w_s are weighting factors to be specified shortly. Assuming that w_s does not depend on $\boldsymbol{\beta}$, differentiation with respect to $\boldsymbol{\beta}$ produces the estimating function

$$\mathbf{q}(\mathbf{y};\boldsymbol{\beta}) = \sum_{s=1}^{S} w_s^{-1}(g_s - \mathbf{x}_s^{\mathrm{T}}\boldsymbol{\beta})\mathbf{x}_s \qquad (10.6)$$

which leads on to the estimator

$$\tilde{\boldsymbol{\beta}} = \left\{ \sum_{s=1}^{S} w_s^{-1}\mathbf{x}_s\mathbf{x}_s^{\mathrm{T}} \right\}^{-1} \sum_{s=1}^{S} w_s^{-1} g_s \mathbf{x}_s$$

as a solution of the estimating equation $\mathbf{q}(\mathbf{y};\boldsymbol{\beta}) = \mathbf{0}$. Note that (10.6) resembles the quasi-score function (9.1): the observed quantity here is g_s, and its mean is approximately $\mathbf{x}_s^{\mathrm{T}}\boldsymbol{\beta}$ for large n_s under a model which specifies

$$\log\{\pi_s/(1-\pi_s)\} = \mathbf{x}_s^{\mathrm{T}}\boldsymbol{\beta}$$

The modified transform

$$g_s = \log\{(r_s + 1/2)/(n_s - r_s + 1/2)\}$$

is often used: extreme values $r_s = 0$ or n_s do not then cause g_s to become infinite, and the approximation of $E(g_s)$ by $\mathbf{x}_s^{\mathrm{T}}\boldsymbol{\beta}$ is improved (Cox, 1970, section 3.2). To complete the correspondence between (10.6) and (9.1), w_s must be taken as var (g_s) which, for large n_s, can be well approximated (applying the delta method) by the sample estimate $\{n_s p_s(1-p_s)\}^{-1}$. Berkson (1953) noted that, with this choice of w_s, and for large n_s,

$$Q(\mathbf{y};\boldsymbol{\beta}) \doteqdot \sum_{s=1}^{S} \left[\frac{(r_s - n_s\pi_s)^2}{n_s\pi_s} + \frac{\{(n_s - r_s) - n_s(1-\pi_s)\}^2}{n_s(1-\pi_s)} \right]$$

the χ^2-statistic for testing goodness of fit in a $2 \times S$ contingency table with sth column containing r_s and $n_s - r_s$. Correspondingly, the method is known as **minimum logit chi-square**.

■ *Illustration 10.3 Failures of metal coating* The table below shows the numbers of failures and numbers tested of coated metal specimens over a range of temperatures.

temp°C	10	30	50	70	90	110
no. fail r	3	5	10	11	12	8
no. tested n	25	25	20	20	15	10

There are six groups here and the regression model relates the failure proportions to temperature. There are no within-groups factors to differentiate specimens: the properties of the estimates actually rely on the group sizes being 'large'. ■

10.2.2 *Longitudinal data*

Each individual is monitored now over p occasions, so we have $p \times 1$ vector observations $\mathbf{y}_i (i = 1, \ldots, n)$. There are 2^p distinct binary vectors of length p, say $\mathbf{b}_1, \ldots, \mathbf{b}_{2^p}$. Let π_{sk} be the probability that $\mathbf{y}_i = \mathbf{b}_k$ when individual i belongs to subpopulation s, so $\pi_{s+} = 1$, and let $\boldsymbol{\pi}_s = [\pi_{s1} \; \pi_{s2} \ldots]$ be the $2^p \times 1$ vector of these probabilities. Suppose that there are n_s individuals in the sample from subpopulation s of whom r_{sk} yield response profile $\mathbf{y} = \mathbf{b}_k$, so $r_{s+} = n_s$. In terms of the $\boldsymbol{\pi}_s$ vectors, the corresponding log-product of multinomial likelihoods is then

$$Q(\mathbf{y}; \boldsymbol{\pi}_1, \ldots, \boldsymbol{\pi}_S) = \log \prod_{s=1}^{S} \left\{ n_s! \prod_{k=1}^{2^p} (\pi_{sk}^{r_{sk}} / r_{sk}!) \right\} \tag{10.7}$$

If the $\boldsymbol{\pi}_s$'s are unconstrained their MLEs are thus given by the sample proportions: $\hat{\boldsymbol{\pi}}_s = \mathbf{p}_s$ with kth component $p_{sk} = r_{sk}/n_s$.

For regression modelling of the π_{sk} simple logits will not suffice because there are 2^p such probabilities per group, instead of just π_{s1} and $1 - \pi_{s1}$ as previously. More detailed models, like the ones given in section 10.1.1, can be used for input to (10.7). Alternatively, the weighted least-squares methodology described below in section 10.4.2 can be applied.

10.3 Regression models for categorical data

10.3.1 *Univariate measures*

Suppose that $y_i = c$ when the ith individual falls into category $c (c = 1, \ldots, C)$. The probability $\pi_{ic} = \mathrm{pr}(y_i = c)$ depends on the explanatory vector $\mathbf{x}_i$ in some way to be specified in context. The log-likelihood function is

$$Q(\mathbf{y}; \boldsymbol{\theta}) = \log \prod_{i=1}^{n} \pi_{iy_i} \tag{10.8}$$

Unordered categories (nominal scale)

Suppose that the C categories are not ordered or structured in any way, as with colours, blood groups, religious denominations, or failure modes of equipment. A standard type of regression model for this case is

$$\pi_{ic} = \phi(\mathbf{x}_i, \boldsymbol{\beta}_c) \Big/ \sum_{l=1}^{C} \phi(x_i, \boldsymbol{\beta}_l)$$

where ϕ is a given positive function. A common choice is $\phi(\mathbf{x}_i, \boldsymbol{\beta}_c) = \exp(\mathbf{x}_i^T \boldsymbol{\beta}_c)$, in which case $\log(\pi_{ic}/\pi_{ic'}) = \mathbf{x}_i^T(\boldsymbol{\beta}_c - \boldsymbol{\beta}_{c'})$, a log-linear relation for ratios of probabilities of different categories. For identifiability of the $\boldsymbol{\beta}_c$ in this case it is usual to take $\boldsymbol{\beta}_C = \mathbf{0}$.

Ordered categories (ordinal scale)

Suppose now that the C categories have a natural ordering, as with colours (ordered by wavelength), classifications (bad/satisfactory/good, low/medium/high). A standard type of regression model for this case is

$$\pi_{ic} = F(\alpha_c + \mathbf{x}_i^T \boldsymbol{\beta}_c) - F(\alpha_{c-1} + \mathbf{x}_i^T \boldsymbol{\beta}_{c-1})$$

where $-\infty = \alpha_0 < \alpha_1 < \cdots < \alpha_C = \infty$ and F is a distribution function on $(-\infty, \infty)$. In terms of the cumulative probabilities $\tau_{ic} = \mathrm{pr}(y_i \leqslant c)$,

$$\tau_{ic} = \pi_{i1} + \cdots + \pi_{ic} = F(\alpha_c + \mathbf{x}_i^T \boldsymbol{\beta}_c)$$

Note that the model is preserved under the merging of adjacent categories. A constraint commonly made is that $\boldsymbol{\beta}_c = \boldsymbol{\beta}$ for all c. Then one just has the cut-points $\alpha_1, \ldots, \alpha_{C-1}$ on the z-scale of $F(z)$ and the sole effect of $\mathbf{x}_i$ is to make a location shift $\mathbf{x}_i^T \boldsymbol{\beta}$.

■ *Illustration 10.4* Proportional odds model (McCullagh, 1980) Take for F the logistic form, $F(z) = (1 + e^{-z})^{-1}$, and denote the *odds ratio* $\tau_{ic}/(1 - \tau_{ic})$ by λ_{ic}. Then $\log \lambda_{ic} = \alpha_c + \mathbf{x}_i^T \boldsymbol{\beta}_c$ and, if $\boldsymbol{\beta}_c = \boldsymbol{\beta}$ for all c, $\lambda_{ic}/\lambda_{ic'} = \exp(\alpha_c - \alpha_{c'})$ is independent of $\mathbf{x}_i$. Again, the odds ratio for category c between two individuals, i and i', is $\lambda_{ic}/\lambda_{i'c} = \exp\{(\mathbf{x}_i - \mathbf{x}_{i'})^T \boldsymbol{\beta}_c\}$; if $\boldsymbol{\beta}_c = \boldsymbol{\beta}$ this is independent of c so the effect on the odds ratio of changing the $\mathbf{x}$ value, for example a treatment indicator, is the same for all categories. ■

Since the y_i are categorical they do not have numerical mean values and so MQLE is not directly applicable. A simple dodge is to convert the data to binary form (see, for example, Liang, Zeger and Qaqish, 1992) so that MQLE can be applied as follows. Let $y_{ic}^* = I(y_i = c)$, where I denotes the indicator function, and let $\mathbf{y}_i^* = (y_{i1}^*, \ldots, y_{i,C-1}^*)^T$. Then $\mathbf{y}_i^*$ is a binary vector of length $C-1$ in which at most one entry can be non-zero. Its mean $\boldsymbol{\mu}_i^*$ has cth component π_{ic} and its covariance matrix $\boldsymbol{\Sigma}_i^*$ has (c, d)th entry

$$\Sigma_{icd}^* = E(y_{ic}^* y_{id}^*) - \pi_{ic}\pi_{id} = \pi_{ic}(\delta_{cd} - \pi_{id})$$

using the Kronecker delta symbol. Thus, we have the ingredients $\mathbf{y}_i^*$, $\boldsymbol{\mu}_i^*$ and $\boldsymbol{\Sigma}_i^*$, for MQLE. Incidentally, had we taken $\mathbf{y}_i^*$ to be of full length C, its components would have had sum 1; this linear constraint would make its covariance matrix singular, thus causing complications for constructing a quasi-score function.

10.3.2 Longitudinal data: GLMs with a working correlation matrix

The data now comprise $p_i \times 1$ vector observations $\mathbf{y}_i$ $(i = 1, \ldots, n)$ in which the components are categorical, y_{ij} having C_j categories. Let $\pi_{ijc} = \text{pr}(y_{ij} = c)$, and let $\boldsymbol{\pi}_{ij} = [\pi_{ij1} \ldots \pi_{ijC_j}]^{\text{T}}$ and $\boldsymbol{\pi}_i = [\boldsymbol{\pi}_{il}^{\text{T}} \ldots \boldsymbol{\pi}_{ip_i}^{\text{T}}]^{\text{T}}$ be the probability vectors associated with $\mathbf{y}_i$.

General regression models can be constructed along the lines of those given in section 10.3.1. For unordered categories,

$$\pi_{ijc} = \phi(\mathbf{x}_{ij}, \boldsymbol{\beta}_{jc}) \Big/ \sum_{l=1}^{C_j} \phi(\boldsymbol{X}_{ij}, \boldsymbol{\beta}_{jl})$$

is a candidate, where $\mathbf{x}_{ij}$ is the covariate vector for individual i at occasion j. For ordered categories, a useful general framework is given by

$$\tau_{ijc} = \pi_{ij1} + \cdots + \pi_{ijc} = F(\alpha_{jc} + \mathbf{x}_{ij}^{\text{T}}\boldsymbol{\beta}_{jc})$$

for some chosen distribution function F.

The categorical components y_{ij} of $\mathbf{y}_i$ can be converted to binary vectors $\mathbf{y}_{ij}^*$ (of length $C_j - 1$) as described in section 10. 3. 1, and then we take $\mathbf{y}_i^{*\text{T}} = [\mathbf{y}_{i1}^{*\text{T}} \cdots \mathbf{y}_{ip_i}^{*\text{T}}]$ to represent the full binary data vector for individual i; $\mathbf{y}_i^*$ is of length $C_+ - p$, where $C_+ = C_1 + \cdots + C_p$, subject to missing components for individual i.

The subvector $\mathbf{y}_{ij}^*$ has mean $\boldsymbol{\mu}_{ij}^*$, with cth component π_{ijc}, and covariance matrix $\boldsymbol{\Sigma}_{ij}^*$, with (c, d)th entry $\pi_{ijc}(\delta_{cd} - \pi_{ijd})$. The $\boldsymbol{\Sigma}_{ij}^*$ form the diagonal blocks of the full covariance matrix $\boldsymbol{\Sigma}_i^*$ of $\mathbf{y}_i^*$, the (j, k)th off-diagonal block $\boldsymbol{\Sigma}_{ijk}^*$ being open to suitable modelling. Thus,

$$\boldsymbol{\Sigma}_{ijkcd}^* = \rho_{i,jk,cd}\sqrt{\pi_{ijc}(1 - \pi_{ijc})\pi_{ikd}(1 - \pi_{ikd})}$$

where the correlation $\rho_{i,jk,cd}$ might be taken, for example, as $\rho^{|t_j - t_k|}$ to reflect decreasing dependence with separation over time as in section 6.2.4. This choice implies that the serial correlation structure can be described by the single parameter ρ: it is a fairly basic choice. Even so, it can lead to trouble. As pointed out by Liang, Zeger and Qaqish (1992), there are hidden constraints on the correlation parameters which involve the π_{ijc}. If not allowed for, this can cause problems in the computations: as the parameter set is explored in the iterative search for a solution of the estimating equations, the constraints can be violated at some points, for example causing numerical overflows as covariance matrices become singular. In the examples below we will play safe, employing the 'independence model' with $\rho_{i,jk,cd} = 0$ (section 9.2.2).

We now have $\mathbf{y}_i^*$, $\boldsymbol{\mu}_i^*$ and $\boldsymbol{\Sigma}_i^*$, so the quasi-score function (9.9) can be constructed. For any additional covariance parameters, perhaps just ρ, we need to specify $\boldsymbol{W}_i$ in order to apply (9.10): the basic choice is $\boldsymbol{W}_i = \mathbf{I}$, as made in section 9. 3.

Example 10.2 Pain assessment

Back pain is a widespread and miserable condition, but not a glamorous, high-profile specialty for medical researchers. Table A.19 summarizes some results of a clinical trial involving two treatments. The rows of the data matrix contain case number and treatment group followed by four sets of five numbers. Each set comprises pain verbal rating scale (VRS), pain visual analogue scale (VAS), anxiety VAS, alertness VAS and time (in minutes) since treatment: the VRS is scored 1–5, and the VAS 1–100.

Let us focus upon the first assessment, pain VRS, which is an ordered categorical variable. The frequencies of the scores 1–5 in the data are 18, 55, 26, 6, 1; the scores 4 and 5 have been pooled with 3 for the present analysis. The model will relate the four repeated measures to treatment group and time after treatment. We will take

$$\pi_{ijc} = F(\alpha_c + \mathbf{x}_{ij}^{\mathrm{T}}\boldsymbol{\beta}) - F(\alpha_{c-1} + \mathbf{x}_{ij}^{\mathrm{T}}\boldsymbol{\beta})$$

where $F(z) = (1 + e^{-z})^{-1}$ has the logit form and $\mathbf{x}_{ij} = [g_i \ \ t_i \ \ g_i t_i]$ with g_i representing the treatment group indicator and t_i the time since treatment (mins/100). The third component of $\mathbf{x}_{ij}$ allows for an interaction term through which the progress of the pain VRS over time may differ between treatments.

The results of MQLE are $\tilde{\boldsymbol{\alpha}} = [-1.72 \ \ 2.47]$ and $\tilde{\boldsymbol{\beta}} = [0.51 \ \ 0.021 \ \ -0.039]$, the standard errors of the $\tilde{\boldsymbol{\beta}}$'s being computed as (0.40, 0.061, 0. 061). The treatments are not well differentiated (normal deviate (nd) 1.28), time does not show up strongly as a factor (nd = 0.34), and the interaction term is likewise less than arresting (nd = − 0.64).

The corresponding results from Gaussian estimation are $\tilde{\alpha} = [-1.65 \ \ 2.46]$ and $\tilde{\boldsymbol{\beta}} = [0.57 \ \ 0.012 \ \ -0.062]$, with $\mathrm{se}(\tilde{\boldsymbol{\beta}}) = [0.37 \ \ 0.053 \ \ 0.053]$. The inferences are similar to those made from MQLE.

Example 10.3 Anxiety scores

The data in Table A.20 relate to patients who suffer from panic attacks (group 1) and a control set (group 2). The case variables are subject number and group indicator. There follow 11 repeated measures on three variables: the first variable is the score on an anxiety scale, increasing from 0 to 8; the second is CO_2 expiration; and the third is pulse rate. The three variables were recorded together at times 4, 6, 8, 10, 11, 14, 16, 17, 18, 19 and 23 (minutes). The subject was spoken to about the topic of anxiety at times 6, 8 and 10; the subject was asked to hyperventilate at times 16, 17 and 18; the remaining times, 4, 11, 14, 19 and 23, were designated as 'rest' times. The scope for relating anxiety scores to explanatory variables here is wide: we will just focus upon the effects of 'group', 'circumstances' (i.e. rest, spoken to, or hyperventilation) and CO_2 expiration.

The frequencies of the scores 0–8 in the data are 43, 53, 58, 50, 54, 37, 57, 42 and 43; for a first analysis these scores have been grouped as (0, 1, 2), (3, 4, 5) and (6, 7, 8) with frequencies 154, 141 and 142. We will use the logit form, as in Example 10.2 with $\mathbf{x}_{ij} = [g_i \ \ c_{ij} \ \ e_{ij}]$, where g_i represents the treatment group indicator, c_{ij} the circumstances indicator (two dummy variables), and e_{ij} CO_2 expiration.

The results of MQLE are $\tilde{\boldsymbol{\alpha}} = [-3.82 \ \ 1.90]$, and $\tilde{\boldsymbol{\beta}} = [-0.84 \ \ 0.95 \ \ -1.20 \ \ 10.40]$. For the treatment effect the ratio of $\tilde{\boldsymbol{\beta}}_1$ to its standard error gives a normal deviate equal to − 3.64, and that for CO_2 expiration gives a normal deviate, based on $\tilde{\beta}_4$, of

3.01; these aspects seem, then, to be well established. For the 'circumstances' a Wald test gives $W = 42.8$ as χ_2^2, so this aspect also seems to be well established.

The corresponding results from Gaussian estimation are $\tilde{\boldsymbol{\alpha}} = [-3.73 \;\; 1.88]$, and $\tilde{\boldsymbol{\beta}} = [-0.81 \;\; 1.05 \;\; -1.34 \;\; 10.40]$, and the following inferences are similar to those made from MQLE.

10.3.3 Longitudinal data: random effects models

The data are as described in section 10.3.2, vector observations $\mathbf{y}_i$ ($p_i \times 1$) of categorical components y_{ij}. However, we now consider a random effects model of the type covered in section 9.4.1. Thus, let $p_{ijc} = \text{pr}(y_{ij} = c|\mathbf{b}_i)$ and take for p_{ijc} a regression model of the type used for π_{ijc} in section 10.3.2 but with $\mathbf{x}_{ij}^{\mathrm{T}}\boldsymbol{\beta}_c$ augmented to $\mathbf{x}_{ij}^{\mathrm{T}}\boldsymbol{\beta}_c + \mathbf{z}_{ij}^{\mathrm{T}}\mathbf{b}_i$. To proceed with estimation, a distribution for $\boldsymbol{b}_i$ can be proposed, then integration of p_{ijc} over this distribution yields the unconditional probability $E(p_{ijc}) = \text{pr}(y_{ij} = c)$, and then MQLE can be applied as in section 10.3.2.

For the particular case $\mathbf{Z}_i = \mathbf{I}$, where $\mathbf{Z}_i$ is the matrix with rows $\mathbf{z}_{ij}$ Conaway (1989) uses a conditional likelihood method, thus avoiding the specification of, and integration over, a distribution for $\mathbf{b}_i$. He assumes conditional independence of the components of $\mathbf{y}_i$, given $\mathbf{b}_i$, and conditions on the minimal sufficient statistics for $\mathbf{b}_i$. The approach is a generalization of the well-known Rasch model. Agresti and Lang (1993) do something similar for ordered categorical repeated measures. They apply a cumulative logit model (Illustration 10.3) with Rasch-type random individual effects. However, they point out that simple sufficient statistics for the random effects are not available and so move on to a rather ingenious conditioning argument.

10.4 Categorical data from groups of individuals

10.4.1 Univariate measures

The situation is as described in section 10.3: the individuals are grouped into S subpopulations. Suppose that in group s there are n_s individuals of whom r_{sc} yield y-response c, so $r_{s1} + \cdots + r_{sC} = n_s$. For this situation the log-likelihood (10.8) reduces to

$$Q(\mathbf{y}; \boldsymbol{\theta}) = \log \prod_{s=1}^{S} \prod_{c=1}^{C} \pi_{sc}^{r_{sc}} \tag{10.9}$$

Where π_{sc} is the probability that y_i takes value c when individual i is in group s. This is a log-product of S multinomial likelihoods, each based on the C catgories. If the π_{sc} are unconstrained, their MLEs are $\hat{\pi}_{sc} = p_{sc}$, p_{sc} being the observed proportion r_{sc}/n_s. Otherwise, Q is to be maximized over $\boldsymbol{\theta}$, which usually requires an iterative computation.

An alternative, non-iterative method of estimation, generalizing Berkson's minimum logit χ^2, is based on empirical transforms of the p_{sc} (Grizzle, Starmer and Koch, 1969). Let $\boldsymbol{\pi}_s = [\pi_{s1} \ldots \pi_{sC}]^{\mathrm{T}}$ be the $C \times 1$ vector of probabilities for group s and let $\boldsymbol{\pi}^{\mathrm{T}} = [\boldsymbol{\pi}_1^{\mathrm{T}} \ldots \boldsymbol{\pi}_S^{\mathrm{T}}]$ contain all SC probabilities. Consider the model defined by the constraints $g_k(\boldsymbol{\pi}) = \mathbf{x}_k^{\mathrm{T}} \boldsymbol{\beta}$, where the g_k are specified link functions, for example $g_k(\boldsymbol{\pi}) = \log(\pi_{k1}/\pi_{k2})$ and $\mathbf{x}_k^{\mathrm{T}}\boldsymbol{\beta} = \beta_1 + \beta_2 k$ for $k = 1, \ldots, S$, which specifies a linear trend over groups of the log-ratio of probabilities of categories 1 and 2. The number of such functions must lie between 0, for a completely unconstrained model, and $S(C-1)$, for a completely specified set of π_{sc}. Obviously, the g_k must be chosen coherently, so that the constraints are not mutually contradictory. Nor should any of them be redundant in the sense of being implied by others. These constraints can be written as a single matrix equation $\mathbf{g}(\boldsymbol{\pi}) = \mathbf{X}\boldsymbol{\beta}$, where $\mathbf{g}(\boldsymbol{\pi})$ is the vector with components $g_k(\boldsymbol{\pi})$ and $\mathbf{X}$ has kth row $\mathbf{x}_k^{\mathrm{T}}$ and is of full rank. Let $\mathbf{g}(\mathbf{p})$ be the empirical transform of the vector $\mathbf{p}$ of sample proportions corresponding to $\boldsymbol{\pi}$. As the n_s all become larger, and under correct model specification, the distribution of $\mathbf{g}(\mathbf{p})$ tends to $N(\mathbf{X}\boldsymbol{\beta}, \mathbf{V}_g(\boldsymbol{\pi}))$. Here $\mathbf{V}_g(\boldsymbol{\pi}) = \mathbf{g}'(\boldsymbol{\pi})\mathbf{V}_p(\boldsymbol{\pi})\mathbf{g}'(\boldsymbol{\pi})^{\mathrm{T}}$ can be obtained by the delta method, $\mathbf{g}'(\boldsymbol{\pi})$ being the Jacobian matrix of derivatives of $\mathbf{g}(\boldsymbol{\pi})$ with respect to $\boldsymbol{\pi}$, and $\mathbf{V}_p(\boldsymbol{\pi}) = \mathrm{var}(\mathbf{p})$ is block-diagonal with sth block

$$\mathbf{V}_s(\boldsymbol{\pi}_s) = n_s^{-1}(\mathrm{diag}[\pi_{sc}] - \boldsymbol{\pi}_s\boldsymbol{\pi}_s^{\mathrm{T}})$$

The formal expression of the strictures above, concerning the choice of the g_k functions, is that $\mathbf{V}_g(\boldsymbol{\pi})$ must be nonsingular.

The method is based on minimizing the quadratic form

$$Q(\mathbf{y};\boldsymbol{\beta}) = \{\mathbf{g}(\mathbf{p}) - \mathbf{X}\boldsymbol{\beta}\}^{\mathrm{T}}\mathbf{W}^{-1}\{\mathbf{g}(\mathbf{p}) - \mathbf{X}\boldsymbol{\beta}\}$$

where $\mathbf{W}$ is a weighting matrix to be specified shortly. Assuming that $\mathbf{W}$ is independent of $\boldsymbol{\beta}$, differentiation of $Q(\mathbf{y};\boldsymbol{\beta})$ with respect to $\boldsymbol{\beta}$ produces the estimating function

$$\mathbf{q}(\mathbf{y};\boldsymbol{\beta}) = \mathbf{X}^{\mathrm{T}}\mathbf{W}^{-1}\{\mathbf{g}(\mathbf{p}) - \mathbf{X}\boldsymbol{\beta}\} \qquad (10.10)$$

The estimator

$$\tilde{\boldsymbol{\beta}} = (\mathbf{X}^{\mathrm{T}}\mathbf{W}^{-1}\mathbf{X})^{-1}\mathbf{X}^{\mathrm{T}}\mathbf{W}^{-1}\mathbf{g}(\mathbf{p})$$

is obtained by solving the estimating equation $\mathbf{q}(\mathbf{y};\boldsymbol{\beta}) = \mathbf{0}$ which, under the model, is asymptotically unbiased since $\mathbf{g}(\mathbf{p})$ has asymptotic mean $\mathbf{X}\boldsymbol{\beta}$. In (10.10) the matrix $\mathbf{W}$ can be taken as $\mathbf{V}_g(\boldsymbol{\pi})$, as in MQLE (equation 9.5), or its estimate $\mathbf{V}_g(\mathbf{p})$. With the second form for $\mathbf{W}$, solving $\mathbf{q}(\mathbf{y};\boldsymbol{\beta}) = \mathbf{0}$ is equivalent to minimizing $\mathbf{Q}(\mathbf{y};\boldsymbol{\beta})$. The asymptotic distribution of the estimator $\tilde{\boldsymbol{\beta}}$ is $N(\boldsymbol{\beta}, \mathbf{V}_\beta(\boldsymbol{\pi}))$, where $\mathbf{V}_\beta(\boldsymbol{\pi}) = \{\mathbf{X}^{\mathrm{T}}\mathbf{V}_g(\boldsymbol{\pi})^{-1}\mathbf{X}\}^{-1}$. Wald-type tests can be performed for linear hypotheses about $\boldsymbol{\beta}$ in the usual way (section 7.1.1), using the estimate $\mathbf{V}_\beta(\mathbf{p})$ for $\mathrm{cov}(\hat{\boldsymbol{\beta}})$. A χ^2 goodness-of-fit test for the model is provided by the statistic

$$T = \mathbf{g}(\mathbf{p})^{\mathrm{T}}\{\mathbf{V}_g(\mathbf{p})^{-1} - \mathbf{V}_g(\mathbf{p})^{-1}\mathbf{H}(\mathbf{p})\mathbf{V}_g(\mathbf{p})^{-1}\}\mathbf{g}(\mathbf{p})$$

where $\mathbf{H}(\mathbf{p}) = \mathbf{X}\{\mathbf{X}^{\mathrm{T}}\mathbf{V}_g(\mathbf{p})^{-1}\mathbf{X}\}^{-1}\mathbf{X}^{\mathrm{T}}$ has the familiar 'hat matrix' form. The particular case $\mathbf{X} = \mathbf{0}$ is allowed, so that the null hypothesis $\mathbf{g}(\boldsymbol{\pi}) = \mathbf{0}$ can be looked at.

10.4.2 Longitudinal data

Suppose now that individual i provides a vector observation $\mathbf{y}_i = [y_{i1} \dots y_{ip}]^{\mathrm{T}}$ in which y_{ij} takes a value from the set $(1, \dots, C)$. The number of such distinct profile vectors is C^p. Let π_{sC} be the probability that a randomly selected individual from subpopulation s has y vector $\mathbf{c} = [c_1 \dots c_p]^{\mathrm{T}}$. The marginal probabilities for occasion j will also figure in the following: define

$$\pi_{sjk} = \sum \pi_{sc}$$

with summation over all profile vectors $\mathbf{c}$ with jth component $c_j = k$; π_{sjk} is the probability that an individual i from subpopulation s has y_{ij} equal to k.

The weighted least-squares approach for this set-up has been presented by Koch *et al.* (1977) and Landis *et al.* (1988), among others. In principle, the analysis is the same as that described in section 10.4.1: the differences are that the former category label c now becomes a vector, and more involved models and hypotheses arise. Some typical examples of the latter are as follows.

1. $\pi_{sc} = \pi_{s'c}$: the probability of profile $\mathbf{c}$ is the same in groups s and s';
2. $\pi_{sc} = 2\pi_{sc'}$: within group s, profile $\mathbf{c}$ is twice as probable as $\mathbf{c}'$;
3. $\pi_{sjk} = \pi_{s'jk}$: the outcome $y_{ij} = k$ is equally probable in groups s and s';
4. $\pi_{sjk} = \pi_{sj'k}$: within group s, the probability of response k is the same at occasions j and j';
5. $\pi_{sjk} = \pi_{sjk'}$: within group s, responses k and k' are equally probable at occasion j.

These constraints can obviously be elaborated. For instance, the first might be proposed for every pair (s, s') to specify homogeneity of the probabilities over all the groups.

Let $\boldsymbol{\pi}_s$ be the $C^p \times 1$ vector of probabilities π_{sc} (listed in some defined order) for subpopulation s, and let $\boldsymbol{\pi}^{\mathrm{T}} = [\boldsymbol{\pi}_1^{\mathrm{T}} \dots \boldsymbol{\pi}_S^{\mathrm{T}}]$. Correspondingly, define $\mathbf{p}^{\mathrm{T}} = [\mathbf{p}_1^{\mathrm{T}} \dots \mathbf{p}_S^{\mathrm{T}}]$, comprising the vectors of observed proportions. Then one can investigate more complex constraints of type $g_k(\boldsymbol{\pi}) = \mathbf{x}_k^{\mathrm{T}}\boldsymbol{\beta}$, involving regression parameters, and the analysis goes through as before (section 10.4.1).

Note that, by working in terms of the π_{sc}, the probabilities of complete profiles of length p, the modelling of dependence between the p occasions is avoided. Such dependencies are simply estimated in the matrix $\mathbf{V}_{\mathrm{p}}(\mathbf{p})$.

The analysis described above is available in some widely used computer packages. Examples are given in Koch *et al.* (1977), Landis *et al.* (1988), Crowder and Hand (1990, section 8.4), among others.

The data could be presented as a $S \times C \times \cdots \times C$ contingency table with SC^p cells, and the S subpopulations could be further cross-classified (for example, gender by blood group) if appropriate. Then standard log-linear modelling could be applied (Bishop, Fienberg and Holland, 1975). The general approach via log-linear models will not be pursued here, however. For longitudinal data there are certain drawbacks, as pointed out by Liang, Zeger and Qaqish (1992). For instance, the log-linear terms represent conditional probabilities, given the other components of the observation vector. Thus, their interpretation depends on the presence or absence of other components, unlike the marginal models of section 10.3.

10.5 Further reading

A useful starting point for repeated categorical measures are issues 1 and 2 of Vol. 7 (1988) of *Statistics in Medicine*, a special issue containing a collection of papers on the general theme. Agresti (1989), Kenward and Jones (1992) and Fitzmaurice, Laird and Rotnitzky (1993) also make very worthwhile reading.

Stram, Wei and Ware (1988) fitted repeated ordered categorical measures at each time point separately before combining the results. This 'separate times' approach also appears in Wei and Stram (1988) and Moulton and Zeger (1989).

Prentice (1988) suggested the additional quasi-score function, of type (10.3), for estimating $\boldsymbol{\tau}$. Actually, his $\mathbf{u}_i$ was based on cross-products of standardized components, $\tilde{y}_{ij} = (y_{ij} - \pi_{ij})/\sqrt{\pi_{ij}(1 - \pi_{ij})}$, and he took $\mathbf{W}_i$ to be diagonal with (l, l)th entry $\text{var}\{(\mathbf{u}_i)_l\}$.

Lipsitz, Laird and Harrington (1991) and Liang, Zeger and Qaqish (1992) proposed the use of (10.2) and (10.3) underpinned by a modelling approach to the odds ratios

$$\psi_{ijk} = \frac{\text{pr}(y_{ij} = 0)|y_{ik} = 0)/\text{pr}(y_{ij} = 1|y_{ik} = 0)}{\text{pr}(y_{ij} = 0|y_{ik} = 1)/\text{pr}(y_{ij} = 1|y_{ik} = 1)} = \frac{\pi_{ijk}(1 - \pi_{ij} - \pi_{ik} + \pi_{ijk})}{(\pi_{ij} - \pi_{ijk})(\pi_{ik} - \pi_{ijk})} \qquad (10.11)$$

where $\pi_{ijk} = \text{pr}(y_{ij} = 1, y_{ik} = 1) = E(y_{ij}y_{ik})$. One can solve (10.11) to obtain π_{ijk}, i.e. the components of $\boldsymbol{\upsilon}_i$, in terms of π_{ij}, π_{ik} and ψ_{ijk} for input to (10.3); ψ_{ijk} is modelled as a function of explanatory variables, for instance log-linearly as $\log\psi_{ijk} = \mathbf{z}_{ijk}^{\text{T}}\boldsymbol{\tau}$. Lipsitz *et al.* (1991) suggested taking $\mathbf{W}_i$ in (10.3) as diag $\{\pi_{ijk}(1 - \pi_{ijk})\}$, which reflects $\text{var}(y_{ij}y_{ik}) = \pi_{ijk}(1 - \pi_{ijk})$, $y_{ij}y_{ik}$ being a binary variate, but replaces the off-diagonal terms $\text{cov}(y_{ij}y_{ik}, y_{ij'}\, y_{ik'})$ by zeros. Fitzmaurice and Lipsitz (1995) have continued the work, proposing serial pattern models for the odds ratio analogous to the Markov correlation structure for continuous measures (section 6.2.4).

Zhao and Prentice (1990) started with the joint probability function for $\mathbf{y}_i$ given by

$$f(\mathbf{y}_i) \propto \exp\{\boldsymbol{\theta}_i^{\mathrm{T}}\mathbf{y}_i + \boldsymbol{\lambda}_i^{\mathrm{T}}\mathbf{u}_i + c_i(\mathbf{y}_i)\} \tag{10.12}$$

where $\mathbf{u}_i$ is the vector of length $p_i(p_i - 1)/2$ whose components are the distinct products of pairs $y_{ij}y_{ik}$, $\boldsymbol{\theta}_i$ and $\boldsymbol{\lambda}_i$ are the canonical parameter vectors of this quadratic exponential-family model, and the $c_i(\mathbf{y}_i)$ are shape functions; the constant of proportionality in (10.12) is determined by $\sum f(\mathbf{y}_i) = 1$, the summation being over all binary vectors $\mathbf{y}_i$ of length p_i. They went on to reparametrize the distribution in terms of vectors $\boldsymbol{\beta}$ and $\boldsymbol{\tau}$, where $\boldsymbol{\beta}$ governs the mean $\boldsymbol{\mu}_i(\boldsymbol{\beta}) = E(\mathbf{y}_i)$ and both govern the covariance matrix $\boldsymbol{\Sigma}_i(\boldsymbol{\beta}, \boldsymbol{\tau}) = \mathrm{cov}(\mathbf{y}_i)$. The score function can then be written as $\mathbf{q}(\mathbf{y}; \boldsymbol{\beta}, \boldsymbol{\tau}) = \sum \mathbf{D}_i^{\mathrm{T}}\mathbf{V}_i^{-1}\mathbf{d}_i$, where

$$\mathbf{D}_i = \begin{bmatrix} \partial\boldsymbol{\mu}_i/\partial\boldsymbol{\beta} & \mathbf{0} \\ \partial\boldsymbol{\upsilon}_i/\partial\boldsymbol{\beta} & \partial\boldsymbol{\upsilon}_i/\partial\boldsymbol{\tau} \end{bmatrix} \quad \mathbf{d}_i = \begin{bmatrix} \mathbf{y}_i - \boldsymbol{\mu}_i \\ \mathbf{u}_i - \boldsymbol{\upsilon}_i \end{bmatrix}$$

$$\mathbf{V}_i = \mathrm{var}(\mathbf{d}_i) = \begin{bmatrix} \mathrm{var}(\mathbf{y}_i) & \mathrm{cov}(\mathbf{y}_i, \mathbf{u}_i) \\ \mathrm{cov}(\mathbf{u}_i, \mathbf{y}_i) & \mathrm{var}(\mathbf{u}_i) \end{bmatrix} \tag{10.13}$$

with $\mathbf{u}_i$ redefined to have mean-centred products $(y_{ij} - \mu_{ij})(y_{ik} - \mu_{ik})$ and $\boldsymbol{\upsilon}_i = E(\mathbf{u}_i)$.

Zhao and Prentice (1990) noted that $\mathrm{cov}(\mathbf{y}_i, \mathbf{u}_i)$ and $\mathrm{var}(\mathbf{u}_i)$ respectively involve third and fourth joint moments of the y_{ij} which are computable by summations like $\sum y_{ij}y_{ik}y_{il}f(\mathbf{y}_i)$. However, these will not be needed for $\mathbf{q}(\mathbf{y}; \boldsymbol{\beta}, \boldsymbol{\tau})$ if $\mathbf{V}_i$ is replaced by a 'working' covariance matrix in which $\mathrm{cov}(\mathbf{y}_i, \mathbf{u}_i)$ and $\mathrm{var}(\mathbf{u}_i)$ are replaced by matrices whose entries are suitable functions of $\boldsymbol{\beta}$ and $\boldsymbol{\tau}$. Zhao and Prentice proposed some candidate forms for such, including the Gaussian values $\mathrm{cov}(\mathbf{y}_i, \mathbf{u}_i) = \mathbf{0}$ and

$$\mathrm{cov}(u_{ijk}, u_{ilm}) = \boldsymbol{\Sigma}_{jl}\boldsymbol{\Sigma}_{km} + \boldsymbol{\Sigma}_{jm}\boldsymbol{\Sigma}_{kl}$$

The general strategy follows Liang and Zeger's (1986) approach (section 9.2.2), and was generalized from the present binary case in Prentice and Zhao (1991) (section 9.2.3).

Carey, Zeger and Diggle (1993) noted that the matrix $\mathbf{V}_i$ in (10.13), which must be inverted to compute $\mathbf{q}(\mathbf{y}; \boldsymbol{\beta}, \boldsymbol{\tau})$, could be large for data sets arising commonly in practice. In addition, the desirable use of regression modelling for the odds ratios ψ_{ijk} leads to the undesirable burden of solving (10.11) for the π's from which the elements of $\mathbf{V}_i$ are constructed. They suggested a neat way round the computational problem by splitting the estimating equation into two: the usual one based on $\mathbf{y}_i$ only; and one based on an offset logistic regression of the conditional probabilities $\mathrm{pr}(y_{ij} = 1 | y_{ik})$ on $\log\psi_{ijk}$.

Zeger, Liang and Self (1985) tackled logistic regression models for binary repeated measures using working models for the dependence structure. For the particular case of stationary binary time series they gave results to show that (a) the regression coefficient estimator is consistent and asymptotically normal for the 'independence' working model (section 9.2, $\bar{\mathbf{R}}_i = \mathbf{I}$), and (b) the working model is a stationary Markov chain with $\operatorname{corr}(y_{ij}, y_{i,j-1}) = \tau$ (to be estimated) and the true structure also has $\operatorname{corr}(y_{ij}, y_{i,j-1})$ homogeneous over j.

In Anderson and Aitkin (1985) y_{ij} in section 10.1.3 was the binary response from interviewee j to interviewee i in a survey. A logit link function was used, together with the normal distribution for $\mathbf{b}_i$. The integral (9.15) was performed by Gaussian quadrature. Stiratelli, Laird and Ware (1984) proposed a strategy for the same model involving both maximization of certain likelihoods and empirical Bayes methods. A major focus of their paper was on computational aspects.

Conaway (1990) noted that $f(\mathbf{y}_i|\mathbf{b}_i)$ in (10.3) can be written as a sum of terms of the form $\Pi_{j \in S} m_{ij}$, where S is a subset of $\{1, 2, \ldots, p_i\}$; for example, with $p_i = 3$ and $\mathbf{y}_i = [1\,0\,1]$, (10.3) becomes

$$m_{i1}(1 - m_{i2})m_{i3} = m_{i1}m_{i3} - m_{i1}m_{i2}m_{i3}$$

Thus, the integral (9.15) will be tractable if $E(\Pi_{j \in S} m_{ij})$ can be evaluated analytically for each S. Conaway suggested, for the case of a one-dimensional random coefficient b_i, the forms

$$\log(-\log m_{ij}) = \mathbf{x}_{ij}^{\mathrm{T}}\boldsymbol{\beta} + b_i, \quad f(b_i) = \Gamma(\alpha)^{-1}\lambda^{\alpha}\exp(\alpha b_i - \lambda e^{b_i})$$

as a compatible pair; the form for m_{ij} is the complementary log-log function (Illustration 10.1 (3)), and $f(b_i)$ is a univariate log-gamma density. Conaway extended the method to certain conditionally dependent models and gave a second tractable density for b_i.

Autoregressive models for binary data have been considered by many authors. A GLM for $\pi_{ij} = \operatorname{pr}(y_{ij} = 1 | H_{ij}, \mathbf{x}_{ij})$, where H_{ij} represents the outcome history for individual i prior to time j, is $g_j(\pi_{ij}) = \mathbf{x}_{ij}^{\mathrm{T}}\boldsymbol{\beta}_1 + \mathbf{h}_{ij}^{\mathrm{T}}\boldsymbol{\beta}_2$, where $\mathbf{h}_{ij}$ is a given vector of functions of H_{ij}. The likelihood function for $\mathbf{y}_i$ is then easily computed as

$$\prod_{j=1}^{p} \{\pi_{ij}^{y_{ij}}(1 - \pi_{ij})^{1 - y_{ij}}\}$$

Bonney (1987) proposed such a framework in which g_j is the logit link function and $\mathbf{h}_{ij}$ has components linear in $y_{i1}, \ldots, y_{i,j-1}$.

Alternative treatments of the situation described in section 10.3.3 have been given by Harville and Mee (1984), Gianola (1980), and Quaas and Van Vleck (1980).

This chapter has covered binary and categorical measures, these being probably the most common types of discrete data encountered in practice. Counts, of binomial, Poisson or other types, have not been addressed. A useful recent reference for Poisson repeated counts is Thall and Vail (1990).

Part Three: Comparisons of Methods

CHAPTER 11

Relationships between methods

11.1 Kronecker product and the vec notation

In this chapter we will make use of the Kronecker product, ⊗, and the vec notation.

Given an $n \times p$ matrix

$$\mathbf{Y} = \begin{bmatrix} y_{11} & y_{12} & \cdots & y_{1p} \\ y_{21} & y_{22} & \cdots & y_{2p} \\ \cdots & \cdots & \cdots & \cdots \\ y_{n1} & y_{n2} & \cdots & y_{np} \end{bmatrix}$$

we define

$$\text{vec}(\mathbf{Y}) = \begin{bmatrix} y_{11} \\ y_{12} \\ \cdots \\ y_{1p} \\ y_{21} \\ \cdots \\ y_{np} \end{bmatrix}$$

That is, vec($\mathbf{Y}$) is the vector formed by running along the rows of $\mathbf{Y}$ sequentially. Some authors (for example, Muirhead, 1982) define vec($\mathbf{Y}$) by running down the columns. Our definitions is more convenient for our purposes.

The **Kronecker product** (sometimes called the **direct product**) of two matrices $\mathbf{A}_{p\times q} = (a_{ij})$ and $\mathbf{B}_{r\times s} = (b_{ij})$ is denoted by $\mathbf{A} \otimes \mathbf{B}$ and is defined as

$$\mathbf{A} \otimes \mathbf{B} = \begin{bmatrix} a_{11}\mathbf{B} & a_{12}\mathbf{B} & \cdots & a_{1q}\mathbf{B} \\ a_{21}\mathbf{B} & a_{22}\mathbf{B} & \cdots & a_{2q}\mathbf{B} \\ \cdots & \cdots & \cdots & \cdots \\ a_{p1}\mathbf{B} & a_{p2}\mathbf{B} & \cdots & a_{pq}\mathbf{B} \end{bmatrix}$$

We shall make use of the property of ⊗ and vec that if $\mathbf{A}$, $\mathbf{B}$ and $\mathbf{C}$ are three conformable matrices, then

$$\mathrm{vec}(\mathbf{ABC}) = (\mathbf{A} \otimes \mathbf{C}^{\mathrm{T}})\mathrm{vec}(\mathbf{B})$$

11.2 Relationship between manova, anova and regression estimators

Now, with n individuals, each measured on p occasions, the (univariate) anova approach described in Chapter 3 treats this as a single $(np \times 1)$ observation vector with possible non-zero correlation within consecutive blocks of p variables. Thus

$$\mathbf{y} = \begin{bmatrix} y_{11} \\ \vdots \\ y_{1p} \\ y_{21} \\ \vdots \\ y_{np} \end{bmatrix} = \mathbf{X}\boldsymbol{\beta} + \mathbf{u} \tag{11.1}$$

where $\mathbf{X}$ is an $np \times q$ design matrix, $\boldsymbol{\beta}$ is a $q \times 1$ vector of parameters, and $\mathbf{u}$ is an $np \times 1$ vector of errors with $E(\mathbf{u}) = \mathbf{0}$. Since different subjects are assumed to be independent, but with possible correlation within subjects, we have $\mathrm{var}\,(\mathbf{u}) = \mathbf{I} \otimes \boldsymbol{\Sigma}_p$ (where $\mathbf{I}$ is $n \times n$ and $\boldsymbol{\Sigma}_p$ is $p \times p$). We will relax the restriction that all individuals have the same covariance matrix below, but we retain it for the time being. As described in Chapter 3, standard anova assumes the observations (all np observations in this case) are independent or have particular covariance structures – which are often unlikely to hold with longitudinal data – so one often has to make an adjustment to the test statistics.

The manova model for longitudinal data, described in Chapter 2, has the general form

$$\mathbf{Y} = \mathbf{X}^{(\mathrm{M})}\boldsymbol{\Gamma}\mathbf{B} + \mathbf{U} \tag{11.2}$$

where we have put the $n \times r$ between-subjects design matrix as $\mathbf{X}^{(\mathrm{M})}$ to distinguish it from the $np \times q$ design matrix in (11.1). We showed in Chapter 2 how this could be reduced to the usual manova formulation $E(\mathbf{Z}) = \mathbf{X}^{(\mathrm{M})}\boldsymbol{\Theta}$ by post-multiplying by a suitable matrix. This could then be analysed by standard manova methods.

What, then, is the relationship between (11.1) and (11.2)? First, of course, $\mathbf{y} = \mathrm{vec}(\mathbf{Y})$. So

$$\begin{aligned} E(\mathbf{y}) &= E(\mathrm{vec}(\mathbf{Y})) = \mathrm{vec}(\mathbf{X}^{(\mathrm{M})}\boldsymbol{\Gamma}\mathbf{B}) \\ &= (\mathbf{X}^{(\mathrm{M})} \otimes \mathbf{B}^{\mathrm{T}})\mathrm{vec}\boldsymbol{\Gamma} \\ &= \mathbf{X}\boldsymbol{\beta}. \end{aligned}$$

if we define $\mathbf{X} = \mathbf{X}^{(M)} \otimes \mathbf{B}^T$ and $\boldsymbol{\beta} = \text{vec}(\boldsymbol{\Gamma})$. Hence (11.1) and (11.2) are alternative parametrizations of each other.

■ *Illustration 11.1* The manova model in Illustration 2.7 had the form

$$E\begin{bmatrix} y_{11} & y_{12} & y_{13} \\ y_{21} & y_{22} & y_{23} \\ y_{31} & y_{32} & y_{33} \\ y_{41} & y_{42} & y_{43} \\ y_{51} & y_{52} & y_{53} \end{bmatrix} = \begin{bmatrix} 1 & 0 \\ 1 & 0 \\ 1 & 0 \\ 0 & 1 \\ 0 & 1 \end{bmatrix} \begin{bmatrix} \mu_{11} & \mu_{12} & \mu_{13} \\ \mu_{21} & \mu_{22} & \mu_{23} \end{bmatrix} \begin{bmatrix} 1 & 1 & 1 \\ 1 & 0 & -1 \\ 0 & 1 & -1 \end{bmatrix}$$

$$= \begin{bmatrix} 1 & 0 \\ 1 & 0 \\ 1 & 0 \\ 0 & 1 \\ 0 & 1 \end{bmatrix} \begin{bmatrix} \mu_{11}+\mu_{12} & \mu_{11}+\mu_{13} & \mu_{11}-\mu_{12}-\mu_{13} \\ \mu_{21}+\mu_{22} & \mu_{21}+\mu_{23} & \mu_{21}-\mu_{22}-\mu_{23} \end{bmatrix}$$

$$= \begin{bmatrix} \mu_{11}+\mu_{12} & \mu_{11}+\mu_{13} & \mu_{11}-\mu_{12}-\mu_{13} \\ \mu_{11}+\mu_{12} & \mu_{11}+\mu_{13} & \mu_{11}-\mu_{12}-\mu_{13} \\ \mu_{11}+\mu_{12} & \mu_{11}+\mu_{13} & \mu_{11}-\mu_{12}-\mu_{13} \\ \mu_{21}+\mu_{22} & \mu_{21}+\mu_{23} & \mu_{21}-\mu_{22}-\mu_{23} \\ \mu_{21}+\mu_{22} & \mu_{21}+\mu_{23} & \mu_{21}-\mu_{22}-\mu_{23} \end{bmatrix}$$

According to the above, this should yield the same expected values as

$$\begin{bmatrix} y_{11} \\ y_{12} \\ y_{13} \\ \vdots \\ y_{53} \end{bmatrix} = \begin{bmatrix} 1 & 0 \\ 1 & 0 \\ 1 & 0 \\ 0 & 1 \\ 0 & 1 \end{bmatrix} \otimes \begin{bmatrix} 1 & 1 & 0 \\ 1 & 0 & 1 \\ 1 & -1 & -1 \end{bmatrix} \begin{bmatrix} \mu_{11} \\ \mu_{12} \\ \mu_{13} \\ \mu_{21} \\ \mu_{22} \\ \mu_{23} \end{bmatrix}$$

$$= \begin{bmatrix} 1 & 1 & 0 & 0 & 0 & 0 \\ 1 & 0 & 1 & 0 & 0 & 0 \\ 1 & -1 & -1 & 0 & 0 & 0 \\ \vdots & \vdots & \vdots & \vdots & \vdots & \vdots \\ 0 & 0 & 0 & 1 & -1 & -1 \end{bmatrix} \begin{bmatrix} \mu_{11} \\ \mu_{12} \\ \mu_{13} \\ \mu_{21} \\ \mu_{22} \\ \mu_{23} \end{bmatrix}$$

$$= \begin{bmatrix} \mu_{11}+\mu_{12} & \mu_{11}+\mu_{13} & \mu_{11}-\mu_{12}-\mu_{13} \\ \mu_{11}+\mu_{12} & \mu_{11}+\mu_{13} & \mu_{11}-\mu_{12}-\mu_{13} \\ \mu_{11}+\mu_{12} & \mu_{11}+\mu_{13} & \mu_{11}-\mu_{12}-\mu_{13} \\ \mu_{21}+\mu_{22} & \mu_{21}+\mu_{23} & \mu_{21}-\mu_{22}-\mu_{23} \\ \mu_{21}+\mu_{22} & \mu_{21}+\mu_{23} & \mu_{21}-\mu_{22}-\mu_{23} \end{bmatrix}$$

And this is the same as Illustration 3.1. ■

Now, by putting

$$\mathbf{X} = \begin{bmatrix} \mathbf{X}_1 \\ \mathbf{X}_2 \\ \vdots \\ \mathbf{X}_n \end{bmatrix}$$

the regression model is derived from the above by putting

$$\mathbf{X}_i = [x_{i1}^{(M)} \dots x_{iq}^{(M)}] \otimes \mathbf{B}^{\mathrm{T}} = \mathbf{X}_i^{(M)} \otimes \mathbf{B}^{\mathrm{T}}$$

(where $\mathbf{X}_i^{(M)}$ is a row vector) yielding

$$E(\mathbf{y}_i) = \mathbf{X}_i \boldsymbol{\beta}$$

where $\mathbf{y}_i = [y_{i1}, y_{i2}, \dots, y_{ip}]^{\mathrm{T}}$.

Since the manova, anova, and regression models above are simply reformulations of each other, we would expect them to yield the same estimators. From Chapter 3, the univariate anova estimate is

$$\hat{\boldsymbol{\beta}} = (\mathbf{X}^{\mathrm{T}} \boldsymbol{\Sigma}^{-1} \mathbf{X})^{-1} \mathbf{X}^{\mathrm{T}} \boldsymbol{\Sigma}^{-1} \mathbf{y} \tag{11.3}$$

where $\boldsymbol{\Sigma}^{-1} = \operatorname{diag}[\boldsymbol{\Sigma}_1^{-1} \boldsymbol{\Sigma}_2^{-1} \dots \boldsymbol{\Sigma}_n^{-1}]$ (with $\boldsymbol{\Sigma}_i$ the covariance matrix of the ith individual, so that $\boldsymbol{\Sigma} = \mathbf{I} \otimes \boldsymbol{\Sigma}_p$), a block diagonal matrix. A little algebra then shows that

$$\hat{\boldsymbol{\beta}} = \left(\sum_{i=1}^{n} \mathbf{X}_i^{\mathrm{T}} \boldsymbol{\Sigma}_i^{-1} \mathbf{X}_i \right)^{-1} \left(\sum_{i=1}^{n} \mathbf{X}_i^{\mathrm{T}} \boldsymbol{\Sigma}_i^{-1} \mathbf{y}_i \right)$$

which is the regression estimator of Chapter 4.

Now, assuming that the $\boldsymbol{\Sigma}_i$ are the same ($= \boldsymbol{\Sigma}_p$, say), then

$$\hat{\boldsymbol{\beta}} = \left(\sum_{i=1}^{n} \mathbf{X}_i^{\mathrm{T}} \boldsymbol{\Sigma}_P^{-1} \mathbf{X}_i \right)^{-1} \left(\sum_{i=1}^{n} \mathbf{X}_i^{\mathrm{T}} \boldsymbol{\Sigma}_P^{-1} \mathbf{y}_i \right)$$

where $\boldsymbol{\Sigma}_p$ is the covariance matrix of each subject's p scores. Now,

$$\mathbf{X}_i^{\mathrm{T}} \boldsymbol{\Sigma}_p^{-1} = (\mathbf{X}_i^{(M)\mathrm{T}} \otimes \mathbf{B}) \boldsymbol{\Sigma}_p^{-1} = \mathbf{X}_i^{(M)\mathrm{T}} \otimes (\mathbf{B} \boldsymbol{\Sigma}_p^{-1})$$

so that

$$\begin{aligned} \mathbf{X}_i^{\mathrm{T}} \boldsymbol{\Sigma}_p^{-1} \mathbf{X}_i &= \{\mathbf{X}_i^{(M)\mathrm{T}} \otimes (\mathbf{B} \boldsymbol{\Sigma}_p^{-1})\} \{\mathbf{X}_i^{(M)} \otimes \mathbf{B}^{\mathrm{T}}\} \\ &= (\mathbf{X}_i^{(M)\mathrm{T}} \mathbf{X}_i^{(M)}) \otimes (\mathbf{B} \boldsymbol{\Sigma}_p^{-1} \mathbf{B}^{\mathrm{T}}) \end{aligned}$$

Hence

$$\begin{aligned} \left(\sum_{i=1}^{n} \mathbf{X}_i^{\mathrm{T}} \boldsymbol{\Sigma}_p^{-1} \mathbf{X}_i \right)^{-1} &= \left\{ \left(\sum_{i=1}^{n} \mathbf{X}_i^{(M)\mathrm{T}} \mathbf{X}_i^{(M)} \right) \otimes (\mathbf{B} \boldsymbol{\Sigma}_p^{-1} \mathbf{B}^{\mathrm{T}}) \right\}^{-1} \\ &= (\mathbf{X}^{(M)\mathrm{T}} \mathbf{X}^{(M)})^{-1} \otimes (\mathbf{B} \boldsymbol{\Sigma}_p^{-1} \mathbf{B}^{\mathrm{T}})^{-1} \end{aligned}$$

and

$$\sum_{i=1}^{n} \mathbf{X}_i^{\mathrm{T}} \boldsymbol{\Sigma}_p^{-1} \mathbf{y}_i = \sum_{i=1}^{n} \{\mathbf{X}_i^{(\mathrm{M})\mathrm{T}} \otimes \mathbf{B}\boldsymbol{\Sigma}_p^{-1}\} \mathbf{y}_i = \sum_{i=1}^{n} (\mathbf{X}_i^{(\mathrm{M})\mathrm{T}} \otimes \mathbf{B}\boldsymbol{\Sigma}_p^{-1} \mathbf{y}_i)$$

$$= \begin{bmatrix} \sum_{i=1}^{n} x_{i1}^{(\mathrm{M})} \mathbf{B}\boldsymbol{\Sigma}_p^{-1} \mathbf{y}_i \\ \cdots \\ \sum_{i=1}^{n} x_{ip}^{(\mathrm{M})} \mathbf{B}\boldsymbol{\Sigma}_p^{-1} \mathbf{y}_i \end{bmatrix}$$

$$= \begin{bmatrix} \mathbf{B}\boldsymbol{\Sigma}_p^{-1} \sum_{i=1}^{n} \mathbf{x}_{i1}^{(\mathrm{M})} \mathbf{y}_i \\ \cdots \\ \mathbf{B}\boldsymbol{\Sigma}_p^{-1} \sum_{i=1}^{n} \mathbf{x}_{ip}^{(\mathrm{M})} \mathbf{y}_i \end{bmatrix}$$

$$= \operatorname{vec}(\mathbf{B}\boldsymbol{\Sigma}_p^{-1} \mathbf{Y}^{\mathrm{T}} \mathbf{X}^{(\mathrm{M})})^{\mathrm{T}}$$

$$= \operatorname{vec}(\mathbf{X}^{(\mathrm{M})\mathrm{T}} \mathbf{Y} \boldsymbol{\Sigma}_p^{-1} \mathbf{B}^{\mathrm{T}})$$

so

$$\hat{\boldsymbol{\beta}} = (\mathbf{X}^{(\mathrm{M})\mathrm{T}} \mathbf{X}^{(\mathrm{M})})^{-1} \otimes (\mathbf{B}\boldsymbol{\Sigma}_p^{-1} \mathbf{B}^{\mathrm{T}})^{-1} \operatorname{vec}(\mathbf{X}^{(\mathrm{M})\mathrm{T}} \mathbf{Y} \boldsymbol{\Sigma}_p^{-1} \mathbf{B}^{\mathrm{T}})$$
$$= \operatorname{vec}((\mathbf{X}^{(\mathrm{M})\mathrm{T}} \mathbf{X}^{(\mathrm{M})})^{-1} \mathbf{X}^{(\mathrm{M})T} \mathbf{Y} \boldsymbol{\Sigma}_p^{-1} \mathbf{B}^{\mathrm{T}} (\mathbf{B}\boldsymbol{\Sigma}_p^{-1} \mathbf{B}^{\mathrm{T}})^{-1})$$

which is simply the vec of the manova estimate given in Chapter 2.

We can also get directly from the univariate anova estimate to the manova estimate using $\boldsymbol{\Sigma} = \mathbf{I} \otimes \boldsymbol{\Sigma}_p$ and $\mathbf{X} = \mathbf{X}^{(\mathrm{M})} \otimes \mathbf{B}^{\mathrm{T}}$. Then, from (11.3)

$$\hat{\boldsymbol{\beta}} = \{[\mathbf{X}((^{\mathrm{M})\mathrm{T}} \mathbf{X}^{(\mathrm{M})})^{-1} \mathbf{X}^{(\mathrm{M})}] \otimes [\mathbf{B}\boldsymbol{\Sigma}^{-1} \mathbf{B}^{\mathrm{T}}]^{-1} \mathbf{B}\boldsymbol{\Sigma}^{-1}\} \mathbf{y}$$
$$= \operatorname{vec}((\mathbf{X}^{(\mathrm{M})\mathrm{T}} \mathbf{X}^{(\mathrm{M})})^{-1} \mathbf{X}^{(\mathrm{M})\mathrm{T}} \mathbf{Y} \boldsymbol{\Sigma}_p^{-1} \mathbf{B}^{\mathrm{T}} (\mathbf{B}\boldsymbol{\Sigma}_p^{-1} \mathbf{B}^{\mathrm{T}})^{-1})$$

using the property of vec introduced earlier.

Some comparative properties immediately emerge from the above:

1. To be valid, univariate anova requires as assumption about the structure of the covariance matrix to be made. This means it is more restricted than the manova approach, which imposes no constraint on the form of the covariance matrix. However, if such an assumption seems unjustified then the sphericity adjustments described in Chapter 3 can be made. Whether these will lead to sensible alternatives to manova will depend on power considerations.
2. Only some $\mathbf{X}$ matrices in the above can be factorized in the form $\mathbf{X}^{(\mathrm{M})} \otimes \mathbf{B}^{\mathrm{T}}$. This means that the univariate approach is more general than the manova approach. In particular, the latter forces each subject to have the same within-subjects design matrix $\mathbf{B}$. This can be especially restricting since it does not allow missing data or observation vectors of different lengths. We

capitalize on this advantage of the univariate approach in the regression and random effects methods described in Chapters 4 and 5.

3. In (11.1) above we specified that var $(\mathbf{u}) = \mathbf{I} \otimes \mathbf{\Sigma}_p$. In fact, of course, there was no reason to be quite so restrictive, and we could merely have required that var ($\mathbf{u}$) be block diagonal, with ith block being $\mathbf{\Sigma}_i$. That is, we could have specified a separate covariance matrix for each individual. The manova approach, however, requires that $\mathbf{\Sigma}_i = \mathbf{\Sigma}_p$ for all i. Again, advantage is taken of this greater flexibility of the univariate approach in the regression and random effects methods.
4. The regression approach requires neither that $\mathbf{\Sigma}_i = \mathbf{\Sigma}_p$ for all i nor that $\mathbf{\Sigma}_i$ satisfies sphericity. That is, arbitrary covariance matrices can be used, based perhaps on additional information. In particular, if one considers that a random effects model is suitable, then the particular class of covariance matrices discussed in Chapter 5 arises.
5. Random effects (Chapter 5) are a natural way of modelling correlations between successive measurements. Coupled with correlations between successive error terms, a flexible model with few parameters is introduced. Although the manova approach, with an unrestricted covariance matrix, will be more flexible, it also, and consequently, runs the risk of over-parametrization.

11.3 Gaussian estimation and maximum quasi-likelihood estimation

In Part Two of this book the featured methods for data analysis are Gaussian estimation (GE) and maximum quasi-likelihood estimation (MQLE). In both GE and MQLE, the estimating functions are based on the mean vectors μ_i and the covariance matrices $\mathbf{\Sigma}_i$ of the observations $\mathbf{y}_i$ ($i = 1, \ldots, n$). In both, the $\boldsymbol{\mu}_i$ appear in a weighted sum of the deviations $\mathbf{y}_i - \boldsymbol{\mu}_i$, in fact, when the parameters governing $\boldsymbol{\mu}_i$ and $\mathbf{\Sigma}_i$ are 'decoupled' (sections 7.1.2 and 7.1.3) the estimating function for $\boldsymbol{\mu}_i$ is exactly the same under both GE and MQLE. On the other hand, $\mathbf{\Sigma}_i$ is handled differently by GE and MQLE. In this light, the results for the examples reported in section 9.3 are not surprising: the estimates for $\boldsymbol{\beta}$, the parameter vector governing the $\boldsymbol{\mu}_i$, come out with very similar values under both GE and MQLE, while those for $\boldsymbol{\tau}$, the parameter vector governing the $\mathbf{\Sigma}_i$, come out with rather less similar values. Further, the inferences concerning the components of $\boldsymbol{\beta}$ are also broadly in accord between GE and MQLE, apparently little affected by any $\boldsymbol{\tau}$ differences. In the examples of Chapter 10, where the repeated measures are categorical, the parameters of $\boldsymbol{\mu}_i$ and $\mathbf{\Sigma}_i$ are not decoupled. Even so, the estimates produced by GE and MQLE are not far apart, though this might just be a lucky accident for these particular data sets.

If, as a result of further and wider experience, it proves to be the case that GE and MQLE generally give similar results, this is good news for the applied scientist, who can run data through a standard normal-models package with a clear conscience without having to ingratiate him/herself with a prickly statistician. However, such low pragmatism cannot entirely sweep aside the need for high expertise. the scientist should be aware that the routine, normal-based standard errors and likelihood ratio tests are not appropriate for non-normal data: the information sandwich and Wald tests (section 7.1) are required. Also, with categorical data the hidden constraints on the covariances, mentioned in section 10.3, are likely to force an early call for proper statistical advice. Recent work, some of which is referred to in section 10.5, is concerned with the problems of modelling correlation strucutres in such situations. More theoretical work is also needed on the question of existence of MQL solutions: it is possible for the MQL equations to have no solution (Crowder, 1995). On the other hand, any function has a minimum over a (limited) parameter space, so GE will generally produce an estimator. For the examples of Chapters 9 and 10 the MQL estimates were computed by minimizing the function $|\mathbf{q}_\beta|^2 + |\mathbf{q}_\tau|^2$, reaching the target zero in all cases there.

APPENDIX A

The data sets used in the examples

The data sets used in the examples are given below (occasionally with extra data which we did not consider). Appendix B gives further data sets not analysed above. These data sets are available on the internet, at

http: //www. chaphall. com/chaphall /stats. html

		Day										
Diet		1	8	15	22	29	36	43	44	50	57	64
1	1	240	250	255	260	262	258	266	266	265	272	278
1	2	225	230	230	232	240	240	243	244	238	247	245
1	3	245	250	250	255	262	265	267	267	264	268	269
1	4	260	255	255	265	265	268	270	272	274	273	275
1	5	255	260	255	270	270	273	274	273	276	278	280
1	6	260	265	270	275	275	277	278	278	284	279	281
1	7	275	275	260	270	273	274	276	271	282	281	284
1	8	245	255	260	268	270	265	265	267	273	274	278
2	9	410	415	425	428	438	443	442	446	456	468	478
2	10	405	420	430	440	448	460	458	464	475	484	496
2	11	445	445	450	452	455	455	451	450	462	466	472
2	12	555	560	565	580	590	597	595	595	612	618	628
3	13	470	465	475	485	487	493	493	504	507	518	525
3	14	535	525	530	533	535	540	525	530	543	544	559
3	15	520	525	530	540	543	546	538	544	553	555	548
3	16	510	510	520	515	530	538	535	542	550	553	569

Table A.1. *Body weights, in grams, of rats on three different diets, measured over* 64 *days (Crowder and Hand, 1990, Example 2.4)*

	Day											
Group	0	2	4	6	8	10	12	14	16	18	20	21
1	42	51	59	64	76	93	106	125	149	171	199	205
1	40	49	58	72	84	103	122	138	162	187	209	215
1	43	39	55	67	84	99	115	138	163	187	198	202
1	42	49	56	67	74	87	102	108	136	154	160	157
1	41	42	48	60	79	106	141	164	197	199	220	223
1	41	49	59	74	97	124	141	148	155	160	160	157
1	41	49	57	71	89	112	146	174	218	250	288	305
1	42	50	61	71	84	93	110	116	126	134	125	−9
1	42	51	59	68	85	96	90	92	93	100	100	98
1	41	44	52	63	74	81	89	96	101	112	120	124
1	43	51	63	84	112	139	168	177	182	184	181	175
1	41	49	56	62	72	88	119	135	162	185	195	205
1	41	48	53	60	65	67	71	70	71	81	91	96
1	41	49	62	79	101	128	164	192	227	248	259	266
1	41	49	56	64	68	68	67	68	−9	−9	−9	−9
1	41	45	49	51	57	51	54	−9	−9	−9	−9	−9
1	42	51	61	72	83	89	98	103	113	123	133	142
1	39	35	−9	−9	−9	−9	−9	−9	−9	−9	−9	−9
1	43	48	55	62	65	71	82	88	106	120	144	157
1	41	47	54	58	65	73	77	89	98	107	115	117
2	40	50	62	86	125	163	217	240	275	307	318	331
2	41	55	64	77	90	95	108	111	131	148	164	167
2	43	52	61	73	90	103	127	135	145	163	170	175
2	42	52	58	74	66	68	70	71	72	72	76	74
2	40	49	62	78	102	124	146	164	197	231	259	265
2	42	48	57	74	93	114	136	147	169	205	236	251
2	39	46	58	73	87	100	115	123	144	163	185	192
2	39	46	58	73	92	114	145	156	184	207	212	233
2	39	48	59	74	87	106	134	150	187	230	279	309
2	42	48	59	72	85	98	115	122	143	151	157	150
3	42	53	62	73	85	102	123	138	170	204	235	256
3	41	49	65	82	107	129	159	179	221	263	291	305
3	39	50	63	77	96	111	137	144	151	146	156	147
3	41	49	63	85	107	134	164	186	235	294	327	341
3	41	53	64	87	123	158	201	238	287	332	361	373
3	39	48	61	76	98	116	145	166	198	227	225	220
3	41	48	56	68	80	83	103	112	135	157	169	178
3	41	49	61	74	98	109	128	154	192	232	280	290
3	42	50	61	78	89	109	130	146	170	214	250	272
3	41	55	66	79	101	120	154	182	215	262	295	321
4	42	51	66	85	103	124	155	153	175	184	199	204
4	42	49	63	84	103	126	160	174	204	234	269	281
4	42	55	69	96	131	157	184	188	197	198	199	200
4	42	51	65	86	103	118	127	138	145	146	−9	−9
4	41	50	61	78	98	117	135	141	147	174	197	196

Group	Day 0	2	4	6	8	10	12	14	16	18	20	21
4	40	52	62	82	101	120	144	156	173	210	231	238
4	41	53	66	79	100	123	148	157	168	185	210	205
4	39	50	62	80	104	125	154	170	222	261	303	322
4	40	53	64	85	108	128	152	166	184	203	233	237
4	41	54	67	84	105	122	155	175	205	234	264	264

Table A.2. *Body weights (in grams) of chicks on a normal diet measured on alternate days (Crowder and Hand, 1990, Example 5.3). There are four groups, on different protein diets, as indicated in the first column*

Time (Minutes) 0	1	45	75	105	135	195	255	315	375	435
4.22	4.92	8.09	6.74	4.30	4.28	4.59	4.49	5.29	4.95	4.34
4.52	4.22	8.46	9.12	7.50	6.02	4.66	4.69	4.26	4.29	4.47
4.47	4.47	7.95	7.21	6.35	5.58	4.57	3.90	3.44	4.18	4.32
4.27	4.33	6.61	6.89	5.64	4.85	4.82	3.82	4.31	3.81	2.92
4.81	4.85	6.08	8.28	5.73	5.68	4.66	4.62	4.85	4.69	4.86
4.61	4.68	6.01	7.35	6.38	6.16	4.41	4.96	4.33	4.54	4.81

Table A.3. *Blood glucose levels of six volunteers after a 6 a.m. meal. The data are extracted from Crowder and Hand (1990, Table 2.3)*

Time (days postpartum) 1	2	3	4	5	5.5	6	6.5	7	7.5	8	8.5	9	9.5	10
516	633	1697	441	596	909	1610	528	493	621	820	1577	1258	692	1270
367	398	458	463	516	390	436	645	505	1195	823	564	692	669	497
471	457	526	594	775	673	446	350	472	398	1394	1137	568	607	635
450	452	465	456	651	595	454	407	717	494	704	1850	580	860	727
709	798	883	984	872	817	894	718	750	820	875	429	913	967	867
273	261	348	537	624	593	354	294	351	2124	471	729	379	676	654
346	375	408	359	614	394	536	598	586	579	1422	834	2346	5927	2608
216	248	354	450	483	377	302	388	357	333	473	496	401	1103	441
433	550	771	566	550	537	466	574	604	2099	616	1360	549	494	826
388	432	287	316	283	256	639	327	463	568	289	246	1111	378	375
477	369	450	437	464	405	548	667	1365	1032	2068	652	1366	1158	857
513	419	1030	525	416	358	479	399	799	1584	1315	662	600	624	790
360	350	377	429	659	471	296	457	2459	1220	674	695	1463	650	685
373	414	609	578	1634	387	562	318	538	392	455	496	651	468	1094
349	450	501	757	678	682	931	516	587	886	896	608	632	618	688
499	491	449	712	642	464	1355	508	413	819	473	565	714	721	965

Table A.4. *Luteinizing hormone levels in non-suckling cows ($ng\,ml^{-1} \times 1000$). From Raz (1989, Table 6)*

Week 1	2	6	10	14	15	16
0.22	0.00	1.03	0.67	0.75	0.65	0.59
0.18	0.00	0.96	0.96	0.98	1.03	0.70
0.73	0.37	1.18	0.76	1.07	0.80	1.10
0.30	0.25	0.74	1.10	1.48	0.39	0.36
0.54	0.42	1.33	1.32	1.30	0.74	0.56
0.16	0.30	1.27	1.06	1.39	0.63	0.40
0.30	1.09	1.17	0.90	1.17	0.75	0.88
0.70	1.30	1.80	1.80	1.60	1.23	0.41
0.31	0.54	1.24	0.56	0.77	0.28	0.40
1.40	1.40	1.64	1.28	1.12	0.66	0.77
0.60	0.80	1.02	1.28	1.16	1.01	0.67
0.73	0.50	1.08	1.26	1.17	0.91	0.87

Table A.5. *Plasma ascorbic acid of 12 patients measured on seven occasions. The data are reproduced from Crowder and Hand (1990 Table 3.3). Weeks 1 and 2 are before treatment, weeks 6, 10, and 14 are during treatment, and weeks 15 and 16 are after treatment*

Proportion dissolved	0.9	0.7	0.5	0.3	0.25	0.1
1	13.0	16.0	19.0	23.0	24.0	28.0
1	14.0	18.0	22.0	26.0	28.0	32.0
1	19.0	24.0	28.0	33.0	33.0	39.0
1	13.0	17.0	21.0	25.0	26.0	29.0
1	14.0	16.0	19.0	23.0	25.0	27.0
1	13.0	16.0	19.0	23.0	24.0	26.0
2	13.0	17.0	21.5	26.0	27.8	33.5
2	11.5	15.5	20.5	24.5	26.7	31.3
2	10.4	14.4	18.4	23.5	25.1	29.6
2	11.1	15.1	19.0	23.6	24.7	28.7
3	11.7	16.9	22.0	27.6	29.2	34.5
3	13.5	18.7	24.9	30.0	31.8	37.5
3	12.0	17.1	22.7	28.4	30.2	35.8
3	12.1	16.7	21.1	26.8	29.0	33.8
4	11.0	14.5	19.0	24.5	26.0	32.0
4	14.4	19.0	24.0	30.0	31.5	39.0
4	11.0	14.0	17.5	22.0	23.5	29.5

Table A.6. *Times at which a specified proportion of each 17 (the rows of the table) pills remain undissolved. Here the 'underlying continuum' from which 'occasions' are chosen at which to take measurement is 'proportion remaining undissolved'. There are four groups of pills*

Hours after injection				
0.0	1.5	3.0	4.5	6.0
77	52	35	56	64
77	56	43	56	64
77	43	54	39	64
81	56	22	35	39
90	73	52	60	64
103	47	22	60	99
99	47	52	47	81
90	35	22	26	22
90	64	47	47	60
90	35	12	26	68
85	68	56	68	73

Table A.7. *Blood sugar levels of 11 rabbits following an injection of insulin. The data are taken from Lindsey (1993, Table A.5), group 1, first A treatment*

1	21.0	20.0	21.5	23.0
1	21.0	21.5	24.0	25.5
1	20.5	24.0	24.5	26.0
1	23.5	24.5	25.0	26.5
1	21.5	23.0	22.5	23.5
1	20.0	21.0	21.0	22.5
1	21.5	22.5	23.0	25.0
1	23.0	23.0	23.5	24.0
1	20.0	21.0	22.0	21.5
1	16.5	19.0	19.0	19.5
1	24.5	25.0	28.0	28.0
2	26.0	25.0	29.0	31.0
2	21.5	22.5	23.0	26.5
2	23.0	22.5	24.0	27.5
2	25.5	27.5	26.5	27.0
2	20.0	23.5	22.5	26.0
2	24.5	25.5	27.0	28.5
2	22.0	22.0	24.5	26.5
2	24.0	21.5	24.5	25.5
2	23.0	20.5	31.0	26.0
2	27.5	28.0	31.0	31.5
2	23.0	23.0	23.5	25.0
2	21.5	23.5	24.0	28.0
2	17.0	24.5	26.0	29.5
2	22.5	25.5	25.5	26.0
2	23.0	24.5	26.0	30.0
2	22.0	21.5	23.5	25.0

Table A.8. *The Potthof and Roy (1964) data on distance from pituitary to pteryomaxillary fissure. The first column gives sex (1 = girl, 2 = boy) and the other four columns the distances (in millimeters) at ages 8, 10, 12 and 14, respectively*

116	119	116	124	120	117	114	122
110	110	114	115	106	112	110	110
117	118	120	120	120	120	120	124
112	116	115	113	115	116	116	119
113	114	114	118	114	117	116	112
119	115	94	116	100	99	94	97
110	110	105	118	105	105	115	115

Table A.9. *Visual response time data, reproduced from Crowder and Hand (1990). The first four variables show response time (in milliseconds) for the left eye with lenses of strengths 6/6, 6/18, 6/36 and 6/60, respectively. The last four variables show the corresponding times for the right eye*

1	20	19	20	20	18	2	9	11	14	11	14
1	14	15	16	9	6	2	6	7	9	12	16
1	7	5	8	8	5	2	13	18	14	20	14
1	6	10	9	10	10	2	9	10	9	8	7
1	9	7	9	4	6	2	6	7	4	5	4
1	9	10	9	11	11	2	11	11	5	10	12
1	7	3	7	6	3	2	7	10	11	8	5
1	18	20	20	23	21	2	8	18	19	15	14
1	6	10	10	13	14	2	3	3	3	1	3
1	10	15	15	15	14	2	4	10	9	17	10
1	5	9	7	3	12	2	11	10	5	15	16
1	11	11	8	10	9	2	1	3	2	2	5
1	10	2	9	3	2	2	6	7	7	6	7
1	17	12	14	15	13	2	0	3	2	0	0
1	16	15	13	7	9	2	18	19	15	17	20
1	7	10	4	10	5	2	15	15	15	14	12
1	5	0	5	0	0	2	14	11	8	10	8
1	16	7	7	6	10	2	6	6	5	5	8
1	5	6	9	5	6	2	10	10	6	10	9
1	2	1	1	2	2	2	4	6	6	4	2
1	7	11	7	5	11	2	4	13	9	8	7
1	9	16	17	10	6	2	14	7	8	10	6
1	2	5	6	7	6						
1	7	3	5	5	5						
1	19	13	19	17	17						
1	7	5	8	8	6						

Table A.10. *Alzheimer's data (from Hand and Taylor, 1987, Table G.1). There are two groups of patients (given by the first column): group 1 received a placebo and group 2 lecithin. The measurements were the number of words correctly recalled at each of five times, spaced at 0, 1, 2, 4 and 6 units after the experiment commenced*

93	121	112	117	121
116	135	114	98	135
125	137	119	105	102
144	173	148	124	122
105	119	125	91	133
109	83	109	80	104
89	95	88	91	116
116	128	122	107	119
151	149	141	126	138
137	139	125	109	107

Table A.11. *Plasma citrate concentrations, from Andersen, Jensen and Schou (1981). Measurements of plasma citrate concentrations in micromoles per litre were taken on ten subjects at five equally spaced times*

1	455	460	510	504	436	466
1	467	565	610	596	542	587
1	445	530	580	597	582	619
1	485	542	594	583	611	612
1	480	500	550	528	562	576
2	514	560	565	524	552	597
2	440	480	536	484	567	569
2	495	570	569	585	576	677
2	520	590	610	637	671	702
2	503	555	591	605	649	675
3	496	560	622	622	632	670
3	498	540	589	557	568	609
3	478	510	568	555	576	605
3	545	565	580	601	633	649
3	472	498	540	524	532	583

Table A.12. *These data show the body weights in grams of guinea pigs at the ends of weeks 1, 3, 4, 5, 6 and 7 of an investigation into the effect of a diet supplement on growth. All animals were given a growth-inhibiting substance during week 1, and the vitamin E supplement was started at the beginning of week 5. There are three treatment groups, indicated in column 1 of the data matrix, differing in the dose levels of vitamin E (low, medium and high). The data are from Crowder and Hand (1990, Table 3.1)*

	Time (minutes)						
Group	0	5	10	20	30	45	60
1	100	90	100	70	36	50	28
1	100	83	97	60	83	83	97
1	70	83	77	21	36	53	53

Table A.13. (Continued)

Group	0	5	Time (minutes) 10	20	30	45	60
1	77	36	36	36	36	36	28
1	100	61	99	83	97	83	100
2	90	28	44	33	36	44	36
2	83	14	28	28	21	28	36
2	69	36	28	21	14	21	21
2	100	77	44	28	44	14	*
2	61	53	53	44	44	28	44

Table A.13. *A medical experiment involving two groups of five subjects measured at seven times (Jones, 1993)*

Male				
66	47.10	31.05	−9.00	32.80
70	44.10	31.50	−9.00	37.00
44	39.70	33.70	−9.00	24.50
70	43.30	18.35	−9.00	36.60
74	37.40	32.25	−9.00	29.05
65	45.70	35.50	−9.00	39.80
54	44.90	34.10	−9.00	32.05
63	42.90	32.05	−9.00	−9.00
71	46.50	28.80	−9.00	37.80
68	42.10	34.40	34.00	36.05
69	38.25	29.40	32.85	30.50
64	43.00	33.70	34.10	36.65
70	37.80	26.60	26.70	30.60
Female				
60	37.25	26.50	−9.00	38.45
52	−9.00	27.95	−9.00	33.95
52	27.00	32.50	−9.00	31.95
75	38.35	32.30	−9.00	37.90
72	38.80	32.55	−9.00	26.85
54	44.65	32.25	−9.00	34.20
71	38.00	27.10	−9.00	37.85
58	34.00	23.20	−9.00	25.95
77	44.80	37.20	−9.00	29.70
66	45.95	29.10	−9.00	26.70
53	41.85	31.95	37.15	37.60
74	38.00	31.65	38.40	35.70
78	42.20	34.00	32.90	33.25
74	39.70	33.45	26.60	32.65
79	37.50	28.20	28.80	30.30
71	34.55	30.95	30.60	28.75
68	35.50	24.70	28.10	28.75

Table A.14. *Ages and four measurements of haematocrit for 13 male and 17 female hip-replacement patients. The first measurement was before the operation. '−9' represents a missing value. From Crowder and Hand (1990, Table 5.2)*

slip

0.00	0.10	0.20	0.30	0.40	0.50	0.60	0.70
0.80	0.90	1.00	1.20	1.40	1.60	1.80	

loads

0.00	2.38	4.34	6.64	8.05	9.78	10.97	12.05
12.98	13.94	14.74	16.13	17.98	19.52	19.97	
0.00	2.69	4.75	7.04	9.20	10.94	12.23	13.19
14.08	14.66	15.37	16.89	17.78	18.41	18.97	
0.00	2.85	4.89	6.61	8.09	9.72	11.03	12.14
13.18	14.12	15.09	16.68	17.94	18.22	19.40	
0.00	2.46	4.28	5.88	7.43	8.32	9.92	11.10
12.23	13.24	14.19	16.07	17.43	18.36	18.93	
0.00	2.97	4.68	6.66	8.11	9.64	11.06	12.25
13.35	14.54	15.53	17.38	18.76	19.81	20.62	
0.00	3.96	6.46	8.14	9.35	10.72	11.84	12.85
13.83	14.85	15.79	17.39	18.44	19.46	20.05	
0.00	3.17	5.33	7.14	8.29	9.86	11.07	12.13
13.15	14.09	15.11	16.69	17.69	18.71	19.54	
0.00	3.36	5.45	7.08	8.32	9.91	11.06	12.21
13.16	14.05	14.96	16.24	17.34	18.23	18.87	

Table A.15. *A timber plank is sandwiched between two concrete surfaces and an increasing load is applied, with the slippage (in millimetres) being recorded at intervals of 2 seconds. The data are pruned by selecting those load–slip pairs in which the slip takes predetermined values. Thus the observed slippage becomes the x variable, playing the role of the underlying continuum. The table shows the data from eight experimental runs*

Subject	Weight (kg)	Energy expenditure w				heart rate y			
1	51.1	1.57	1.31	4.67	4.23	50	60	125	135
2	58.1	1.46	1.41	4.49	5.56	80	78	120	131
3	61.5	1.20	1.22	3.65	7.83	85	81	127	135
4	60.3	1.20	0.94	3.31	7.85	65	74	100	122
5	61.5	1.16	1.35	4.13	10.6	70	74	84	125
6	50.9	1.44	2.78	4.18	6.45	86	92	121	130
7	50.1	1.29	1.33	3.03	5.06	78	95	110	128

Table A.16. *The weight, energy expenditure and heart rate of seven female volunteers when performing four physical tasks (Crowder, 1978). A detailed description is given in Example 8.3*

times −1.0 0.0 2.0 4.0 6.0 8.0 10.0 12.0
15.0 18.0 21.0 24.0 27.0 30.0

glucose date 1

3.0	3.0	4.7	6.0	6.3	4.3	3.0	2.0	4.5	3.8	3.2	2.6	2.6	2.6
4.0	3.6	6.0	8.6	8.8	7.2	5.0	3.8	4.2	4.0	2.6	2.5	2.6	3.8
3.5	3.5	6.0	7.3	7.5	6.2	5.0	4.1	4.3	4.1	3.4	3.8	3.8	3.9
3.8	3.8	4.4	6.0	6.8	5.7	4.6	3.8	4.3	4.5	4.5	4.2	3.8	4.2
3.7	4.0	5.2	7.0	6.6	6.0	5.2	4.7	4.5	4.5	4.2	3.5	3.7	3.8
3.5	3.1	3.1	3.6	4.2	3.8	3.5	4.2	3.7	4.1	3.2	3.4	3.4	3.2
3.0	2.9	5.0	6.2	7.7	5.9	3.9	5.8	5.0	5.2	4.3	4.0	3.5	3.5

glucose date 2

2.2	2.8	4.4	5.6	5.8	4.5	3.6	3.3	3.2	3.3	3.0	3.0	3.2	3.1
4.1	4.2	6.3	7.0	8.3	5.7	2.9	3.0	3.4	3.5	3.2	3.8	4.3	3.7
3.8	3.8	5.0	5.5	7.0	5.0	3.8	3.6	3.6	3.5	3.5	3.5	3.5	4.0
3.6	3.6	4.3	5.5	6.3	5.7	5.3	4.7	4.0	3.5	3.6	3.7	4.0	3.7
3.8	3.8	4.7	7.0	7.7	6.0	5.0	4.7	4.3	4.2	3.7	3.4	3.7	3.8
3.6	3.5	4.4	6.2	7.0	5.9	4.8	3.9	3.9	4.0	3.7	3.5	3.8	3.8
3.3	2.9	4.2	5.8	5.8	5.8	4.4	4.0	3.8	3.7	3.4	3.6	3.6	3.6

insulin date 1

<5	<5	23	53	67	43	38	11	45	28	21	10	7	7
<5	<5	32	58	78	92	58	42	60	50	9	10	<5	<5
<5	9	53	58	95	95	58	42	58	38	13	11	15	9
13	19	20	62	99	102	77	48	51	57	67	35	21	36
10	14	39	80	97	105	77	71	65	61	46	29	23	15
13	15	20	26	51	30	42	66	35	58	28	26	20	24
21	16	68	99	150	122	−1	168	121	111	131	88	63	32

insulin date 2

<5	<5	21	40	33	20	15	11	15	7	7	7	7	7
<5	<5	25	48	75	70	40	25	17	17	13	13	12	5
5	7	17	30	38	37	13	17	9	7	7	6	9	11
13	14	28	50	73	66	65	57	36	20	13	29	31	24
13	13	29	49	74	51	50	36	28	30	20	13	21	15
15	14	21	37	52	47	34	24	36	15	18	21	18	16
18	16	32	86	100	121	81	44	41	38	40	26	30	−1

alcohol date 1

−1	−1	24	47	73	85	77	82	73	75	64	53	47	40
−1	−1	25	67	100	103	103	95	92	82	68	57	45	37
−1	−1	22	47	66	61	45	39	36	22	20	17	10	<5
−1	−1	10	42	70	57	53	53	47	42	37	30	25	15
−1	−1	30	53	50	70	62	45	48	38	24	20	10	4
−1	−1	5	19	46	42	50	46	41	32	30	25	18	10
−1	−1	17	40	66	75	65	75	47	60	41	40	35	28

alcohol date 2

−1	−1	33	77	105	92	90	80	72	67	65	57	50	42
−1	−1	54	74	125	122	124	108	103	92	65	75	68	54

−1	−1	23	48	66	62	54	48	47	37	29	26	18	10
−1	−1	18	50	70	75	67	64	60	46	40	35	26	23
−1	−1	12	45	60	48	48	51	35	28	20	22	12	6
−1	−1	5	25	55	58	50	42	32	35	26	23	15	15
−1	−1	19	50	61	91	80	64	82	61	55	48	46	−1

Table A.17. *Blood glucose levels for seven volunteers. The subject took alcohol at time 0 and gave a blood sample at 14 times over a period of 5 hours. The whole procedure was repeated at a later date but with a dietary additive. Insulin and alcohol levels were also measured in the same blood samples. Times shown are minutes/10, and −1 represents a missing value (Crowder and Tredger, 1981; Crowder, 1983)*

1	1	DSH	6.0	15.0	4	5	1	4	3	5	2	1	2	4	3	5	1	5	4	5
2	1	DSH	−9.	7.0	2	4	3	5	1	1	4	3	4	4	1	1	3	4	2	2
3	0	DSH	4.0	15.0	3	4	4	4	2	4	1	4	1	4	2	4	4	4	3	4
4	0	DSH	5.0	1.0	1	3	2	4	4	4	3	2	3	1	4	1	2	1	1	1
5	1	DSH	5.0	5.0	4	1	1	4	3	1	2	1	2	1	3	1	1	1	4	1
6	0	DSH	3.0	.7	2	5	3	5	1	1	4	5	4	5	1	1	3	1	2	5
7	1	BBL	6.0	7.0	3	1	4	1	2	1	1	1	1	1	2	2	4	1	3	1
8	1	DSH	4.0	7.0	1	5	2	5	4	1	3	1	3	1	4	1	2	3	1	3
9	1	DSH	3.5	6.0	4	4	1	2	3	2	2	2	2	4	3	2	1	2	4	2
10	1	DSH	5.0	7.0	2	1	3	1	1	1	4	1	4	1	1	1	3	1	2	1
11	0	DLH	3.5	4.0	3	4	4	4	2	5	1	5	1	2	2	5	4	5	3	5
12	1	DSH	9.0	8.0	1	5	2	5	4	5	3	5	3	5	4	5	2	5	1	5
13	0	DSH	3.0	2.0	4	5	1	5	3	5	2	5	2	5	3	5	1	2	4	5
14	1	DSH	4.0	.5	2	3	3	1	1	1	4	1	4	1	1	1	3	1	2	1
15	1	DSH	−9.	6.0	3	4	4	4	2	5	1	4	1	4	2	4	4	4	3	4
16	1	DSH	4.0	.7	1	4	2	4	4	4	3	4	3	4	4	4	2	4	1	4
17	1	DSH	8.0	3.5	4	1	1	1	3	1	2	1	2	1	3	1	1	1	4	1
18	0	DSH	4.0	5.0	2	4	3	5	1	4	4	5	4	4	1	4	3	4	2	4
19	1	SIA	−9.	−9.	3	1	4	1	2	1	1	5	1	5	2	5	4	4	3	1
20	1	DSH	4.0	.5	1	4	2	4	4	4	3	4	3	4	4	2	2	4	1	2
21	0	DSH	4.0	.5	4	5	1	4	3	4	2	5	2	4	3	5	1	3	4	4
22	0	DSH	4.0	1.5	2	4	3	4	1	5	4	4	4	5	1	1	3	4	2	1
23	0	DSH	4.0	4.0	3	1	4	1	2	1	1	1	1	1	2	1	4	1	3	1
24	0	DSH	3.0	3.0	1	1	2	4	4	1	3	1	3	1	4	1	2	1	1	1
25	0	DSH	2.0	.5	4	4	1	4	3	4	2	4	2	4	3	4	1	4	4	4
26	1	DSH	−9.	15.0	2	5	3	5	1	4	4	5	4	5	1	5	3	4	2	5
27	1	DSH	3.0	4.0	3	5	4	4	2	5	1	5	1	5	2	5	4	5	3	5

Table A.18. *Cat medicines. There are five case variables (cat number, sex, breed, weight and age) and eight pairs of measures showing formulation and palatability on days 1, 2, 3, 4, 8, 9, 10 and 11. Palatability is scored as 1 = readily accepted, 2 = accepted, 3 = reluctantly accepted, 4 = accepted in food, and 5 = refused. Missing values are coded as −9*

1	1	2	29	9	27	65	2	22	12	48	298	2	33	6	9	545	2	11	32	8	785
2	1	2	31	13	91	90	1	0	14	12	342	1	1	6	29	575	1	6	3	81	855
3	1	2	10	15	53	270	2	20	35	46	374	2	25	64	50	631	4	86	78	17	870
4	1	2	37	26	41	67	1	6	6	6	337	1	0	2	2	550	2	7	8	2	808
5	1	2	27	77	85	45	2	23	40	45	297	2	23	25	12	535	3	23	60	52	765
6	1	3	71	66	61	125	3	56	76	32	319	2	27	20	13	540	2	20	25	9	865
7	1	2	32	33	52	75	2	29	14	42	325	2	18	10	63	565	2	12	9	15	790
8	1	1	0	22	54	220	−9	−9	−9	−9	−9	1	0	21	91	525	2	22	19	77	740
9	1	3	48	38	41	85	2	36	46	85	335	2	30	27	45	560	2	35	22	37	830
10	1	1	7	5	24	75	1	2	13	67	360	2	14	2	50	555	2	21	10	64	795
11	1	4	73	67	80	114	4	72	99	70	295	4	66	99	76	583	4	80	84	87	720
12	1	1	14	69	82	85	2	25	45	35	330	3	67	82	15	560	2	15	21	6	790
13	1	2	13	4	44	110	3	50	31	53	395	2	24	56	85	578	2	12	20	84	835
14	2	2	23	1	20	57	2	11	0	14	297	3	17	0	47	552	2	30	0	27	796
15	2	3	53	36	70	25	3	49	−9	55	271	3	61	12	73	515	3	45	22	47	745
16	2	2	48	34	76	85	3	69	72	71	325	2	19	33	20	565	2	22	29	27	805
17	2	3	58	9	46	45	1	2	2	89	285	3	24	9	88	537	2	1	2	97	785
18	2	1	12	8	75	165	1	4	14	97	415	1	10	6	67	660	1	1	9	46	890
19	2	3	55	22	55	56	2	69	22	73	300	2	28	22	25	556	3	22	3	17	793
20	2	3	−9	−9	−9	42	3	50	17	91	283	2	8	12	33	530	2	12	5	28	735
21	2	2	28	−9	−9	86	5	46	20	80	315	3	10	51	88	569	3	38	75	69	858
22	2	3	63	26	90	92	2	43	13	74	352	2	36	16	74	572	2	31	15	66	782
23	2	2	41	48	84	60	3	41	17	75	320	2	14	23	53	560	3	31	43	19	795
24	2	3	47	53	82	109	3	28	45	63	350	3	31	30	54	570	2	21	27	30	825
25	2	2	24	22	78	60	2	36	7	84	300	2	27	14	70	540	2	23	8	58	780
26	2	2	28	21	92	70	2	20	18	78	315	2	33	14	97	540	4	74	14	97	860
27	2	2	35	38	73	70	1	0	0	56	235	1	2	2	75	475	1	12	5	73	715

Table A.19. *Back pain clinical trial with two treatment groups, 13 intercostal/epidural and 14 morphine infusion. Patients are asked to give an assessment of their state of disturbance or discomfort on four different scales, one verbal rating scale (VRS) and three visual analogue scales (VASs). The variables appearing are case number and treatment group followed by four sets (for four equally spaced times after treatment) each comprising the five values: pain VRS (scores 1 to 5), pain VAS (scores 0 to 100), anxiety VAS (scores 0 to 100), alertness VAS (scores 0 to 100) and time since treatment (minutes). Missing values are coded as* −9

1	1	4	5	8	6	4	4	5	6	6	7	4
		33	30	21	26	27	28	18	15	14	19	22
		108	108	128	116	112	108	124	124	120	108	100
2	1	5	6	6	5	3	3	7	8	8	8	4
		40	39	40	40	39	40	23	23	20	25	31
		94	104	106	104	92					108	84
3	1	5	5	5	5	5	5	6	6	7	6	4
		30	23	25	25	30	30	18	15	13	18	24
		96	96	96	84	96	96	120	108	108	96	84
4	1	2	3	3	3	1	1	7	8	7	7	2
		29	29	28	28	28	29	19	17	15	19	22
		90	90	90	90	84	84	96	96	90	84	84

5	1	4	5	7	8	4	2	2	3	4	2	1
		35	31	31	30	32	34	19	17	16	27	33
		78	78	84	78	78	78	78	78	84	84	78
6	1	3	5	6	7	5	3	7	7	7	4	2
		38	34	28	22	27	33	21	17	18	24	31
		84	108	114	108	90	84	96	108	96	80	78
7	1	1	5	6	6	2	1	5	7	8	5	1
		36	38	36	35	35	35	21	18	17	25	30
		86	94	88	82	84	80	104	104	104	96	74
8	1	1	4	6	6	4	2	4	6	7	4	1
		32	32	30	32	32	31	19	13	10	10	31
		70	74	72	72	70	68	86	84	82	68	70
9	1	1	1	5	2	1	2	2	2	2	1	0
		37	36	35	36	37	37	19	17	16	23	29
		80	80	80	78	78	72	96	96	90	80	76
10	1	4	8	8	8	3	4	7	7	8	8	5
		22	21	19	19	21	19	14	11	11	14	17
		78	78	78	78	72	78	90	90	90	84	74
11	1	2	2	4	6	7	4	7	7	7	7	7
		31	31	31	29	29	30	21	18	17	22	24
		90	90	96	96	96	90	90	90	90	84	78
12	1	8	8	8	8	7	6	4	3	2	4	2
		30	29	29	29	30	34	17	16	15	23	29
		124	124	132	130	118	124	140	152	148	130	102
13	1	3	4	8	8	7	4	6	8	8	7	8
		26	25	23	23	23	24	19	17	16	19	20
		80	82	86	82	80	86	92	90	84	80	80
14	1	2	2	2	3	2	1	4	5	6	3	1
		28	29	28	26	26	27	18	15	14	18	20
		98	110	110	110	108	102	132	134	140	114	78
15	1	3	6	8	8	6	8					
		35	37	34	34	33	28					
		96	98	104	104	98	122					
16	1	4	7	8	8	3	2	7	8		7	3
		34	33	32	32	34	34	19	15		23	30
		66	76	74	70	62	64	94	92		72	62
17	1	4	6	6	6	4	6	8	8		7	7
		32	32	30	30	28	26	19	18		19	19
		86	86	86	82	78	78	120	122		112	94
18	1	5	5	8	6	2	3	6	8		5	3
		34	34	32	33	33	34	23	20		25	32
		96	108	118	116	100	100	124	130		102	90
19	1	4	6	6	6	4	3	4	6	6	5	2
		32	30	28	27	31	32	16	15	15	19	27
		86	88	90	88	82	92	100	98	92	84	82
20	1	2	3	4	4	2	2	7	8		4	2
		36	36	37	36	35	35	20	18		27	36
		80	86	86	82	78	78	114	122		86	80

Table A.20. (Continued)

21	1	7	7	6	7	6	5	7			6	6
		31	31	32	31	31	32	15			28	30
		104	100	98	98	98	96				98	96
22	1	6	6	8	8	5	3	6	6	6	4	1
		30	31	31	33	30	30	25	21	21	23	28
		114	108	108	114	108	112	120	120	122	108	108
23	2	1	6	7	7	6	3	3	4	5	6	4
		40	39	37	36	39	39	18	14	14	17	21
		92	104	102	102	92	84	146	166	166	132	102
24	2	2	4	6	7	2	1	5	5	7	3	1
		34	34	34	33	34	33	22	19	18	25	31
		88	100	102	96	90	84	96	96	102	84	84
25	2	1	3	3	4	1	0	3	3	3	2	0
		37	38	37	37	38	36	20	16	13	22	32
		84	86	84	88	74	72	108	108	108	84	74
26	2	0	0	0	0	0	0	0	1	1	0	0
		34	34	34	33	33	33	19	18	17	27	34
		84	78	88	84	84	84	90	90	90	84	82
27	2	1	4	8	8	2	1	2	3	4	2	1
		37	36	36	36	37	37	20	17	15	24	30
		80	100	98	96	92	84	120	126	132	90	80
28	2	0	3	6	6	0	0	0	1	1	0	0
		37	37	37	36	36	36	21	20	18	25	31
		64	68	66	66	66	66	72	72	74	66	62
29	2	0	0	0	0	0	0	1	4	4	0	0
		35	34	35	35	35	35	25	22	20	28	35
		98	104	106	106	94	94	110	112	118	94	90
30	2	1	3	6	7	3	1	6	8	8	3	0
		40	38	37	39	40	41	18	13	10	18	36
		92	104	108	108	92	90	134	160	172	114	90
31	2	1	1	4	8	3	0	6	8	8	6	2
		34	34	25	20	23	28	15	14	13	17	20
		74	80	84	88	70	68	122	132	126	92	68
32	2	1	1	1	1	1	0	1	2	2	1	0
		32	31	31	32	32	32	21	19	18	29	32
		84	84	84	82	82	82	82	84	84	82	80
33	2	0	0	2	3	0	0	3	4	5	0	0
		35	33	33	34	34	35	19	18	17	26	32
		70	78	76	78	74	72	80	80	80	74	70
34	2	1	2	2	2	1	1	2	3	3	4	2
		40	40	40	40	40	40	18	16	14	27	38
		64	66	64	62	64	64	98	84	80	64	66
35	2	0	2	2	2	0	0	1	2	2	1	0
		36	36	36	35	35	35	18	15	13	20	28
		90	98	98	98	90	84	108	112	114	98	84

36	2	1	1	2	4	2	1	2	5		3	1
		31	31	30	27	28	31	15	13		23	27
		90	94	88	86	82	78	104	116		82	82
37	2	1	4	6	7	0	0	5	6	7	5	0
		39	38	37	36	37	40	21	16	14	20	28
		78	84	80	80	70	68	108	110	114	84	68
38	2	2	4	4	5	3	1	5	5	6	3	3
		37	35	35	35	35	34	20	15	13	25	32
		90	96	92	94	94	92	102	122	130	86	82
39	2	2	4	6	7	2	2	3	5	5	4	2
		40	40	40	40	40	40	17	15	13	23	33
		78	86	80	80	76	80	100	106	106	78	86
40	2	1	4	8	8	2	1	2	6	6	4	1
		38	38	38	38	39	40	23	19	18	26	38
		98	96	98	94	98	94	108	120	122	98	98
41	2	3	3	3	3	3	3	4			4	6
		27	26	26	25	25	25	22			23	24
		134	130	126	124	124	118				118	140

Table A.20. *Panic attacks. The first two variables are case variables: subject number and panic attacks (1 = yes, 2 = no). The next 11 show repeated measures on an anxiety scale, measured at the following times (in minutes): 4, 6, 8, 10, 11, 14, 16, 17, 18, 19, 23. The next 11 show repeated measures on a* CO_2 *expiration scale, measured at the same times. The final 11 show pulse rates measured at the same times. Times 4, 11, 14, 19 and 23 are rest times. Times 6, 8 and 10 are times at which subjects are spoken to on the topic about which they are anxious. Times 16, 17 and 18 are times at which the subjects are asked to hyperventilate. Blanks signify missing values*

APPENDIX B

Extra data sets

1	22	0	103	67	75	65	59	46	16	37	45	29	35	36
	76	27	114	100	112	96	81	35	28	0	0	11	0	36
	6	4	5	5	4	2	4	984	1595	1984	1098	1727	1300	1224
	38	29	39	110	59	66	54	104	67	24	66	50	91	48
	−9	−9	375	536	542	260	440							
2	18	0	96	96	98	103	70	34	13	39	38	33	32	44
	47	41	103	122	145	112	126	48	32	0	0	0	21	4
	14	11	18	20	23	22	15	1029	1562	1380	1160	1691	1339	1633
	−9	−9	70	118	92	129	151	75	61	59	74	10	139	101
	−9	−9	232	455	396	280	360							
3	73	37	118	76	107	80	110	−9	51	53	44	−9	58	90
	131	114	150	106	140	112	134	27	82	29	0	0	38	0
	9	8	13	13	11	9	4	1282	1088	1200	1695	2854	1467	1502
	56	126	55	71	54	64	60	57	58	176	205	116	156	82
	−9	−9	260	620	496	650	540							
4	30	25	74	110	148	39	36	15	25	53	47	58	36	47
	64	51	122	141	169	83	75	0	16	44	0	0	32	39
	14	12	18	15	16	13	17	2002	1923	1176	1933	1952	1193	1633
	44	48	43	47	42	39	44	170	189	166	−9	200	200	98
	−9	−9	500	−9	650	650	650							
5	54	42	133	132	130	74	56	43	58	57	32	114	84	47
	92	76	172	124	145	116	77	77	66	26	41	28	21	99
	19	15	13	12	19	4	13	1073	1107	1168	886	1691	1273	714
	95	88	211	109	138	69	82	75	60	58	205	91	150	79
	−9	−9	425	620	650	650	650							
6	16	30	127	106	139	63	40	20	33	68	54	51	−9	35
	76	40	156	128	173	84	71	110	45	32	0	9	10	0
	9	10	12	15	16	9	8	1247	881	1027	1497	973	843	1413
	69	85	59	83	63	44	49	48	40	35	200	118	164	200
	−9	−9	210	275	550	650	561							
7	30	109	117	90	117	75	88	37	36	37	83	84	47	38
	65	55	145	106	134	92	92	5	13	3	48	0	0	25
	24	17	25	28	29	25	22	1700	1587	1500	1198	1433	1099	991
	50	55	40	58	38	49	36	60	52	−9	47	110	149	96
	−9	−9	−9	200	480	650	465							

Table B.1. (Continued)

8	70	130	180	180	160	123	41	62	84	55	37	43	30	32
	60	54	149	98	98	72	56	42	39	131	0	0	16	69
	0	0	16	21	25	18	18	769	1379	767	3045	2923	1667	1143
	−9	141	47	59	83	56	54	84	77	100	200	145	133	130
	−9	−9	195	195	400	580	352							
9	31	54	124	56	77	28	40	63	70	71	−9	57	47	57
	140	140	200	330	190	173	108	10	0	0	0	0	7	21
	17	18	19	30	30	24	25	1050	1010	1200	2100	710	1610	1260
	80	80	80	80	−9	−9	80	−9	−9	85	94	−9	53	73
	320	208	248	216	490	265	240							
10	140	140	164	128	112	66	77	35	63	40	46	53	33	49
	160	168	113	189	134	117	106	5	49	12	0	11	24	13
	35	30	27	26	−9	20	20	1380	660	770	1480	1420	1020	860
	140	200	190	−9	170	−9	180	−9	50	125	69	176	10	19
	267	155	163	346	501	350	440							
11	60	80	102	128	116	101	67	48	39	48	55	57	35	48
	95	98	155	114	140	123	110	18	−9	7	19	9	7	10
	19	19	20	23	28	19	24	2300	300	1270	2500	1190	1560	1680
	180	70	80	60	70	−9	100	56	32	93	128	149	111	93
	415	170	107	240	650	265	175							
12	73	50	108	126	117	91	87	26	43	48	46	68	45	33
	120	130	108	142	135	132	110	−9	−9	−9	−9	−9	−9	−9
	38	38	28	33	36	36	35	830	1200	1000	1960	900	1190	1490
	100	130	70	−9	170	−9	70	−9	−9	96	180	126	106	117
	−9	−9	−9	390	650	425	442							

Table B.1. *Twelve patients measured on seven occasions (weeks 1, 2, 6, 10, 14, 15, 16) on nine variables. The first row contains seven repeated measurements on variable 1, followed by seven on variable 2, etc. Plasma ascorbic acid (× 100), leucocyte ascorbic acid (× 100), whole blood ascorbic acid (× 100), thiamin status-TPP value (× 100), grip strength (× 100), red cell transketolase (× 1000), reaction time (× 100), folate serum (× 100), folate red cell (× 100). Missing values are coded −9*

1	1	2	60.	162.5	57.6	3725	2650	−9	3845	118	406	−9	148
						91	47	−9	58	680	174	−9	741
						176	186	−9	299	1125	275	−9	1425
						3243	4491	−9	2094	1426	1975	−9	920
						165	39	−9	45	140	53	−9	110
						2150	1750	−9	2050	267	162	−9	373
2	1	1	66.	184.2	73.0	4710	3105	−9	3280	153	184	−9	35
						114	104	−9	87	916	296	−9	785
						55	134	−9	159	875	550	−9	850
						2102	6198	−9	1318	1172	3454	−9	733
						154	83	−9	58	130	78	−9	109
						2250	1825	−9	2075	234	219	−9	373

3	1	1	70.	178.0	93.9	4410	3150	−9	3700	174	179	−9	97
						152	61	−9	108	916	166	−9	767
						245	221	−9	189	1250	625	−9	600
						1414	2336	−9	1879	1013	1674	− 9	1339
						130	16	−9	133	180	72	−9	104
						2075	1775	−9	2050	334	218	−9	430
4	1	2	52.	144.0	57.6	−9	2795	−9	3395	181	222	−9	104
						88	66	−9	118	916	262	−9	793
						376	156	−9	102	700	300	−9	600
						2466	3159	−9	2800	1804	1389	−9	1228
						182	44	−9	66	185	83	−9	108
						2000	1675	−9	1975	396	341	−9	423
5	1	2	52.	160.0	63.5	2700	3250	−9	3195	76	83	−9	75
						85	38	−9	61	846	366	−9	1020
						109	1564	−9	221	1025	400	−9	925
						1729	9405	−9	3463	859	4559	−9	1675
						109	13	−9	58	63	33	−9	70
						2075	1825	−9	2150	361	300	−9	491
6	1	2	75.	161.0	49.9	3835	3230	−9	3790	0	390	−9	0
						131	96	−9	96	671	244	−9	898
						1143	569	−9	482	1000	550	−9	1175
						2807	6550	−9	2376	1069	2495	−9	905
						196	22	−9	51	96	83	−9	143
						2175	1925	−9	2000	464	261	−9	505
7	1	2	72.	161.0	47.2	3880	3255	−9	2685	106	271	−9	119
						103	37	−9	69	678	174	−9	794
						1100	1500	−9	357	1550	600	−9	1775
						2183	6070	−9	3687	787	2188	−9	1328
						56	40	−9	104	103	67	−9	68
						2050	1850	−9	1775	432	228	−9	464
8	1	1	44.	170.0	73.0	3970	3370	−9	2450	147	111	−9	184
						131	75	−9	110	1142	375	−9	1020
						68	42	−9	291	1500	575	−9	1400
						578	1333	−9	1498	321	743	−9	840
						101	37	−9	37	99	67	−9	88
						2250	1875	−9	2050	363	166	−9	438
9	1	2	54.	166.0	66.2	4465	3225	−9	3420	107	250	−9	88
						70	59	−9	95	1308	698	−9	802
						661	281	−9	702	1500	625	−9	1925
						5038	1407	−9	2844	2456	711	−9	1437
						48	35	−9	31	108	79	−9	83
						2325	1850	−9	2400	413	178	−9	394
10	1	2	71.	163.0	59.0	3800	2710	−9	3785	182	111	−9	63
						126	98	−9	81	794	384	−9	1203
						566	309	−9	375	1325	600	−9	1450
						730	3274	−9	5075	331	1475	−9	2286
						67	43	−9	107	124	72	−9	131
						2175	1950	−9	2250	402	286	−9	464

Table B.2. (Continued)

11	1	2	58.	−9.0	−9.0	3400	2320	−9	2595	115	−90	−9	70
						96	74	−9	71	1160	340	−9	1003
						860	193	−9	100	850	725	−9	1000
						1314	1690	−9	1659	−9	−9	−9	−9
						120	43	−9	131	154	−9	−9	111
						2200	2000	−9	2275	440	342	−9	513
12	1	1	70.	177.0	78.0	4330	1835	−9	3660	81	388	−9	187
						66	58	−9	39	1020	279	−9	654
						234	210	−9	927	1200	550	−9	975
						557	1477	−9	3515	332	862	−9	2093
						51	35	−9	60	88	66	−9	39
						2000	1775	−9	2150	314	266	−9	−9
13	1	1	74.	168.0	73.6	3740	3225	−9	2905	0	211	−9	0
						66	76	−9	132	977	392	−9	776
						303	268	−9	318	1025	1125	−9	1450
						546	1446	−9	1043	307	812	−9	586
						120	57	−9	70	86	48	−9	71
						2275	2000	−9	2350	321	225	−9	348
14	1	1	65.	176.0	80.7	4570	3550	−9	3980	0	117	−9	125
						115	108	−9	81	1107	419	−9	1020
						379	465	−9	516	1550	800	−9	1325
						1704	1994	−9	2067	1050	1229	−9	1274
						77	61	−9	100	102	37	−9	34
						2575	2200	−9	2225	329	295	−9	337
15	1	1	54.	179.0	78.0	4490	3410	−9	3203	45	63	−9	42
						102	38	−9	9	1102	375	−9	1134
						582	273	−9	679	1700	875	−9	1650
						1330	1270	−9	1928	792	756	−9	1148
						96	35	−9	87	112	65	−9	98
						2350	2050	−9	2225	276	235	−9	324
16	1	1	63.	180.0	81.6	4290	3205	−9	−9	80	45	−9	−9
						121	44	−9	−9	1107	305	−9	−9
						310	756	−9	817	1750	700	−9	−9
						876	1363	−9	1468	546	849	−9	914
						106	69	−9	−9	113	55	−9	−9
						2500	2100	−9	−9	359	298	−9	−9
17	1	2	77.	157.0	59.4	4480	3720	−9	2970	0	105	−9	222
						151	76	−9	133	1395	479	−9	916
						385	326	−9	785	1300	625	−9	1175
						1294	2455	−9	2591	586	1113	−9	1174
						123	77	−9	90	130	124	−9	126
						2325	1800	−9	2300	415	264	−9	317
18	1	2	66.	168.0	47.2	4595	2910	−9	2670	0	316	−9	26
						66	28	−9	103	776	305	−9	837
						206	982	−9	270	1550	575	−9	1275
						1543	1930	−9	2162	556	695	−9	779
						85	107	−9	62	96	42	−9	49
						2350	2075	−9	2425	464	269	−9	409

19	1	1	71.	176.0	85.3	4605	2880	−9	3780	100	261	−9	62
						95	42	−9	105	575	200	−9	689
						323	400	−9	765	1525	700	−9	1625
						1285	1573	−9	1819	836	1024	−9	1184
						75	65	−9	87	140	58	−9	85
						2575	2050	−9	2200	321	264	−9	374
20	2	2	53.	−9.0	−9.0	4185	3195	3715	3760	0	69	0	0
						103	56	54	75	637	296	541	732
						−9	64	4592	876	1350	550	1150	1350
						−9	5150	4983	4646	−9	−9	−9	−9
						55	25	35	97	65	36	46	57
						2225	1925	2100	2400	379	276	586	604
21	2	1	68.	−9.0	−9.0	4210	3440	3400	3605	61	130	0	42
						196	124	128	98	576	244	453	480
						329	−9	1390	1071	1200	650	1150	1400
						1494	−9	2723	3258	−9	−9	−9	−9
						82	29	54	36	51	31	33	44
						2200	2000	2125	2300	336	319	388	440
22	2	2	74.	−9.0	−9.0	3800	3165	3840	3570	64	74	0	0
						134	62	72	68	628	122	366	497
						866	568	1889	1694	1200	550	1225	1525
						1202	2798	4874	5531	−9	−9	−9	−9
						140	44	44	32	77	37	44	47
						2400	1800	2175	2300	423	216	431	535
23	2	1	69.	−9.0	−9.0	3825	2940	3285	3050	104	172	102	364
						108	72	75	68	802	270	602	907
						517	426	1723	1951	975	475	975	1250
						1581	2183	1637	2556	−9	−9	−9	−9
						48	26	31	44	83	25	39	64
						2325	1925	2125	2300	414	256	302	500
24	2	1	64.	−9.0	−9.0	4300	3370	3410	3665	147	419	108	115
						116	79	84	95	532	244	436	506
						363	234	685	784	1150	625	1150	1275
						1154	723	2075	2005	−9	−9	−9	−9
						53	25	33	29	64	29	41	44
						2200	2125	2200	2275	181	164	362	352
25	2	2	78.	−9.0	−9.0	4220	3400	3290	3325	51	278	68	108
						184	69	103	142	541	139	244	375
						424	408	1622	1943	1400	600	1000	1250
						2405	2583	4889	3817	−9	−9	−9	−9
						69	33	62	39	−9	49	49	51
						2450	2250	2300	2350	328	164	423	454
26	2	2	74.	−9.0	−9.0	3970	3345	2660	3265	193	194	396	111
						203	76	105	105	828	209	453	689
						325	353	974	699	1225	725	975	1725
						777	1993	1861	2886	−9	−9	−9	−9
						49	32	9	19	145	75	99	125
						2375	2025	2125	2225	340	247	371	367

Table B.2. (Continued)

27	2	2	79.	−9.0	−9.0	3750	2820	2880	3030	131	150	44	128
						156	82	78	78	724	279	541	602
						675	1333	600	1133	1150	350	875	1050
						458	8405	2563	3787	−9	−9	−9	−9
						59	21	49	27	145	78	82	78
						2225	1975	2100	2175	432	272	376	374
28	2	1	70.	−9.0	−9.0	3780	2660	2670	3060	83	235	174	317
						135	43	76	85	349	174	357	532
						462	787	1906	2350	1175	550	1275	1575
						1015	1758	1883	1996	−9	−9	−9	−9
						84	18	13	20	105	46	70	93
						2200	2000	2125	2175	321	213	395	383
29	2	2	71.	−9.0	−9.0	3455	3095	3060	2875	380	130	148	127
						158	57	99	72	706	166	427	645
						2670	1569	2292	2235	1150	600	1125	1300
						2009	1755	2364	1687	−9	−9	−9	−9
						15	20	24	78	33	6	18	31
						2250	1925	2075	2150	386	340	435	469
30	2	2	68.	−9.0	−9.0	3550	2470	2810	2975	0	138	219	0
						122	60	84	83	794	235	698	724
						1411	700	1870	3125	1350	600	1205	1325
						2690	2096	1126	2129	−9	−9	−9	−9
						55	20	31	68	74	18	38	64
						2300	1850	2150	2200	340	241	407	358

Table B.2. *Hip-replacement data. For convenience a subset of these data has been separated out as Table A.14 (p. 178). For each of 30 patients there are six case variables (case number, group, sex, age, height, and weight) and twelve repeated measures variables (haematocrit, TPP, vitamin E, vitamin A, urinary zinc, plasma zinc, hydroxyprolene (in milligrams), hydroxyprolene (index), ascorbic acid, carotine, calcium, and plasma phosphate) ordered row-wise (row 1 holds haematocrit and TPP values, row 2 holds vitamin E and vitamin A values, etc). Each variable is measured on four equally spaced occasions. Missing values are coded as −9*

slips

0.00	0.10	0.20	0.30	0.40	0.50	0.60	0.70	0.80	0.90	1.00	1.20	1.40	1.60
	1.80	2.00	2.20	2.40	2.60	2.80	3.00						

loads

0.00	4.61	6.17	7.53	8.34	9.79	10.97	12.01	13.01	14.03	14.96	16.71
18.21	19.53	20.60	21.40	22.07	22.57	22.98	23.21	23.46			
0.00	5.24	6.73	8.25	9.72	11.26	12.38	13.69	15.01	15.86	16.90	18.46
19.65	20.53	21.32	21.99	22.48	22.76	22.96	23.06	23.16			
0.00	4.40	6.00	7.27	8.21	9.51	10.57	11.62	12.53	13.51	14.40	16.07
17.62	18.99	20.26	21.28	22.07	22.78	23.40	23.88	24.30			
0.00	3.96	5.36	6.61	7.68	8.51	9.68	10.61	11.47	12.42	13.25	14.96
16.63	18.13	19.45	20.66	21.71	22.55	23.36	24.05	24.66			

0.00	4.28	5.67	6.80	7.86	8.84	9.77	10.21	11.31	12.24	13.11	14.72
16.21	17.52	18.79	19.90	20.86	21.71	22.48	23.15	23.73			
0.00	3.62	5.04	6.12	7.22	8.14	8.99	9.86	10.23	11.32	12.17	13.74
15.19	16.43	17.50	18.40	19.05	19.58	20.05	20.40	20.63			
0.00	2.14	3.53	4.86	6.09	7.31	8.39	9.15	10.37	11.45	12.45	14.25
15.85	17.45	18.81	20.00	20.99	21.76	22.32	22.93	23.46			
0.00	3.40	4.85	6.05	7.09	8.01	9.01	9.88	10.24	11.34	12.25	13.93
15.41	16.72	17.77	18.78	19.53	20.14	20.66	20.94	–9.00			
0.00	4.60	5.68	6.50	7.11	7.67	8.20	8.73	9.22	9.75	9.99	11.34
12.69	14.47	15.29	16.44	17.67	19.03	20.23	21.23	22.13			
0.00	3.16	4.27	5.21	5.92	6.59	7.18	7.72	8.27	8.84	9.46	10.27
12.06	13.71	15.21	16.66	18.04	19.30	20.40	21.33	22.17			
0.00	5.20	6.51	7.49	8.13	8.85	9.50	10.11	10.37	10.86	11.62	12.81
13.98	14.99	16.02	17.03	17.91	18.78	19.49	20.10	20.64			
0.00	3.47	4.82	6.02	7.21	8.36	9.54	10.13	11.34	12.40	13.37	14.91
16.43	17.76	19.06	20.45	21.01	21.70	22.25	22.71	22.98			
0.00	2.19	3.28	4.38	5.42	6.48	7.49	8.42	9.33	10.16	10.92	13.06
14.85	16.45	17.71	18.92	19.92	20.76	21.48	22.15	22.76			
0.00	2.69	3.97	5.09	6.07	7.05	7.89	8.66	9.46	10.21	10.36	12.54
14.16	15.66	16.94	18.19	19.23	20.01	20.72	21.25	21.79			
0.00	2.70	4.35	5.73	7.03	8.31	9.59	10.33	12.03	13.19	14.10	15.77
17.32	18.74	19.89	20.96	21.82	22.55	23.18	23.70	24.13			
0.00	3.04	4.72	6.15	7.45	8.83	10.13	11.40	12.83	14.06	15.29	17.21
18.81	20.25	21.38	22.40	23.25	23.99	24.65	25.22	25.72			
0.00	2.01	4.58	6.33	7.69	8.80	9.81	10.82	12.17	13.23	14.50	16.38
18.17	18.86	20.76	21.73	22.65	23.45	24.18	24.79	25.28			
0.00	1.65	2.68	3.73	4.71	5.60	6.51	7.36	8.21	9.00	9.81	11.29
13.16	14.79	16.29	17.70	18.96	20.12	21.10	21.93	22.67			
0.00	2.34	3.84	5.27	6.53	7.76	8.57	9.72	10.20	11.36	12.29	14.02
15.55	16.96	18.22	19.41	20.41	21.29	22.06	22.73	23.29			
0.00	0.31	0.67	1.25	2.11	3.30	5.07	6.71	8.18	9.14	9.88	11.69
13.41	14.85	16.20	17.45	18.61	19.63	20.56	21.35	22.07			
0.00	2.38	3.48	4.51	5.31	6.08	6.84	7.52	8.33	9.10	9.90	11.31
13.57	15.55	17.38	19.08	20.60	21.83	22.95	23.91	24.72			
0.00	2.15	3.43	4.55	5.53	6.62	7.72	8.91	10.10	11.54	13.14	15.78
18.12	20.14	21.75	23.09	24.30	25.25	26.15	26.87	27.25			
0.00	2.64	3.98	5.30	6.54	7.62	8.72	9.89	10.71	12.27	13.48	15.79
17.94	18.89	21.66	23.02	24.21	25.21	25.98	26.63	27.15			
0.00	3.03	4.44	5.62	6.81	7.89	8.91	9.85	10.53	11.98	13.22	15.45
17.57	19.64	21.49	22.94	23.97	24.89	25.68	26.39	26.87			
0.00	3.62	5.05	6.25	7.38	8.43	9.65	10.19	11.63	12.90	14.14	16.37
18.45	20.22	21.89	23.16	24.22	25.19	25.98	26.63	27.23			
0.00	1.16	2.02	3.14	4.42	6.01	7.18	10.36	13.25	16.04	18.23	20.70
22.31	23.59	24.66	25.56	26.29	26.95	27.56	28.04	28.47			
0.00	3.28	4.55	5.65	6.79	7.99	9.25	10.07	11.46	12.92	14.00	16.21
18.06	19.83	21.35	22.66	23.80	24.73	25.54	26.20	26.70			
0.00	1.45	4.69	5.80	6.79	7.77	8.76	9.75	10.35	11.79	12.97	15.28
17.35	19.14	20.82	22.23	23.36	24.27	24.95	25.52	26.00			

Table B.3. (Continued)

0.00	3.78	5.33	6.48	7.52	8.49	9.40	10.18	10.81	12.03	13.04	14.98
16.90	18.67	20.39	22.08	23.42	24.54	25.55	26.33	26.98			
0.00	2.70	3.99	5.20	6.20	7.14	8.13	9.11	9.93	10.70	12.02	14.43
16.66	18.66	20.66	22.32	23.78	25.03	25.93	26.66	27.15			
0.00	2.07	3.34	4.60	5.78	6.87	8.04	9.36	10.15	11.51	12.77	15.23
17.43	19.58	21.22	22.75	24.08	25.25	26.23	27.08	27.76			
0.00	2.56	3.99	5.41	6.74	8.00	9.30	10.26	11.98	13.47	15.02	17.39
19.78	21.64	23.21	24.63	25.72	26.72	27.49	28.11	28.65			
0.00	5.32	7.08	8.75	10.09	11.92	13.50	14.97	16.32	17.46	18.80	20.98
22.48	24.60	26.04	27.29	28.27	29.08	29.76	30.28	30.69			
0.00	3.50	4.79	5.91	6.93	8.01	9.01	9.98	10.78	12.08	13.24	15.47
17.71	19.76	21.61	23.25	24.62	25.97	27.10	27.99	28.75			

Table B.3. *Timber slip data. These are similar to those of Table A. 15. There are 34 cases and missing values are coded as − 9.0*

Subj.	Age	Sex	Weight (kg)	Height (cm)	Pulse (× 10/min)			kcal (× 1000/min)			Pulvent (litres/min)		
1	21	1	70.9	176.5	1032	1169	1293	2765	3970	5814	10.81	14.11	20.44
2	22	1	70.0	173.9	777	910	1096	3570	5110	7420	13.11	17.03	25.19
3	22	1	61.0	172.0	1080	1894	2400	2928	4026	5246	13.52	19.33	22.59
4	22	1	63.6	169.0	913	1000	1210	2798	3625	5851	14.08	17.81	25.26
5	24	1	70.0	173.9	692	843	1354	3500	4130	6510	13.10	15.00	19.09
6	19	2	54.2	159.0	1190	1227	1378	2060	2385	3035	13.33	12.88	14.67
7	20	2	66.8	172.7	909	1277	1677	2939	5210	6279	12.84	19.33	23.28
8	19	2	52.7	162.5	1083	1363	1567	2688	4269	5323	12.81	17.11	20.52
9	19	2	66.7	167.0	1026	1326	1758	2201	3001	3869	15.60	17.61	25.57
10	19	2	54.5	159.5	1269	1486	1918	3270	3706	5014	13.80	14.61	19.57

Table B.4. *Energy consumption and heart rate, similar to Table A.16. Three exercises ten student volunteers*

1	1	la	32.0	6.0	4	1	1	1	2	1	3	1	3	1	2	1	1	1	4	1
2	0	cb	−9.	6.0	4	1	1	1	2	1	3	1	3	4	2	1	1	1	4	1
3	0	cs	−9.	11.0	3	4	4	2	1	1	2	1	2	2	1	4	3	4	4	1
4	0	sp	22.0	10.0	1	1	2	1	3	1	4	2	4	1	3	1	2	1	1	1
5	0	la	20.0	10.0	2	1	1	1	4	1	3	1	3	1	4	1	1	2	2	1
6	0	sc	17.0	10.0	2	4	3	4	4	4	1	4	1	4	4	4	3	4	2	4
7	0	sc	18.0	12.0	1	2	2	3	3	3	4	2	4	3	3	3	2	3	1	3
8	0	cb	6.0	13.0	3	4	4	1	2	4	1	5	1	2	2	2	4	1	3	5
9	0	cd	15.0	3.0	2	2	3	3	4	1	1	2	1	1	4	1	3	3	2	5
10	1	bco	15.0	3.0	1	2	2	5	3	4	4	4	4	4	3	5	2	4	1	5
11	1	jte	6.0	12.0	1	1	2	1	3	1	4	1	4	1	3	1	2	1	1	1
12	1	gs	−9.	4.0	4	1	3	1	1	1	2	1	2	1	1	1	3	1	4	1
13	1	jte	5.0	8.0	2	4	1	4	4	4	3	4	3	4	4	2	1	5	2	4
14	0	bco	−9.	5.0	4	2	3	3	1	4	2	4	2	4	1	3	3	3	4	3
15	0	bco	−9.	7.0	4	1	3	2	1	1	2	1	2	1	1	2	3	2	4	1

16	0	bco	−9.	12.0	1	4	2	4	3	4	4	3	4	1	3	5	2	4	1	4
17	0	bco	−9.	.5	2	4	1	4	4	2	3	2	3	4	4	1	2	2	1	4
18	0	bco	−9.	1.0	3	2	4	1	2	2	1	2	1	2	2	3	4	1	3	2
19	0	bco	20.0	10.0	2	3	4	1	1	1	3	1	3	1	1	1	4	1	2	1
20	0	gr	25.0	4.0	3	1	4	1	1	1	2	1	2	1	1	1	4	1	3	1
21	1	bco	20.0	1.0	3	4	4	2	2	2	1	1	1	1	2	2	4	1	3	1
22	0	gs	−9.	5.0	3	1	4	1	2	1	1	3	1	3	2	1	4	1	3	1
23	0	ct	−9.	14.0	4	1	3	1	1	1	2	3	2	2	1	1	3	1	4	1
24	1	la	35.0	5.0	4	1	3	4	2	3	1	1	1	1	2	4	3	4	4	1
25	0	cb	22.0	14.0	1	4	2	4	3	4	4	4	4	1	3	5	2	4	1	4
26	1	cb	18.0	3.0	3	4	1	4	4	1	2	4	2	4	4	1	1	4	3	4
27	1	sp	25.0	6.0	2	4	4	1	1	4	3	4	3	4	1	4	4	1	2	4
28	0	jte	−9.	4.0	4	1	3	4	2	4	1	1	1	1	2	5	3	4	4	1
29	1	cor	−9.	8.0	1	5	2	5	3	5	4	2	4	4	3	4	2	5	1	4
30	0	oe	49.0	9.0	2	1	1	1	4	1	3	1	3	1	4	1	2	1	1	1
31	0	ro	45.0	4.0	1	3	2	2	3	2	4	3	4	2	3	3	2	2	1	2
32	0	cb	27.0	3.0	3	3	1	1	4	1	2	2	2	1	4	1	1	1	3	1

Table B.5. *Dog medicines. There are five case variables (dog number, sex, breed, weight, and age) and eight pairs of measures showing formulation and palatability on days 1, 2, 3, 4, 8, 9, 10 and 11. Palatability is scored as 1 = readily accepted, 2 = accepted, 3 = reluctantly accepted, 4 = accepted in food, and 5 = refused. Missing values are coded as −9*

1104	1236	1215	1208	1201	1181	1	1
1208	1201	1215	1299	1229	1236	2	1
1118	1146	1194	1188	1215	1215	3	1
1167	1201	1215	1236	1194	1264	4	1
1236	1181	1229	1215	1208	1236	5	1
1146	1188	1181	1188	1174	1160	6	1
1174	1181	1181	1194	1194	1160	7	1
1208	1160	1208	1194	1222	1201	8	1
1250	1243	1236	1306	1264	1278	1	2
1264	1222	1250	1222	1201	1285	2	2
1313	1292	1215	1340	1264	1167	3	2
1222	1264	1215	1167	1208	1271	4	2
1285	1167	1285	1236	1271	1299	5	2
1201	1201	1208	1222	1250	1243	6	2
1243	1257	1299	1243	1306	1306	7	2
1264	1313	1236	1299	1250	1174	8	2
938	1076	951	958	944	965	1	3
861	833	889	938	903	861	2	3
938	965	903	910	986	1042	3	3
1000	1000	1014	944	1063	1042	4	3
944	924	944	972	938	847	5	3
889	847	924	938	944	1000	6	3
917	993	1000	882	1014	944	7	3
813	924	931	938	993	993	8	3
1215	1146	1215	1181	1257	1194	1	4

Table B.6. (Continued)

1111	1146	1188	1132	1146	1188	2	4
1111	1250	1208	1243	1285	1215	3	4
1125	1250	1188	1264	1229	1132	4	4
1160	1125	1181	1208	1264	1236	5	4
1208	1125	1229	1111	1208	1299	6	4
1229	1167	1167	1250	1236	1215	7	4
1181	1208	1139	1201	1229	1250	8	4
1215	1264	1243	1306	1215	1313	1	5
1215	1257	1257	1264	1250	1299	2	5
1222	1167	1250	1264	1243	1271	3	5
1104	1222	1208	1208	1257	1222	4	5
1069	1236	1236	1299	1257	1264	5	5
1174	1278	1257	1271	1292	1278	6	5
1222	1194	1236	1236	1250	1306	7	5
1215	1243	1278	1125	1257	1257	8	5
931	972	826	938	951	1014	1	6
924	917	1042	986	986	1007	2	6
958	917	944	931	854	1021	3	6
917	944	938	965	965	1028	4	6
903	882	994	951	938	972	5	6
875	896	972	944	1028	979	6	6
924	924	896	951	938	1042	7	6
951	944	951	958	993	1007	8	6

Table B.6. *Parachute rigging lines data. Eight lines on each of six parachutes were tested: the breaking strength was measured at six equally spaced positions on each line. Thus, $n = 6$ and there is an 8×6 array of measurements on each individual parachute. In the data matrix the first six columns correspond to the positions, and the last two are line number and parachute number, respectively. The data were analysed using SPSS in Crowder et al. (1991, Example 7.1)*

cyclic GMP

1	4.1	6.1	7.6	7.5	8.9	9.5	8.7	8.8	−9.0	7.0	−9.0	6.5
1	5.8	7.5	10.1	10.4	10.4	8.9	8.9	8.4	9.9	8.6	−9.0	6.9
1	7.0	8.4	11.2	12.8	10.0	10.3	9.5	9.2	9.0	9.4	−9.0	8.4
1	9.0	7.8	10.8	10.3	9.3	10.3	11.5	12.3	10.0	11.4	−9.0	5.9
1	3.6	4.3	3.9	3.9	4.5	3.2	4.1	4.0	3.5	3.7	3.0	2.8
1	7.7	7.0	6.7	7.0	7.9	7.4	7.3	7.2	6.6	6.6	8.3	7.9
1	3.4	2.1	2.2	2.0	2.2	2.2	2.5	2.3	2.5	2.4	2.0	2.2
1	1.8	1.4	2.1	2.4	2.5	2.3	2.0	2.0	1.9	2.0	2.0	1.4
2	7.6	8.9	8.5	8.4	8.5	8.2	5.6	8.8	8.8	8.4	8.0	8.2
2	4.2	6.5	7.5	7.1	7.2	7.0	5.0	4.2	6.9	9.5	−9.0	−9.0
2	6.9	13.3	12.9	13.5	13.4	13.1	13.6	13.1	14.8	15.3	16.1	16.9
2	8.1	7.4	8.8	9.2	8.4	9.2	7.9	7.9	7.9	7.3	−9.0	7.2
2	4.5	4.9	5.5	5.6	5.2	5.3	6.4	6.0	6.4	6.4	−9.0	6.9
2	4.2	3.2	3.2	4.0	3.2	3.4	3.4	3.2	3.2	3.2	2.8	2.8
3	5.9	5.5	5.5	5.5	5.3	5.0	4.5	4.1	4.3	3.9	3.7	3.5

3	−9.0	0.8	0.4	0.6	0.4	0.4	0.5	0.6	0.5	0.5	0.8	0.7
3	10.1	6.5	6.2	6.3	6.6	5.9	6.5	5.5	5.7	5.1	4.4	4.9
3	5.7	4.3	4.6	3.8	3.9	3.6	3.0	3.7	3.2	3.1	2.7	2.4
3	2.1	2.9	3.2	3.2	2.7	2.7	2.4	2.2	1.8	1.7	1.7	1.5
3	5.5	11.1	10.8	8.7	9.3	10.5	12.7	11.3	19.1	18.9	37.0	39.0
3	0.9	4.9	5.7	7.0	7.0	5.8	6.9	7.7	7.5	8.8	8.1	9.9
4	5.0	5.2	3.4	3.0	3.1	3.6	3.2	2.6	4.6	3.8	4.9	2.7
4	4.2	4.3	4.1	3.5	2.8	2.8	4.7	3.7	3.7	4.2	−9.0	4.4
4	3.2	3.0	3.3	3.5	3.4	3.3	3.3	3.3	3.4	3.2	3.1	3.2
4	5.0	6.9	7.5	5.9	−9.0	7.7	7.3	7.6	7.5	7.5	7.0	7.5
4	2.5	12.0	12.2	11.4	11.6	11.7	12.6	10.1	11.4	12.8	11.5	10.7
4	1.6	1.6	2.1	1.9	1.7	2.5	1.6	1.3	3.5	0.6	−9.0	−9.0

cyclic AMP

1	15.6	14.2	12.7	13.7	13.9	14.6	17.5	17.4	−9.0	14.1	−9.0	13.6
1	19.2	18.8	20.3	19.7	20.4	23.8	20.0	21.0	18.8	19.8	−9.0	18.9
1	15.0	12.3	11.8	13.5	10.0	14.2	13.5	17.4	−9.0	16.8	−9.0	13.0
1	12.6	12.6	11.2	10.0	8.9	9.8	10.5	10.4	10.8	12.3	−9.0	11.3
1	8.1	6.2	7.2	7.1	7.6	6.7	9.6	9.5	7.2	8.0	8.2	8.8
1	4.7	3.7	4.5	3.5	3.4	4.5	4.3	4.0	3.8	5.2	4.2	4.7
1	8.3	8.8	8.1	7.6	7.9	9.0	8.3	9.8	9.8	9.4	10.0	8.8
1	14.2	12.2	12.5	11.6	12.2	12.5	12.9	13.0	12.0	12.4	−9.0	13.3
2	17.2	18.8	20.5	19.6	22.1	20.3	23.2	21.7	23.9	21.7	22.3	24.4
2	16.0	15.7	16.1	14.9	16.2	14.3	16.4	14.6	8.9	14.6	−9.0	−9.0
2	9.0	14.1	14.2	14.5	14.1	12.4	14.2	12.0	13.2	14.7	16.4	16.1
2	12.7	11.0	11.1	11.3	11.6	10.5	10.0	10.5	13.9	13.7	−9.0	13.6
2	13.2	13.4	12.8	12.1	12.9	11.6	13.1	12.8	13.2	13.4	−9.0	13.1
2	−9.0	4.0	7.6	5.9	4.3	4.6	5.3	6.5	6.1	6.4	5.1	5.0
3	5.7	5.6	7.6	6.5	6.9	6.6	6.4	6.3	7.6	6.7	6.7	6.4
3	−9.0	4.7	5.8	5.7	5.7	6.0	5.7	5.5	6.1	5.9	5.4	6.2
3	8.0	11.5	9.4	10.0	6.3	6.8	7.6	7.2	8.6	7.0	5.9	7.3
3	9.5	12.0	12.0	7.8	10.9	10.9	7.2	7.6	11.1	8.8	12.5	11.6
3	−9.0	5.5	5.6	5.3	5.3	5.3	4.8	5.3	4.7	6.6	5.9	6.0
3	12.1	11.7	16.5	11.7	12.1	14.1	13.1	13.2	13.6	13.8	10.5	13.5
3	11.3	10.6	11.0	11.1	13.0	12.3	10.4	12.9	11.7	12.6	13.3	12.8
4	18.8	14.7	18.0	18.3	14.3	16.7	17.6	15.8	11.9	16.3	18.7	12.4
4	10.9	10.2	10.7	11.1	10.2	10.1	9.8	10.0	11.6	11.8	−9.0	11.6
4	7.4	8.9	9.3	10.0	10.5	11.0	12.2	11.1	12.0	12.3	12.5	12.4
4	7.2	7.7	6.8	8.6	−9.0	9.6	10.7	10.0	8.5	10.4	11.4	12.2
4	26.2	24.9	28.6	29.8	37.9	30.4	31.6	27.5	25.9	28.6	36.2	33.8
4	6.4	3.5	7.3	5.5	3.7	5.7	3.4	3.3	12.7	2.2	−9.0	−9.0

Table B.7. *Twenty-seven hospital patients in four groups (8 normal controls, 6 control diabetic, 6 diabetic with postural hypertension, 7 diabetic with hypertension). Repeated measures on two variables: cyclic guanosine monophosate (GMP) and cyclic adenosine monophosphate (AMP), each recorded at times* $-30, -1, 1, 2, 3, 4, 5, 6, 8, 10, 12, 15$ *(mins) relative to a certain activity. Missing values are coded as* -9. *The data are reproduced from Crowder and Hand (1990, Table 2.1)*

1	1.08	1.99	1.46	1.21	1.48	2.50	2.62	1.95
1	1.19	2.10	1.21	0.96	0.62	0.88	0.68	0.48
1	1.22	1.91	1.36	0.90	0.65	1.52	1.32	0.95
1	0.60	1.10	1.03	0.61	0.32	2.12	1.48	1.09
1	0.55	1.00	0.82	0.52	1.48	0.90	0.75	0.44
2	1.20	1.54	1.28	0.79	1.28	2.25	1.95	1.24
2	0.96	2.05	1.65	1.05	0.00	1.36	1.24	0.60
2	1.89	2.55	2.35	1.30	1.42	1.70	1.42	1.05
2	0.00	1.04	0.94	0.57	0.64	0.74	0.50	0.00
2	0.00	0.94	0.96	0.68	0.92	1.72	1.65	1.25

Table B.8. *Antibiotic serum levels ($IU\ ml^{-1}$) for two groups of five human subjects: four repeated measures (1, 2, 3 and 6 hours after administration of medicine) in each of two periods. The first group (column 1) was administered medicine A in period 1 and medicine B in period 2. The second group was administered medicine B in period 1 and medicine A in period 2. From Crowder and Hand (1990, Example 2.2)*

19.9	18.9	11.9	34.4	28.5	38.4	43.3	48.2
29.1	30.8	38.3	66.8	5.28	48.6	29.4	53.2
40.2	50.7	54.5	53.7	78.8	73.8	66.5	65.7
15.3	19.8	26.6	41.7	44.5	46.2	54.0	63.6
95.0	81.6	82.2	74.9	8.80	27.9	25.5	25.2
69.7	76.9	75.6	69.1	62.5	70.3	80.2	57.5
40.4	46.1	54.9	68.9	18.5	28.8	36.8	47.8
64.4	59.5	40.9	56.2	4.30	37.3	49.7	52.7

Table B.9. *Data on eight dogs. A two-period crossover design with four repeated measures of blood levels of chlorpropamide (in $\mu g\ ml^{-1}$) in each period. Measurements are at 0.75, 1.50, 2.25 and 6.00 hours after drug administration for each drug. For dogs 1–4, period 1 is with drug A and period 2 is with drug B. For dogs 5–8 the order of administration is the reverse*

Drug A				Drug B				
68	76	67	28	−9	68	50	42	fem
24	48	48	38	14	48	46	36	fem
34	44	31	20	46	59	44	27	male
27	40	30	20	24	56	63	20	fem
33	48	38	24	52	86	48	46	male

Table B.10. *Data on eight dogs. A two-period crossover design with four repeated measures in each period. The measured variable is serum concentration of Tolbutamide ($mg\ ml^{-1}$) at four times (1, 2, 3, and 6 hours after treatment) under each of the two drugs. Also given is the sex of the dog*

1	26	26	44	43	67	69	79	81	83	83	84	84
1	26	27	44	45	70	72	79	81	84	83	84	84
1	26	27	46	47	70	73	80	83	84	84	84	84
1	27	27	44	44	68	69	79	79	84	81	84	84
1	25	25	44	45	70	71	83	82	84	84	84	84
2	28	29	50	50	83	83	84	84	84	84	84	84
2	28	27	49	49	80	78	84	84	84	84	84	84
2	28	28	48	49	78	80	84	84	84	84	84	84
2	27	27	48	49	78	81	84	84	84	84	84	84
2	28	27	48	49	81	83	84	84	84	84	84	84
3	24	24	42	41	70	71	83	84	84	84	84	84
3	24	24	40	42	68	69	84	84	84	84	84	84
3	24	25	43	43	71	71	84	84	84	84	84	84
3	25	27	43	44	72	75	84	84	84	84	84	84
3	25	25	43	43	72	75	84	84	84	84	84	84

Table B.11. *Growth of fungus on a culture dish. Three pH values, five plates at each. Two diameters of the growth, at right angles, each measured on days 7, 11, 17, 21, 29 and 37. (The order of the data is: both diameters for day 1, both diameters for day 2, etc.). Recorded values (cm × 10) are censored at 84(= diameter of dish)*

times	−15	0	30	60	90	120	180	240	300	360	
	−15	0	30	60	90	120	180	240	300	360	
	−15	0	30	60	90	120	180	240	300	360	420
	−15	0	30	60	90	120	180	240	300	360	
	−15	0	30	60	90	120	180	240	300	360	420
	−15	0	30	60	90	120	180	240	300	360	420

4.90	4.50	7.84	5.46	5.08	4.32	3.91	3.99	4.15	4.41	
4.91	4.18	9.00	9.74	6.95	6.92	4.66	3.45	4.20	4.63	
4.22	4.92	8.09	6.74	4.30	4.28	4.59	4.49	5.29	4.95	4.34
4.05	3.78	8.71	7.12	6.17	4.22	4.31	3.15	3.64	3.88	
5.03	4.99	9.10	10.03	9.20	8.31	7.92	4.86	4.63	3.52	4.17
4.60	4.72	9.53	10.02	10.25	9.29	5.45	4.82	4.09	3.52	3.92
4.61	4.65	7.90	6.13	4.45	4.17	4.96	4.36	4.26	4.13	
4.16	3.42	7.09	6.98	6.13	5.36	6.13	3.67	4.37	4.31	
4.52	4.22	8.46	9.12	7.50	6.02	4.66	4.69	4.26	4.29	4.47
3.94	4.14	7.82	8.68	6.22	5.10	5.16	4.38	4.22	4.27	
4.51	4.50	8.74	8.80	7.10	8.20	7.42	5.79	4.85	4.94	2.07
4.33	4.10	4.36	6.92	9.06	8.11	5.69	5.91	5.65	4.58	3.79
5.37	5.35	7.94	5.64	5.06	5.49	4.77	4.48	4.39	4.45	
4.95	4.40	7.00	7.80	7.78	7.30	5.82	5.14	3.59	4.00	
4.47	4.47	7.95	7.21	6.35	5.58	4.57	3.90	3.44	4.18	4.32
4.19	4.22	7.45	8.07	6.84	6.86	4.79	3.87	3.60	4.92	
4.87	5.12	6.32	9.48	9.88	6.28	5.58	5.26	4.10	4.25	4.55
4.42	4.07	5.48	9.05	8.04	7.19	4.87	5.40	4.35	4.51	4.66
5.10	5.22	7.20	4.95	4.45	3.88	3.65	4.21	−1.00	4.44	
3.82	4.00	6.56	6.48	5.66	7.74	4.45	4.07	3.73	3.58	

Table B.12. (Continued)

4.27	4.33	6.61	6.89	5.64	4.85	4.82	3.82	4.31	3.81	2.92
4.31	4.45	7.34	6.75	7.55	6.42	5.75	4.56	4.30	3.92	
4.55	4.44	5.56	8.39	7.85	7.40	6.23	4.59	4.31	3.96	4.05
4.38	4.54	8.86	10.01	10.47	9.91	6.11	4.37	3.38	4.02	3.84
5.34	4.91	5.69	8.21	2.97	4.30	4.18	4.93	5.16	5.54	
3.76	4.70	6.76	4.98	5.02	5.95	4.90	4.79	5.25	5.42	
4.81	4.85	6.08	8.28	5.73	5.68	4.66	4.62	4.85	4.69	4.86
4.30	4.71	7.44	7.08	6.30	6.50	4.50	4.36	4.83	4.50	
4.79	4.82	9.29	8.99	8.15	5.71	5.24	4.95	5.06	5.24	4.74
5.06	5.04	8.86	9.97	8.45	6.58	4.74	4.28	4.04	4.34	4.35
5.24	5.04	8.72	4.85	5.57	6.33	4.81	4.55	4.48	5.15	
4.13	3.95	5.53	8.55	7.09	5.34	5.56	4.23	3.95	4.29	
4.61	4.68	6.01	7.35	6.38	6.16	4.41	4.96	4.33	4.54	4.81
4.45	4.12	7.14	5.68	6.07	5.96	5.20	4.83	4.50	4.71	
4.33	4.48	8.06	8.49	4.50	7.15	5.91	4.27	4.78	−1.00	5.72
4.43	4.75	6.95	6.64	7.72	7.03	6.38	5.17	4.71	5.14	5.19

Table B.12. *Blood glucose levels for six volunteer student subjects. Six separate occasions with meals at 10 a.m., 2 p.m., 6 a.m., 6 p.m., 2 a.m., 10 p.m. Blood glucose levels recorded at 15 and 0 mins before the meal, and at times 30, 60, 90, 120, 180, 240, 300, 360 and 420 following it. Missing values are coded as −1. (Crowder and Hand, 1990, Example 2.3)*

9.5	10.0	11.0	11.0	11.0	10.5	10.0	9.5	9.5	9.5	9.5
8.0	12.0	13.0	13.5	10.5	9.5	9.5	8.0	8.0	8.0	7.5
7.0	12.0	13.0	12.0	12.0	11.0	10.0	10.0	8.5	8.0	7.0
6.0	10.0	13.0	13.0	12.0	12.0	12.0	11.0	11.0	10.0	10.0
6.0	7.0	8.5	9.0	10.0	10.0	10.0	9.0	10.0	9.5	9.0
5.0	6.0	8.0	9.0	8.0	7.5	6.0	5.5	5.0	5.0	5.0
3.0	6.0	6.0	7.0	5.0	5.0	5.0	4.0	3.0	3.5	3.0
7.0	9.0	15.0	15.0	13.0	8.0	6.0	7.0	7.0	7.0	7.0
5.0	9.0	10.0	10.0	9.5	9.5	9.0	9.0	8.5	9.0	9.0
6.0	8.0	8.0	9.0	9.0	9.0	8.0	8.0	8.0	8.5	8.0
8.0	11.5	13.0	13.0	11.0	9.5	9.5	8.0	8.0	8.0	7.5
6.0	11.0	12.0	12.0	12.0	11.0	11.0	10.5	11.0	11.0	10.5
7.0	11.0	13.0	13.0	10.0	9.5	9.5	8.0	8.0	8.0	8.0

Table B.13. *Electrical activity in the skin. Thirteen student volunteers subjected to a stimulus (an aural bleep). Electrical activity in the skin (potential difference between two electrodes on leg) recorded pre-stimulus and at times 50, 100,... 500 milliseconds post-stimulus*

19°C				24°C				29°C			
0.348	−9.	−2.991	−9.	1.208	1.452	2.384	2.677	−9.	−9.	−9.	−9.
−9.	2.021	−9.	3.462	0.745	2.098	−9.	3.425	−9.	−9.	−9.	−9.
0.500	2.366	−9.	−9.	0.500	−9.	1.970	3.771	−9.	−9.	−9.	−9.
0.500	1.385	4.824	−9.	1.429	2.311	−9.	2.907	−9.	−9.	−9.	−9.
0.401	−9.	−9.	4.646	−9.	−9.	−9.	−9.	0.661	−9.	2.580	2.829
0.500	−9.	−9.	3.934	−9.	−9.	−9.	−9.	1.055	1.293	−9.	2.378
0.500	−9.	2.050	3.623	−9.	−9.	−9.	−9.	0.793	1.721	2.244	−9.
0.500	−9.	2.204	−9.	−9.	−9.	−9.	−9.	1.149	−9.	−9.	1.894
1.028	−9.	3.987	−9.	−9.	−9.	−9.	−9.	0.989	1.549	−9.	−9.
0.301	−9.	3.023	−9.	−9.	−9.	−9.	−9.	−9.	−9.	−9.	−9.
−9.	−9.	−9.	−9.	1.584	−9.	−9.	2.252	−9.	−9.	−9.	−9.

Table B.14. *Carboxyhaemoglobin from blood samples. Eleven student volunteers in a carbon monoxide environment. The CO in carboxy haemoglobin from blood samples is measured at four times (0, 20, 40, 60 mins) on each of three occasions: first occasion at 19°C, second at 24°C, third at 29°C. Missing values are coded as −9*

1	42	51	59	64	76	93	106	125	149	171	199	205
1	40	49	58	72	84	103	122	138	162	187	209	215
1	43	39	55	67	84	99	115	138	163	187	198	202
1	42	49	56	67	74	87	102	108	136	154	160	157
1	41	42	48	60	79	106	141	164	197	199	220	223
1	41	49	59	74	97	124	141	148	155	160	160	157
1	41	49	57	71	89	112	146	174	218	250	288	305
1	42	50	61	71	84	93	110	116	126	134	125	−9
1	42	51	59	68	85	96	90	92	93	100	100	98
1	41	44	52	63	74	81	89	96	101	112	120	124
1	43	51	63	84	112	139	168	177	182	184	181	175
1	41	49	56	62	72	88	119	135	162	185	195	205
1	41	48	53	60	65	67	71	70	71	81	91	96
1	41	49	62	79	101	128	164	192	227	248	259	266
1	41	49	56	64	68	68	67	68	−9	−9	−9	−9
1	41	45	49	51	57	51	54	−9	−9	−9	−9	−9
1	42	51	61	72	83	89	98	103	113	123	133	142
1	39	35	−9	−9	−9	−9	−9	−9	−9	−9	−9	−9
1	43	48	55	62	65	71	82	88	106	120	144	157
1	41	47	54	58	65	73	77	89	98	107	115	117
2	40	50	62	86	125	163	217	240	275	307	318	331
2	41	55	64	77	90	95	108	111	131	148	164	167
2	43	52	61	73	90	103	127	135	145	163	170	175
2	42	52	58	74	66	68	70	71	72	72	76	74
2	40	49	62	78	102	124	146	164	197	231	259	265
2	42	48	57	74	93	114	136	147	169	205	236	251
2	39	46	58	73	87	100	115	123	144	163	185	192
2	39	46	58	73	92	114	145	156	184	207	212	233
2	39	48	59	74	87	106	134	150	187	230	279	309
2	42	48	59	72	85	98	115	122	143	151	157	150
3	42	53	62	73	85	102	123	138	170	204	235	256
3	41	49	65	82	107	129	159	179	221	263	291	305
3	39	50	63	77	96	111	137	144	151	146	156	147
3	41	49	63	85	107	134	164	186	235	294	327	341
3	41	53	64	87	123	158	201	238	287	332	361	373
3	39	48	61	76	98	116	145	166	198	227	225	220
3	41	48	56	68	80	83	103	112	135	157	169	178
3	41	49	61	74	98	109	128	154	192	232	280	290
3	42	50	61	78	89	109	130	146	170	214	250	272
3	41	55	66	79	101	120	154	182	215	262	295	321
4	42	51	66	85	103	124	155	153	175	184	199	204
4	42	49	63	84	103	126	160	174	204	234	269	281
4	42	55	69	96	131	157	184	188	197	198	199	200
4	42	51	65	86	103	118	127	138	145	146	−9	−9
4	41	50	61	78	98	117	135	141	147	174	197	196
4	40	52	62	82	101	120	144	156	173	210	231	238
4	41	53	66	79	100	123	148	157	168	185	210	205
4	39	50	62	80	104	125	154	170	222	261	303	322
4	40	53	64	85	108	128	152	166	184	203	233	237
4	41	54	67	84	105	122	155	175	205	234	264	264

Table B.15. *Body weights, in grams, of chicks. There are four groups, on different protein diets, indicated by column 1 in the data matrix. Measurements are taken on days 0, 2, 4, 6, 8, 10, 12, 14, 16, 18, 20 and 21. Missing values are coded as −9 From Crowder and Hand (1990, Example 5.3)*

Glucose (sucrose-saccharin-caloreen)																		
1	68	100	97	83	76	68	62	56	60	56	68	53	72	106	94	68	52	47
2	73	94	88	98	90	65	65	66	67	68	67	65	71	140	138	−9	88	83
3	75	123	123	123	88	78	76	76	94	−9	76	71	71	135	158	153	121	93
4	59	98	104	70	−9	55	60	58	40	58	52	58	55	68	124	88	52	44
5	56	80	122	120	58	44	−9	−9	−9	−9	−9	−9	55	144	190	130	96	62
6	63	109	101	78	58	60	57	58	55	47	55	58	59	106	147	98	73	44
Insulin																		
1	7	48	13	6	6	3	3	3	1	1	1	1	3	68	36	17	4	3
2	10	53	25	35	11	12	6	9	7	8	6	7	5	77	40	30	36	22
3	4	28	30	23	16	11	4	5	3	5	5	4	5	34	47	55	34	17
4	3	24	20	17	4	4	4	3	3	4	4	4	2	30	41	25	6	4
5	3	26	22	27	6	4	−9	−9	−9	−9	−9	−9	3	36	50	35	25	9
6	6	44	45	22	8	4	3	3	3	3	4	4	6	30	49	37	23	7
Total tryptophan × 10																		
1	129	117	123	110	137	95	74	60	68	57	60	74	132	173	106	78	118	100
2	60	70	74	68	82	73	106	151	131	95	129	95	63	32	24	14	45	73
3	97	94	73	82	97	87	24	50	72	48	58	54	118	163	95	88	100	135
4	82	93	100	98	100	95	106	94	86	94	86	108	113	148	70	120	122	100
5	93	90	114	93	88	108	−9	−9	−9	−9	−9	−9	98	100	100	168	150	100
6	129	126	126	126	109	107	148	84	−9	−9	206	173	119	108	104	110	106	114

Table B.16. *Biochemical multivariate repeated measures. There are six subjects, with three sets of three variables (sucrose, saccharin and caloreen for each of total tryptophan, glucose and insulin determinations). Six repeated measures were taken on each at times 0, 30, 60, 90, 120 and 150 minutes, yielding $3 \times 3 \times 6$ recorded numbers per subject. Missing values are coded as -9*

	Condition 1			Condition 2		
1	2	3	1	2	4	4
1	1	2	3	2	2	4
1	1	1	1	2	2	4
1	4	1	3	1	4	3
1	0	1	1	2	2	3
1	3	4	3	3	4	3
1	4	3	2	4	4	4
1	1	1	3	2	3	4
1	1	1	1	2	2	3
1	1	1	0	3	4	4
1	4	4	3	3	4	3
1	2	2	1	3	3	4
1	2	3	1	3	4	3
1	1	2	2	2	2	2
1	4	4	2	3	4	4
2	3	2	0	3	4	3
2	4	4	1	3	4	3
2	3	3	3	2	4	4
2	4	4	3	4	3	4
2	4	2	2	4	3	3
2	3	3	0	4	4	4
2	0	1	1	0	0	0
2	4	4	3	0	3	4
2	3	2	2	4	3	3
2	1	0	0	2	1	1
2	3	1	0	1	0	0
2	1	2	1	3	2	0
2	2	0	1	1	1	3
2	2	1	0	2	2	2
2	3	1	1	2	2	2

Table B.17. *Neuropsychological measurements. Two groups of subjects, each measured at three under each of two conditions. Each measurement is a count out of 4. Interest lies in differences between the groups and between the conditions (and, presumably, interaction)*

0.37	0.44	0.52	0.25	0.24	0.64	0.51	0.65	0.70	0.65	0.40	0.65	0.37
0.87	1.31	0.61	0.87	0.32	0.82	1.20	0.91	1.08	0.85	0.85	0.80	0.93
1.12	0.28	0.46	0.55	1.04	0.99	0.41	0.72	0.97	1.10	0.98	0.57	0.55
1.05	3.64	2.82	1.97	2.16	1.57	0.75	0.21	0.30	0.62	0.97	1.43	2.55
1.11	0.77	1.31	1.09	0.61	1.05	0.35	0.61	1.41	1.50	1.10	0.78	0.48
0.98	1.24	0.79	1.81	1.09	1.47	1.42	0.93	1.51	1.42	1.06	1.46	0.63
0.62	0.36	0.34	0.25	0.35	0.62	0.46	0.45	0.65	0.55	0.57	0.26	0.28
1.31	0.35	1.24	0.91	1.06	1.61	1.36	0.76	2.26	0.63	1.24	0.62	0.77
1.05	1.43	0.72	1.33	1.48	1.31	0.82	1.61	1.65	0.91	0.74	1.16	0.59
0.92	0.52	0.61	0.70	0.51	0.44	0.34	0.64	0.74	0.51	0.51	0.28	0.47
0.34	0.52	0.70	0.48	0.40	0.58	0.56	0.73	0.74	0.92	0.48	0.29	0.53
0.74	2.76	3.28	0.74	0.49	0.52	0.56	0.13	0.64	0.41	0.34	1.05	0.89
0.78	0.97	0.89	0.90	1.00	1.23	1.61	4.21	2.16	1.60	0.59	1.09	0.59
0.60	0.29	0.49	0.45	0.22	0.25	0.64	0.70	0.90	0.41	0.52	0.34	0.42
0.47	0.42	0.77	0.41	0.47	0.58	0.53	0.62	0.86	0.39	0.64	0.62	0.44
0.61	0.55	0.76	0.91	0.82	0.31	0.42	0.29	0.61	0.35	0.76	0.22	0.53

Table B.18 *Melatonin measurements. Sixteen subjects, 13 recordings each (one per month) of melatonin. The hypothesis is of sinusoidal variation over the year. Measurements were taken at the following times (months): 1, 2, 3, 4, 5, 6, 7, 8, 9, 10, 11, 12 and 13*

1	1	4	15	28	4	4	17	29	4
2	1	4	13	26	4	4	15	29	4
3	1	4	12	19	1	4	14	26	5
4	1	3	11	16	3	3	17	24	5
5	1	5	12	27	3	5	18	29	5
6	1	4	11	17	2	4	15	26	4
7	1	4	12	22	3	5	17	28	5
8	1	4	14	16	5	4	14	25	5
9	1	5	14	20	3	5	16	29	5
10	1	4	13	25	2	4	14	25	4
11	1	3	10	21	3	3	14	26	5
12	1	5	15	25	2	5	16	29	4
13	2	0	5	12	2	0	11	20	1
14	2	4	12	20	2	3	14	26	4
15	2	4	11	17	3	4	12	24	5
16	2	5	9	22	2	5	13	27	3
17	2	4	12	22	5	5	16	26	5
18	2	1	4	9	2	2	9	17	5
19	2	5	16	28	4	5	18	28	4
20	2	4	14	23	2	4	16	27	3
21	2	5	16	26	3	5	16	29	4

Table B.19. *An ante-natal study with a control group of 12 subjects and a treatment group of nine who attended a course. The questionnaire yielded four responses, completed both before and after the experiment. The four responses measure knowledge on scales with levels respectively 0–5, 0–20, 0–30 and 0–5. Rows of the data matrix comprise case number, group, four responses before, four responses after*

1	1	5	5	12	13	14	14
2	1	7	9	9	13	13	19
3	1	1	5	5	9	15	16
4	1	0	0	1	3	9	11
5	1	6	6	8	12	12	18
6	2	0	0	0	2	5	5
7	2	0	0	0	0	0	7
8	2	3	5	5	6	8	12
9	2	6	9	9	9	14	17
10	2	0	0	3	10	10	10
11	3	1	1	1	4	5	5
12	3	7	9	9	9	9	11
13	3	11	11	11	11	11	15
14	3	5	9	14	14	14	14
15	3	5	5	5	5	5	5
16	4	3	3	3	4	4	7
17	4	0	3	3	3	3	7
18	4	6	12	12	16	16	16
19	4	2	5	5	6	6	6
20	4	7	7	9	12	14	14
21	5	9	11	11	24	24	25
22	5	2	5	9	9	9	13
23	5	3	6	11	12	16	17
24	5	9	13	13	13	15	15
25	5	13	18	20	20	23	23

Table B.20. *Effect of a drug on constipation in rats. Five rats in each of five treatment groups, one group being saline control. The measures are cumulative pellet counts at times 1.5, 3, 4, 5, 6 and 7 hours after administration. Do the drugs have an effect and, if so, when?*

1	1	1	2	14	50	3.1	2.6	9.0	9.0	9.0	1 1 3 3 3	1 1 2 2 2	1 1 2 2 2	2 1 3 3 3
1	2	1	1	8	50	2.3	.3	2.0	2.0	1.9	1 1 1 1 0	1 1 1 1 0	0 0 0 0 0	1 0 1 1 0
1	3	0	2	10	50	2.9	2.6	2.4	2.4	2.0	1 1 1 0 0	1 0 0 1 0	0 0 0 0 0	2 0 1 1 0
2	1	1	2	5	35	2.2	3.0	.4	2.2	1.2	1 0 0 0 0	1 0 0 0 0	1 0 0 0 0	2 0 0 0 0
2	2	1	2	5	35	1.8	1.8	1.8	1.8	1.8	1 0 0 0 0	1 1 0 0 0	1 0 0 0 0	2 1 1 1 1
2	3	1	2	10	–9	2.4	.8	.6	2.0	1.7	1 0 0 0 0	1 1 0 0 0	1 0 0 0 0	2 1 1 1 0
2	4	0	2	4	25	.2	1.8	9.0	9.0	9.0	1 2 3 3 3	1 1 2 2 2	1 1 3 3 3	2 2 3 3 3
3	1	1	2	5	50	2.4	2.1	2.2	2.4	1.0	0 0 0 0 0	1 0 0 0 0	0 0 0 0 0	2 1 0 0 0
3	2	0	2	5	40	2.9	2.9	2.6	2.4	2.0	1 1 0 0 0	1 1 0 0 0	0 0 0 0 0	2 2 2 1 0
4	1	1	2	2	50	.4	–.3	.7	2.3	2.1	1 0 0 0 0	1 0 0 0 0	0 0 0 0 0	1 1 0 0 0
4	2	1	1	7	36	2.0	2.8	2.5	1.6	2.0	1 1 0 0 0	1 1 0 0 0	0 0 0 0 0	1 1 1 0 0
4	3	0	2	7	40	2.2	1.8	.8	–9	–9	0 0 0 0 0	0 0 0 0 0	0 0 0 0 0	1 0 0 0 0
5	1	1	2	12	50	3.4	2.2	1.8	1.8	1.8	1 0 0 0 0	1 0 0 0 0	0 0 0 0 0	1 0 0 0 0
5	2	1	2	10	50	2.6	2.2	2.6	2.0	1.8	2 1 0 0 0	1 0 0 0 0	0 0 0 0 0	2 2 1 0 0
5	3	1	1	5	40	4.0	4.0	3.0	3.4	1.8	1 0 0 0 0	0 0 0 0 0	0 0 0 0 0	2 1 0 0 0
5	4	0	2	7	50	2.6	.6	3.2	3.0	3.4	1 0 0 0 1	1 0 0 0 1	0 0 0 0 0	1 0 0 2 1

6	1	1	1	11	50	3.5	2.4	2.4	3.0	1.0	1	1	0	0	1	1	1	0	0	0	0	0	0	0	0	2	1	1	1	0
6	2	1	2	14	50	1.1	3.0	1.8	2.4	2.8	1	1	0	0	0	1	1	0	0	0	0	0	0	0	0	1	1	0	0	1
6	3	0	2	11	50	2.6	2.5	2.5	1.6	1.8	0	0	0	0	0	0	0	0	0	0	0	0	0	0	0	1	0	0	0	0
7	1	1	1	18	–9	3.0	–9	2.0	–9	2.0	1	9	0	9	0	0	9	0	9	0	2	9	0	9	0	1	9	0	9	0
7	2	1	2	21	–9	1.0	–9	2.0	–9	1.0	1	9	0	9	0	0	9	0	9	0	1	9	1	9	0	0	9	0	9	0
7	3	1	1	18	–9	4.0	–9	1.0	–9	1.5	1	9	0	9	0	1	9	0	9	0	2	9	1	9	0	2	9	0	9	0
7	4	1	2	21	–9	2.0	–9	1.0	–9	1.5	1	9	0	9	0	1	9	0	9	0	2	9	1	9	0	2	9	0	9	0
7	5	1	1	18	–9	3.5	–9	1.0	–9	1.0	1	9	0	9	0	1	9	0	9	0	2	9	0	9	0	2	9	0	9	0
7	6	1	2	21	–9	1.5	–9	2.0	–9	1.0	0	9	0	9	0	0	9	0	9	0	1	9	0	9	0	1	9	0	9	0
7	7	0	2	21	–9	4.0	–9	2.0	–9	1.0	1	9	0	9	0	1	9	0	9	0	1	9	0	9	0	0	9	0	9	0
7	8	0	1	21	–9	3.5	–9	1.5	–9	1.0	0	9	0	9	0	1	9	0	9	0	2	9	0	9	0	1	9	0	9	0
8	1	1	1	35	–9	4.0	2.2	2.2	.6	2.0	1	0	0	0	0	1	1	1	1	0	9	9	9	9	9	0	0	0	0	0
8	2	1	2	35	–9	4.0	5.1	3.0	2.9	2.0	1	0	0	0	0	1	1	1	1	0	9	9	9	9	9	0	1	2	2	2
8	3	1	2	35	–9	3.6	2.0	2.7	2.2	2.2	0	0	0	0	0	0	0	0	0	0	9	9	9	9	9	0	0	0	0	0
8	4	1	2	35	–9	2.6	2.6	3.3	2.0	2.4	0	0	0	0	0	0	0	0	0	0	9	9	9	9	9	0	0	0	0	0
8	5	1	1	35	–9	2.6	2.2	.6	1.8	3.5	0	0	0	0	0	0	0	0	0	0	9	9	9	9	9	0	0	0	0	0
8	6	1	1	35	–9	3.6	.9	2.9	2.2	1.8	0	0	0	0	0	0	0	0	0	0	9	9	9	9	9	0	0	0	0	0
8	7	1	1	35	–9	2.3	1.3	.4	1.1	1.1	0	0	0	0	0	0	0	0	0	0	9	9	9	9	9	0	0	0	0	0
8	8	1	2	35	–9	2.2	2.2	2.0	2.2	2.0	0	0	0	0	0	0	0	0	0	0	9	9	9	9	9	0	0	0	0	0
8	9	1	2	35	–9	2.3	1.5	2.7	1.7	2.9	0	0	0	0	0	0	0	0	0	0	9	9	9	9	9	0	0	0	0	0
8	10	1	2	35	–9	2.7	2.4	1.7	2.2	2.0	0	0	0	0	0	0	0	0	0	0	9	9	9	9	9	0	0	0	0	0
8	11	0	1	35	–9	2.7	1.3	1.5	2.2	2.0	0	0	0	0	0	0	0	0	0	0	9	9	9	9	9	0	0	0	0	0
8	12	0	2	35	–9	2.7	1.5	2.0	2.0	2.0	0	0	0	0	0	0	0	0	0	0	9	9	9	9	9	0	0	0	0	0
8	13	0	1	35	–9	1.8	2.0	1.3	2.0	2.0	0	0	0	0	0	0	0	0	0	0	9	9	9	9	9	0	0	0	0	0
8	14	0	2	35	–9	1.8	2.2	1.7	2.2	2.2	0	0	0	0	0	0	0	0	0	0	9	9	9	9	9	0	0	0	0	0
8	15	0	2	35	–9	2.2	2.2	1.7	2.0	2.2	0	0	0	0	0	0	0	0	0	0	9	9	9	9	9	0	0	0	0	0
8	16	0	2	35	–9	2.9	2.7	2.2	2.2	2.7	0	0	0	0	0	0	0	0	0	0	9	9	9	9	9	0	0	0	0	0
8	17	0	1	35	–9	6.0	5.8	4.7	3.3	2.0	1	1	1	0	0	1	1	1	1	0	9	9	9	9	9	0	0	1	2	1
8	18	0	1	35	–9	2.9	5.6	6.3	6.5	6.0	0	0	0	1	1	0	0	0	1	0	9	9	9	9	9	0	0	0	1	2
8	19	0	2	35	–9	2.7	3.8	4.2	3.6	2.2	0	0	0	0	0	0	0	0	0	0	9	9	9	9	9	0	0	1	0	0
9	1	1	1	10	–9	4.8	2.6	1.8	1.6	2.0	1	1	1	1	0	1	1	1	0	0	1	1	1	1	1	1	1	1	0	0
9	2	1	1	25	–9	–3.0	3.1	–.4	9.0	9.0	2	1	1	3	3	1	1	1	2	2	2	2	2	3	3	2	1	1	3	3
9	3	1	1	10	–9	5.0	1.6	1.4	1.4	1.4	1	1	1	0	0	1	1	0	0	0	9	9	9	9	9	1	1	0	0	0
9	4	1	1	10	–9	4.8	1.6	2.0	1.4	1.4	1	1	1	0	0	1	1	0	0	0	1	1	0	0	0	1	1	0	0	0
9	5	1	1	10	–9	4.6	2.4	2.2	1.8	1.4	1	1	1	0	0	1	1	0	0	0	9	9	9	9	9	2	1	1	0	0
9	6	0	1	10	–9	2.8	1.6	1.6	1.6	1.6	1	1	1	1	1	1	1	1	0	0	9	9	9	9	9	2	2	1	1	0

Table B.21. *Measures on 52 calves. The first block contains case variables: farm, calf, treatment (0 = control, 1 = test drug), sex, age (days), weight (kg). The other five blocks contain repeated measures on five consecutive days for rectal temperature (excess over 102°F, 9 = dead); morbidity (0 = normal, 1 = depressed, 2 = moribund/comatose, 3 = dead); appetite (0 = normal, 1 = inappetant, 2 = dead); dehydration (0 = nil, 1 = slight, 2 = severe, 3 = dead); and scour (0 = none, 1 = slight, 2 = severe, 3 = dead). Missing values are coded* – 9

times 0 0 0 0 0 0 15 30 45 60 0 0 0 15 30 45 60
0 0 0 15 30 45 60 0 0 0 15 30 45 60

1. 1. 2.	2. 26.	61.0	87.5	118.5	118.5	118.5
	3. 12.	71.0	102.5	103.5	108.0	112.5
	5. 3.	83.5	−9.0	103.5	118.5	127.5
	7. 26.	62.5	90.0	107.5	115.0	124.5
1. 1. 2.	2. 12.	70.0	93.0	107.5	111.0	110.0
	3. 3.	86.0	106.0	101.0	107.0	97.5
	5. 12.	86.0	108.0	127.5	126.5	130.0
	7. 12.	72.5	125.0	112.0	116.0	106.0
1. 1. 2.	2. 3.	84.0	112.5	106.0	106.0	100.0
	3. 26.	52.5	88.0	97.5	105.0	108.0
	5. 12.	91.0	119.0	118.5	110.0	112.5
	7. 3.	77.5	115.0	107.0	120.0	119.0
2. 1. 2.	2. 26.	67.5	110.0	192.5	118.5	115.0
	3. 12.	71.0	101.0	110.0	114.0	122.5
	5. 3.	74.0	117.5	127.5	132.5	121.5
	7. 26.	50.0	106.0	108.5	131.0	117.5
2. 1. 2.	2. 12.	72.5	135.0	138.5	128.5	127.5
	3. 3.	83.0	102.5	133.0	126.0	122.0
	5. 12.	70.0	121.5	115.5	134.5	124.0
	7. 12.	54.0	109.5	106.5	118.0	117.5
2. 1. 2.	2. 3.	88.5	127.5	137.5	167.5	128.5
	3. 26.	51.0	96.5	122.0	114.0	125.0
	5. 12.	91.0	141.0	137.5	145.0	130.0
	7. 3.	84.0	125.5	132.0	135.0	127.5
1. 1. 1.	2. 26.	51.0	90.0	117.5	126.0	113.5
	3. 12.	62.5	82.5	76.0	97.5	85.0
	5. 3.	77.5	94.0	112.5	114.0	101.0
	7. 26.	65.0	100.0	111.0	174.5	108.5
1. 1. 1.	2. 12.	78.5	133.5	120.5	127.5	112.5
	3. 3.	75.0	94.0	96.0	97.0	107.5
	5. 26.	58.5	91.0	132.5	133.5	112.5
	7. 12.	82.5	120.0	108.5	124.0	110.0
1. 1. 1.	2. 3.	72.5	92.0	105.0	112.0	110.0
	3. 26.	41.0	100.0	127.5	122.5	101.0
	5. 12.	76.0	116.0	117.5	115.0	102.5
	7. 3.	86.0	120.0	124.0	112.0	117.5
2. 1. 1.	2. 26.	71.0	127.5	161.0	135.0	132.5
	3. 12.	67.5	105.0	101.0	111.0	126.0
	5. 3.	76.0	100.0	110.0	102.5	97.5
	7. 26.	62.5	135.5	159.0	120.0	118.0
2. 1. 1.	2. 12.	67.5	110.0	136.0	127.5	127.5
	3. 3.	86.0	102.0	122.0	117.5	117.0
	5. 26.	65.0	120.0	113.5	119.0	111.0
	7. 12.	86.0	159.0	117.5	102.0	118.5
2. 1. 1.	2. 3.	86.0	148.0	135.0	125.0	118.5
	3. 26.	88.0	162.5	166.0	161.0	137.5

	5. 12.	73.5	125.0	125.5	127.5	115.0
	7. 3.	85.0	128.0	119.5	122.5	132.5
1. 2. 2.	2. 26.	85.0	118.0	141.0	152.0	141.0
	3. 12.	75.0	110.0	119.0	135.0	145.5
	5. 3.	85.0	113.5	122.0	125.0	113.5
	7. 26.	90.0	160.0	185.0	153.5	111.0
2. 2. 2.	2. 12.	76.0	100.0	105.0	116.0	116.0
	3. 3.	81.0	106.0	112.5	120.0	137.5
	5. 26.	66.0	105.0	113.5	138.5	117.5
	7. 12.	120.0	152.5	150.0	180.0	126.0
1. 2. 2.	2. 3.	76.0	87.5	104.0	107.5	104.0
	3. 26.	65.0	87.5	116.5	135.0	137.5
	5. 12.	66.0	87.5	110.0	146.5	147.0
	7. 3.	74.0	100.0	177.5	172.0	176.0
2. 2. 2.	2. 26.	70.0	105.0	107.5	111.0	116.0
	3. 12.	95.0	132.0	116.5	116.0	130.0
	5. 3.	101.0	126.0	137.5	142.5	138.5
	7. 26.	95.0	130.0	139.0	160.0	139.0
2. 2. 2.	2. 12.	82.0	95.0	94.0	100.5	100.0
	3. 3.	80.0	106.0	105.0	120.0	137.5
	5. 26.	92.5	116.0	127.5	132.5	135.0
	7. 12.	97.5	117.5	136.0	132.0	133.5
2. 2. 2.	2. 3.	76.5	87.5	95.0	85.0	115.0
	3. 26.	77.5	106.0	138.5	115.0	118.5
	5. 12.	80.0	106.0	113.5	110.0	121.5
	7. 3.	83.5	102.5	113.0	116.0	117.0
1. 2. 1.	2. 26.	63.5	112.5	106.0	106.0	95.5
	3. 12.	60.0	102.5	100.0	76.5	85.0
	5. 3.	63.5	101.0	94.0	60.5	70.0
	7. 26.	118.0	160.0	172.0	167.5	170.0
1. 2. 1.	2. 12.	72.5	95.0	72.5	56.0	67.5
	3. 3.	52.5	76.0	107.5	75.0	62.5
	5. 26.	62.5	93.5	125.0	86.0	88.0
	7. 12.	103.5	172.0	143.5	116.0	113.0
1. 2. 1.	2. 3.	81.5	127.5	150.0	161.0	141.5
	3. 26.	73.0	138.5	169.0	157.5	138.5
	5. 12.	98.0	132.5	164.0	138.5	122.5
	7. 3.	122.5	172.5	169.0	161.0	161.0
2. 2. 1.	2. 26.	63.5	127.5	152.5	127.5	141.5
	3. 12.	74.0	119.0	102.0	94.0	95.0
	5. 3.	72.0	87.5	−9.0	97.5	103.5
	7. 26.	82.5	147.5	149.0	230.0	138.0
2. 2. 1.	2. 12.	63.5	105.0	94.0	72.5	68.0
	3. 3.	75.0	92.5	98.5	94.0	83.0
	5. 26.	66.0	109.0	106.0	137.5	132.5
	7. 12.	74.0	114.0	124.5	97.5	86.0

Table B.22. (Continued)

2. 2. 1.	2.	3.	95.0	125.0	136.0	105.0	105.0
	3.	26.	77.5	124.0	138.5	155.0	120.0
	5.	12.	105.0	122.5	117.5	107.5	131.5
	7.	3.	137.5	139.0	127.5	93.5	102.0

Table B.22. *Rat diets. Twenty-four rats in a 2^3 factorial design, three rats per cell. Factors: diet regime (control, test), fatness (lean, obese), sex (male, female). The rats were subjected to a treatment, and then blood levels of a substance were recorded at five times (0, 15, 30, 45 and 60 mins). Each rat underwent the procedure at weeks 2, 3, 5 and 7, and on each occasion it was preceded by a period of fasting (3, 12 or 26 hours). Apart from the fasting periods, the design has the appearance of perfect balance. However, there are a couple of missing observations (coded as −9) and, more seriously, rat 14 turned out to have been wrongly sexed in a post-experimental check. The data matrix contains sex, fatness, diet (first line), followed by four lines each containing week number, fasting period, and five repeated measures*

1	1	2	210	215	230	244	259	266	277	292	292	290	264
1	2	2	230	240	258	277	277	293	300	323	327	340	343
1	3	2	226	233	248	277	297	313	322	340	354	365	362
1	4	2	233	239	253	277	292	310	318	333	336	353	338
1	5	2	238	241	262	282	300	314	319	331	338	348	338
1	6	2	225	228	237	261	271	288	300	316	319	333	330
1	7	2	224	225	239	257	268	290	304	313	310	318	318
1	8	2	237	241	255	276	293	307	312	336	336	344	328
1	9	2	237	224	234	239	256	266	276	300	302	293	269
1	10	2	233	239	259	283	294	313	320	347	348	362	352
1	11	2	217	222	235	256	267	285	295	317	315	308	301
1	12	2	228	223	246	266	277	287	300	312	308	328	333
1	13	2	241	247	268	290	309	323	336	348	359	372	370
1	14	2	221	221	240	253	273	282	292	307	306	317	318
1	15	2	217	220	235	259	262	276	284	305	303	315	317
1	16	2	214	221	237	256	271	283	287	314	316	320	298
1	17	2	224	231	241	256	265	283	295	314	313	328	334
1	18	2	200	203	221	236	248	262	276	294	291	311	310
1	19	2	238	232	252	268	285	298	303	320	324	320	327
1	20	2	230	222	243	253	268	284	290	316	314	330	330
1	21	2	217	224	242	265	284	302	309	324	328	338	334
1	22	2	209	209	221	238	256	267	281	295	301	309	289
1	23	2	224	227	245	267	279	294	312	328	329	297	297
1	24	2	230	231	244	261	272	283	294	318	320	333	338
1	25	2	216	218	223	243	259	270	270	290	301	314	297

1	26	2 231	239	254	276	294	304	317	335	333	319	307
1	27	2 207	216	228	255	275	285	296	314	319	330	330
1	28	2 227	236	251	264	276	287	297	315	309	313	294
1	29	2 221	232	251	274	284	295	300	323	319	333	322
1	30	2 233	238	254	266	282	294	295	310	320	327	326
2	31	1 233	224	245	258	271	287	287	287	290	293	297
2	32	1 231	238	260	273	290	300	311	313	317	321	326
2	33	1 232	237	245	265	285	298	304	319	317	334	329
2	34	1 239	246	268	288	308	309	327	324	327	336	341
2	35	1 215	216	239	264	282	299	307	321	328	332	337
2	36	1 236	226	242	255	263	277	290	299	300	308	310
2	37	1 219	229	246	265	279	292	299	299	298	300	290
2	38	1 231	245	270	292	302	321	322	334	323	337	337
2	39	1 230	228	243	255	272	276	277	289	289	300	303
2	40	1 232	240	247	263	275	286	294	302	308	319	326
2	41	1 234	237	259	289	311	324	342	347	355	368	368
2	42	1 237	235	258	263	282	304	318	327	336	349	353
2	43	1 229	234	254	276	294	315	323	341	346	352	357
2	44	1 220	227	248	273	290	308	322	326	330	342	343
2	45	1 232	241	255	276	293	309	310	330	326	329	330
2	46	1 210	225	242	260	272	277	273	295	292	305	306
2	47	1 229	241	252	265	274	285	303	308	315	328	328
2	48	1 204	198	217	233	251	258	272	283	279	295	298
2	49	1 220	221	236	260	274	295	300	301	310	318	316
2	50	1 233	234	250	268	280	298	308	319	318	336	333
2	51	1 234	234	254	274	294	306	318	334	343	349	350
2	52	1 200	207	217	238	252	267	284	282	282	284	288
2	53	1 220	213	229	252	254	273	293	289	294	292	298
2	54	1 225	239	254	269	289	308	313	324	327	347	344
2	55	1 236	245	257	271	294	307	317	327	328	328	325
2	56	1 231	231	237	261	274	285	291	301	307	315	320
2	57	1 208	211	238	254	267	287	306	312	320	337	338
2	58	1 232	248	261	285	292	307	312	323	318	328	329
2	59	1 233	241	252	273	301	316	332	336	339	348	345
2	60	1 221	219	231	251	270	272	287	294	292	292	299

Table B.23. *Weights of calves in a trial on the control of intestinal parasites. The variables are: group (two groups), calf number (30 in each group), and then 11 measurements of weight on the following days from the start of the experiment: 0, 14, 28, 42, 56, 70, 84, 98, 112, 126, 133. From Kenward (1987)*

1	2	−1.00	1.55	0.84	0.99	1.01	1.07	0.57
2	1	2.11	3.22	2.11	1.95	1.76	−1.00	−1.00
3	1	−1.00	1.34	1.37	1.34	1.95	2.14	−1.00
4	2	−1.00	−1.00	−1.00	1.22	2.11	3.08	−1.00
5	1	−1.00	1.67	1.02	1.71	1.02	1.32	−1.00
6	2	−1.00	−1.00	−1.00	1.32	1.32	−1.00	1.32
7	1	−1.00	1.31	1.99	1.73	1.40	1.89	2.66
8	2	−1.00	1.00	1.36	1.42	1.10	−1.00	−1.00
9	1	−1.00	2.17	1.67	4.18	2.23	−1.00	−1.00
10	1	1.01	2.36	1.51	−1.00	4.82	−1.00	−1.00
11	2	−1.00	−1.00	2.73	1.32	−1.00	−1.00	−1.00
12	2	2.45	−1.00	1.43	1.40	1.41	−1.00	−1.00
13	2	0.93	−1.00	−1.00	3.74	2.12	1.62	1.83
14	1	2.31	2.25	−1.00	1.16	2.25	1.64	1.71
15	1	3.34	1.66	2.93	1.98	2.46	−1.00	1.63
16	2	−1.00	2.26	1.90	2.20	2.29	1.12	−1.00
17	1	−1.00	−1.00	1.21	1.95	1.62	1.25	−1.00
18	1	−1.00	−1.00	−1.00	1.16	−1.00	−1.00	−1.00
19	1	−1.00	1.98	−1.00	−1.00	1.71	−1.00	−1.00
20	2	−1.00	−1.00	1.42	1.04	−1.00	−1.00	−1.00
21	2	0.77	−1.00	1.67	1.21	1.08	1.38	−1.00
22	2	1.20	1.10	1.13	3.49	1.57	1.54	−1.00
23	1	2.70	3.50	−1.00	−1.00	−1.00	−1.00	−1.00
24	2	−1.00	−1.00	1.40	−1.00	−1.00	−1.00	−1.00
25	2	2.71	2.04	2.61	2.17	2.15	1.81	1.27
26	2	2.01	−1.00	2.03	−1.00	1.52	−1.00	−1.00
27	1	0.60	0.67	2.84	2.10	2.00	1.60	1.27
28	2	1.02	1.43	1.61	1.70	2.82	1.55	−1.00
29	2	1.71	1.71	1.21	0.90	0.61	1.66	−1.00
30	1	1.17	1.63	1.83	2.43	1.84	−1.00	−1.00
31	2	2.77	1.13	0.94	1.39	1.19	−1.00	−1.00
32	1	1.26	0.73	−1.00	−1.00	−1.00	−1.00	−1.00
33	2	1.22	1.55	1.86	−1.00	−1.00	−1.00	−1.00
34	1	1.05	1.06	1.47	1.91	2.74	−1.00	−1.00
35	1	1.51	1.38	1.44	−1.00	−1.00	−1.00	−1.00
36	2	−1.00	1.13	1.03	1.60	1.98	−1.00	−1.00
37	2	1.16	0.78	0.51	0.85	0.88	0.49	0.63
38	1	1.51	1.26	1.47	1.41	−1.00	−1.00	−1.00
39	1	−1.00	1.20	2.40	−1.00	−1.00	−1.00	−1.00
40	1	1.10	1.36	2.31	2.06	4.24	−1.00	−1.00
41	2	−1.00	2.09	1.43	1.43	1.17	1.31	−1.00
42	2	−1.00	1.00	1.44	−1.00	−1.00	−1.00	−1.00
43	2	−1.00	3.14	1.75	1.31	−1.00	−1.00	−1.00
44	1	−1.00	1.71	2.07	2.25	3.15	2.23	3.20
45	1	3.07	2.90	2.96	3.12	−1.00	3.27	−1.00
46	2	0.85	1.25	1.66	2.13	1.04	0.62	0.83
47	2	1.33	2.32	−1.00	1.72	2.04	−1.00	−1.00
48	1	−1.00	1.39	2.09	2.25	2.22	2.81	−1.00

49	1	1.42	3.40	4.10	2.92	2.65	3.40	2.25
50	2	−1.00	0.81	1.34	1.53	1.34	0.92	0.78
51	1	−1.00	1.21	1.20	−1.00	−1.00	−1.00	−1.00
52	2	0.72	0.74	−1.00	−1.00	−1.00	−1.00	−1.00
53	1	−1.00	1.11	3.36	2.25	−1.00	−1.00	−1.00
54	2	0.60	−1.00	1.30	−1.00	−1.00	−1.00	−1.00
55	1	−1.00	2.19	1.46	3.09	−1.00	−1.00	−1.00
56	2	1.43	2.28	2.40	−1.00	−1.00	−1.00	−1.00
57	2	−1.00	1.81	2.10	2.10	2.03	1.95	−1.00
58	1	−1.00	−1.00	1.31	−1.00	−1.00	−1.00	−1.00
59	1	1.08	2.16	4.00	−1.00	−1.00	−1.00	−1.00
60	1	0.61	−1.00	0.59	0.99	1.02	0.83	0.55
61	2	−1.00	−1.00	0.74	0.90	1.00	−1.00	−1.00
62	2	−1.00	1.89	2.74	2.61	−1.00	−1.00	−1.00
63	1	−1.00	−1.00	1.82	5.30	3.09	1.42	2.31
64	2	−1.00	4.49	3.26	2.06	1.53	−1.00	−1.00
65	1	1.10	1.00	−1.00	−1.00	−1.00	−1.00	−1.00
66	2	0.60	−1.00	−1.00	−1.00	−1.00	−1.00	−1.00
67	1	1.00	1.10	5.40	2.10	−1.00	−1.00	−1.00
68	2	0.60	2.50	2.20	1.20	1.10	1.00	1.00
69	2	2.80	2.80	2.10	2.60	2.20	−1.00	−1.00
70	1	0.90	2.30	2.70	1.70	1.10	1.30	−1.00
71	2	0.90	0.80	0.70	1.00	0.80	0.60	0.70
72	1	2.60	3.60	2.60	2.90	−1.00	3.70	−1.00
73	1	−1.00	1.40	4.40	3.00	2.30	2.30	−1.00
74	2	3.40	3.30	3.40	3.40	2.10	1.50	3.10
75	2	1.10	1.20	1.50	2.40	1.50	3.20	−1.00
76	1	1.10	−1.00	3.20	3.40	3.40	2.70	2.20
77	2	4.60	1.20	3.20	2.30	2.30	1.50	−1.00
78	1	−1.00	1.90	3.90	2.20	1.40	1.60	3.40
79	2	0.60	1.00	−1.00	1.50	1.00	−1.00	1.50
80	2	1.20	1.30	−1.00	1.40	1.50	−1.00	−1.00
81	1	2.00	1.80	3.20	2.50	−1.00	−1.00	−1.00
82	1	−1.00	2.00	−1.00	−1.00	−1.00	−1.00	−1.00
83	2	1.60	0.70	1.80	2.10	1.30	1.10	1.30
84	1	1.10	1.40	1.00	2.60	0.90	2.10	1.50
85	2	−1.00	1.20	1.60	1.90	1.20	1.60	1.60
86	1	−1.00	1.40	2.90	2.90	2.40	4.10	−1.00
87	2	0.80	1.70	2.30	1.80	1.80	−1.00	−1.00
88	1	2.30	2.20	3.80	3.50	2.50	1.80	3.10
89	2	−1.00	1.58	1.56	1.35	1.96	1.33	−1.00
90	1	−1.00	2.95	0.88	2.26	−1.00	0.65	0.48
91	2	0.40	0.96	1.01	0.71	0.59	0.60	0.48
92	2	−1.00	2.05	2.03	−1.00	1.08	1.09	−1.00
93	2	1.80	1.40	1.00	1.30	2.40	2.40	−1.00
94	1	−1.00	1.60	1.40	3.40	2.60	−1.00	−1.00

Table B.24. (Continued)

95	1	2.80	2.70	−1.00	−1.00	−1.00	−1.00	−1.00
96	1	0.81	1.20	1.12	1.61	1.49	1.61	−1.00
97	1	−1.00	−1.00	3.17	2.16	3.04	2.18	−1.00
98	1	−1.00	−1.00	5.99	5.22	3.56	−1.00	−1.00

Table B.24. *These data arose from a two-group clinical trial of improvement after an operation. There are two groups (of 98 subjects in all), and seven measurements of the response, one pre-treatment and then at 10 days, 2 months, 4 months, 6 months, 9 months and 12 months after the operation. Missing values are coded as −1. From Crowder and Hand (1990, Table 9.1)*

References

Agresti, A. (1989) A survey of models for repeated ordered categorical response data. *Statist. in Medicine* **8**, 1209–24.

Agresti, A. and Lang, R. (1993) A proportional odds model with subject-specific effects for repeated ordered categorical responses. *Biometrika* **80**, 527–34.

Aitkin, M., Anderson, D., Francis, B. and Hinde, J. (1989) *Statistical Modelling with GLIM*. Oxford University Press, Oxford.

Andersen, A. H., Jensen, E. B. and Schou, G. (1981) Two-way analysis of variance with correlated errors. *Internat. Statist. Rev.* **49**, 153–67.

Anderson, A. J. B. (1991) Repeated measures: groups × occasions designs. In *New Developments in Statistics for Psychology and the Social Sciences*. The British Psychological Society and Routledge, London.

Anderson, D. A. and Aitkin, M. (1985) Variance component models with binary response: interviewer variability. *J. Roy. Statist. Soc. B* **47**, 203–10.

Anderson, T. W. (1969) Statistical inference for covariance matrices with linear structure. In *Multivariate Analysis II*, ed. P. Krishnaiah. Academic Press, New York.

Anderson, T. W. (1970) Estimation of covariance matrices which are linear combinations or whose inverses are linear combinations of given matrices. In *Essays in Probability and Statistics*, ed. R. C. Bose, I. M. Chakravarti, P. C. Mahalanobis, C. R. Rao and K. J. C. Smith pp. 1–24. University of North Carolina Press, Chapel Hill.

Anderson, T. W. (1973) Asymptotically efficient estimation of covariance matrices with linear structure *Ann. Statist.* **1**, 135–41.

Andrade, D. F. and Helms, R. W. (1984) Maximum likelihood estimates in the multivariate normal with patterned mean and covariance via the EM algorithm. *Comm. Statist. Theory Methods* **13**, 2239–51.

Aranda-Ordaz, F. J. (1981) On two families of transformations to additivity for binary response data. *Biometrika* **68**, 357–63.

Atkinson, A. C. (1985) *Plots, Transformations, and Regression*. Oxford University Press, Oxford.

Belsley, D. A., Kuh, E. and Welsch, R. E. (1980) *Regression Diagnostics*. Wiley, New York.

Berkey, C. S. (1982) Bayesian approach for a nonlinear growth model. *Biometrics* **38**, 953–61.
Berkey, C. S. and Laird, N. M. (1986) Nonlinear growth curve analysis: estimating the population parameters. *Ann. Hum. Biol.* **13**, 111–28.
Berkson, J. (1953) A statistically precise and relatively simple method of estimating the bioassay with quantal response, based on the logistic function. *J. Amer. Statist. Assoc.* **48**, 565–99.
Bickel, P. and Doksum, K. (1981) An analysis of transformations revisited. *J. Amer. Statist. Assoc.* **76**, 296–11.
Bishop, Y. M. M., Fienberg, S. E. and Holland, P. W. (1975) *Discrete Multivariate Analysis: Theory and Practice.* MIT Press, Cambridge, Mass.
Bock, R. D. (1975) *Multivariate statistical methods in behavioural research.* McGraw-Hill, New York.
Bonney, G. E. (1987) Logistic regression for dependent binary observations. *Biometrics* **43**, 951–73.
Box, G. E. P. (1949) A general distribution theory for a class of likelihood criteria. *Biometrika* **36**, 317–46.
Box, G. E. P. and Cox, D. R. (1964) An analysis of transformations (with discussion) *J. Roy, Statist. Soc. B.* **26**, 211–52.
Box, G. E. P. and Cox, D. R. (1982) An analysis of transformations revisited, rebutted. *J. Amer. Statist. Assoc.* **77**, 209–10.
Breslow, N. E. and Clayton, D. G. (1993) Approximate inference in generalized linear mixed models. *J. Amer. Statist. Assoc.* **88**, 9–25.
Byrne, P. J. and Arnold, S. F. (1983) Inference about multivariate means for a nonstationary autoregressive model. *J. Amer. Statist. Assoc.* **78**, 850–5.
Carey, V., Zeger, S. L. and Diggle, P. (1993) Modelling multivariate binary data with alternating logistic regressions. *Biometrika* **80**, 517–26.
Carroll, R. J. and Ruppert, D. (1984) Power transformations when fitting theoretical models to data *J. Amer. Statist. Assoc.* **79**, 321–8.
Collett, D. (1991) *Modelling Binary Data.* Chapman & Hall, London.
Conaway, M. R. (1989) Analysis of repeated categorical measurements with conditional likelihood methods. *J. Amer. Statist. Assoc.* **84**, 53–62.
Conaway, M. R. (1990) A random effects model for binary data. *Biometrics* **46**, 317–28.
Cook, R. D, and Weisberg, S. (1982) *Residuals and Influence in Regression.* Chapman & Hall, London.
Cox, D. R. (1970) *Analysis of Binary Data.* Methuen, London.
Cox, D. R. and Reid, N. (1987) Parameter orthogonality and approximate conditional inference (with discussion) *J. Roy. Statist. Soc. B* **49**, 1–39.
Crowder, M. J. (1978) On concurrent regression lines *Appl. Statist.* **27**, 310–18.
Crowder, M. J. (1983) A growth curve analysis for EDP curves. *Appl. Statist.* **30**, 147–52.
Crowder, M. J. (1985) Gaussian estimation for correlated binomial data. *J. Roy. Statist. Soc. B* **47**, 229–37.

Crowder, M. J. (1986) Consistency and inconsistency of estimating equations. *Econ. Theory* **2**, 305–30.

Crowder, M. J. (1992) Interlaboratory comparisons: round robins with random effects. *Appl. Statist.* **41**, 409–25.

Crowder, M. J. (1995) On the use of a working correlation matrix in using generalized linear models for repeated measures. *Biometrika.* **82**, 407–10.

Crowder, M. J. and Hand, D. J. (1990) *Analysis of Repeated Measures.* Chapman & Hall, London.

Crowder, M. J. and Tredger, J. A. (1981) The use of exponentially damped polynomials for biological recovery data. *Appl. Statist.* **32**, 15–18.

Crowder, M. J., Kimber, A. C., Smith, R. L. and Sweeting, T. J. (1991) *Statistical Analysis of Reliability Data.* Chapman & Hall, London.

Davidian, M. and Gallant, A. R. (1993) The nonlinear mixed effects model with a smooth random effects density. *Biometrika* **80**, 475–88.

Davidian, M. and Giltinan, D. M. (1993) Some simple methods for estimating the intra-individual variability in nonlinear mixed effects models. *Biometrics* **49**, 59–73.

Davison, A. C. and Snell, E. J. (1991) Residuals and diagnostics. In *Statistical Theory and Modelling: In Honour of Sir David Cox*, ed. D. V. Hinkley, N. Reid and E. J. Snell, Chapter 14. Chapman & Hall, London.

Dempster, A. P. (1972) Covariance selection. *Biometrics* **28**, 157–75.

Diggle, P. J. (1988) An approach to the analysis of repeated measurements. *Biometrics* **44**, 959–71.

Diggle, P. J. (1990) *Time Series: A Biostatistical Introduction.* Clarendon Press, Oxford.

Diggle, P. J., Liang, K.-Y. and Zeger, S. L. (1994) *Analysis of Longitudinal Data.* Oxford University Press, Oxford.

Dobson, A. J. (1990) *An Introduction to Generalized Linear Models.* Chapman & Hall, London.

Efron, B. and Hinkley, D. V. (1978) Assessing the accuracy of the maximum likelihood estimator: observed versus expected Fisher information (with discussion) *Biometrika* **65**, 457–87.

Finney, D. J. (1952) *Probit Analysis*, 2nd edn. Cambridge University Press, Cambridge.

Firth, D. (1991) Generalized linear models. In *Statistical Theory and Modelling: In Honour of Sir David Cox*, ed. D. V. Hinkley, N. Reid and E. J. Snell, Chapter 3. Chapman & Hall, London.

Fisk, P. R. (1967) Models of the second kind in regression analysis. *J. Roy. Statist. Soc. B* **29**, 266–81.

Fitzmaurice, G. M., Laird, N. M. and Rotnitzky, A. G. (1993) Regression models for discrete longitudinal responses (with discussion) *Statist. Sci.* **8**, 284–309.

Fitzmaurice, G. M. and Lipsitz, S. R. (1995) A model for binary time series data with serial odds ratio patterns *Appl. Statist.* **44**, 51–61.

Gabriel, K. R. (1961) The model of ante-dependence for data of biological growth. *Bull. Internat. Statist. Inst.*, 33rd session, Paris, 253–64.

Gabriel, K. R. (1962) Ante-dependence analysis of an ordered set of variables. *Ann. Math. Statist.* **33**, 201–12.

Gay, D. M. and Welsch, R. E. (1988) Maximum likelihood and quasi-likelihood for nonlinear exponential family regression models. *J. Amer. Statist. Assoc*, **83**, 990–8.

Gianola, D. (1980) Genetic evaluation of animals for traits with categorical responses. *J. Animal Sci.* **41**, 1272–6.

Gilchrist, R., Francis, B. and Whittaker, J. (1985) *Generalized Linear Models.* Springer-Verlag, Berlin.

Goldstein, H. (1979) *The Design and Analysis of Longitudinal Studies.* Academic Press, New York.

Goldstein, H. (1986a) Efficient statistical modelling of longitudinal data, *Ann. Hum. Biol.* **13**, 129–41.

Goldstein, H. (1986b) Multilevel mixed linear model analysis using iterative generalized least-squares. *Biometrika* **36**, 721–7.

Goldstein, H. (1987) *Multilevel Models in Educational and Social Research.* Griffin, London.

Goyan, J. E. (1965) Dissolution rate studies III. Penetration model for describing dissolution of a multiparticulate system. *J. Pharm. Sci.* **54**, 645–7.

Greenhouse, S. W. and Geisser, S. (1959) On the methods in the analysis of profile data. *Psychometrika* **24**, 95–112.

Grieve, A. P. (1984) Tests of sphericity of normal distributions and the analysis of repeated measures designs. *Psychometrika*, **49**, 257–67.

Grizzle, J. E. and Allen, D. M. (1969) Analysis of growth and dose response curves. *Biometrics* **25**, 357–81.

Grizzle, J. E., Starmer, C. F. and Koch, G. G. (1969) Analysis of categorical data by linear models. *Biometrics* **25**, 489–504.

Halperin, M. (1963) Confidence interval estimation in nonlinear regression. *J. Roy. Statist. Soc. B* **25**, 330–3.

Hand, D. J. (1994) Deconstructing statistical questions (with discussion) *J. Roy. Statist. Soc. A* **157**, 317–56.

Hand, D. J. and Taylor, C. C. (1987) *Multivariate Analysis of Variance and Repeated Measures.* Chapman & Hall, London.

Hand, D. J., Daly, F., Lunn, A. D., McConway, K. J. and Ostrowski, E. (1994) *A Handbook of Small Data Sets.* Chapman & Hall, London.

Harris, P. (1984) An alternative test for multisample sphericity. *Psychometrika*, **49**, 273–5.

Hartley, H. O. and Searle, S. R. (1969) A discontinuity in mixed model analysis. *Biometrics* **25**, 573–6.

Harville, D. A. (1977) Maximum likelihood approaches to variance component estimation and to related problems. *J. Amer. Statist. Assoc.* **72**, 320–40.

Harville, D. A. and Mee, R. W. (1984) A mixed-model procedure for analysing ordered categorical data. *Biometrics* **40**, 393–408.

Hertzog, C. and Rovine, M. (1985) Repeated-measures analysis of variance in developmental research: selected issues. *Child Development* **56**, 787–809.

Hinkley, D. V. and Runger, G. (1984) The analysis of transformed data (with discussion) *J. Amer. Statist. Assoc.* **79**, 302–20.

Hixson, A. W. and Crowell, J. H. (1931) Dependence of reaction velocity upon surface and agitation. I Theoretical consideration. *Indust. Engr. Chem.* **23**, 923–31.

Hocking, R. R. (1973) A discussion of the two-way mixed model. *Amer. Statist.* **27**, 148–52.

Huber, P. J. (1967) The behaviour of maximum likelihood estimators under nonstandard conditions. *Proceedings of the Fifth Berkeley Symposium on Mathematical Statistics and Probability*. University of California Press, Berkeley.

Huynh, H. and Feldt, L. S. (1976) Estimation of the Box correction for degrees of freedom for sample data in randomised block and split-plot designs. *J. Educational Statist.* **1**, 69–82.

John, S. (1971) Some optimal multivariate tests. *Biometrika* **58**, 123–7.

John, S. (1972) The distribution of a statistic used for testing sphericity of normal distributions. *Biometrika* **59**, 169–73.

Jones, R. H. (1990) Serial correlation or random subject effects? *Comm. Statist. Simula.* **19**, 1105–23.

Jones, R. H. (1993) *Longitudinal Data with Serial Correlation: A State-Space Approach.* Chapman & Hall, London.

Jorgensen, B. (1987) Exponential dispersion models. *J. Roy. Statist. Soc. B* **49**, 127–62.

Kenward, M. G. (1986) The distribution of a generalized least squares estimator with covariance adjustment. *J. Multivariate Anal.* **20**, 244–50.

Kenward, M. G. (1987) A method for comparing profiles of repeated measurements. *Appl. Statist.* **36**, 296–308.

Kenward, M. G. and Jones, B. (1992) Alternative approaches to the analysis of binary and categorical repeated measurements. *J. Biopharm. Statist.* **2**, 137–70.

Khatri, C. G. (1966) A note on a Manova model applied to problems in growth curve *Ann. Inst. Statist. Math.* **18**, 75–86.

Koch, G. G., Landis, J. R., Freeman, J. L., Freeman, D. H. and Lehnen, R. G. (1977) A general methodology for the analysis of experiments with repeated measurements of categorical data. *Biometrics* **33**, 133–58.

Laird, N. M. and Ware, J. H. (1982) Random-effects models for longitudinal data. *Biometrics* **38**, 963–74.

Landis, J. R., Miller, M. E., Davis, C. S. and Koch, G. G. (1988) Some general methods for the analysis of categorical data in longitudinal studies. *Statist. in Medicine* **7**, 109–37.

Liang, K.-Y. and Zeger, S. L. (1986) Longitudinal data analysis using generalized linear models. *Biometrika* **73**, 13–22.

Liang, K.-Y., Zeger, S. L. and Qaqish, B. (1992) Multivariate regression analysis for categorical data (with discussion) *J. Roy. Statist. Soc. B* **54**, 3–40.

Lindley, D. V. and Smith, A. F. M. (1972) Bayes estimates for the linear model (with discussion) *J. Roy. Statist. Soc. B*, **34**, 1–41.

Lindsey, J. K. (1993) *Models for Repeated Measurements*. Clarendon Press, Oxford.

Lindstrom, M. J. and Bates, D. M. (1990) Nonlinear mixed effects models for repeated measures data. *Biometrics* **46**, 673–87.

Lipsitz, S. R., Laird, N. M. and Harrington, D. P. (1991) Generalized estimating equations for correlated binary data: using the odds ratio as a measure of association. *Biometrika* **78**, 153–60.

Longford, N. T. (1993) *Random Coefficient Models*. Clarendon Press, Oxford.

McCabe, B. P. M. and Leybourne, S. J. (1993) Testing for parameter variation in non-linear regression models. *J. Roy. Statist. Soc. B* **55**, 133–44.

McCall, R. B. and Appelbaum, M. I. (1973) Bias in the analysis of repeated-measures designs: some alternative methods. *Child Development* **44**, 401–15.

McCullagh, P. (1980) Regression models for ordinal data (with discussion) *J. Roy. Statist. Soc. B* **42**, 109–42.

McCullagh, P. (1983) Quasi-likelihood functions. *Ann. Statist.* **11**, 59–67.

McCullagh, P. and Nelder, J. A. (1989) *Generalized Linear Models*, 2nd edn. Chapman & Hall, London.

Mardia, K. V. (1975) Assessment of multinormality and the robustness of Hotelling's T^2 test. *Appl. Statist.* **24**, 163–71.

Matthews, J. N. S., Altman, D. G., Campbell, M. J. and Royston, P. (1990) Analysis of serial measurements in medical research. *British Med. J.* **300**, 230–5.

Mead, R. (1970) Plant density and crop yield. *Appl. Statist.* **19**, 64–81.

Morrison, D. F. (1976) *Multivariate Statistical Methods*, 2nd edn. McGraw-Hill, New York.

Moulton, L. H. and Zeger, S. L. (1989) Analyzing repeated measures on generalized linear models via the bootstrap. *Biometrics* **45**, 381–94.

Muirhead, R. J. (1982) *Aspects of Multivariate Statistical Theory*. Wiley, New York.

Nelder, J. A. and Wedderburn, R. W. M. (1972) Generalized linear models. *J. Roy. Statist. Soc. A* **135**, 370–84.

O'Brien, R. G. and Kaiser, M. K. (1985) MANOVA method for analyzing repeated measures designs: an extensive primer. *Psych. Bull.* **97**, 316–33.

Olson, C. L. (1974) Comparative robustness of six tests in multivariate analysis of variance. *J. Amer. Statist. Assoc.* **69**, 894–908.

Olson, C. L. (1976) On choosing a test statistic in multivariate analysis of variance. *Psych. Bull.* **83**, 579–85.

Olson, C. L. (1979) Practical considerations in choosing a MANOVA test statistic: a rejoinder to Stevens. *Psych. Bull.* **86**, 1350–2.

Palmer, M. J., Phillips, B. F. and Smith, G. T. (1991) Application of nonlinear models with random coefficients to growth data. *Biometrics* **47**, 623–35.

Patterson, H. D. and Thompson, R. (1971) Recovery of inter-block information when block sizes are unequal. *Biometrika* **58**, 545–54.

Pierce, D. A. and Schafer, D. W. (1986) Residuals in generalized linear models. *J. Amer. Statist. Soc.* **81**, 977–86.

Potthof, R. F. and Roy, S. N. (1964) A generalized multivariate analysis of variance model useful especially for growth curve problems. *Biometrika* **51**, 313–26.

Prentice, R. L. (1988) Correlated binary regression with covariate specific to each binary observation. *Biometrics* **44**, 1033–48.

Prentice, R. L. and Zhao, C. (1991) Estimating equations for parameters in means and covariances of multivariate discrete and continuous responses. *Biometrics* **47**, 825–39.

Quass, R. L. and Van Vleck, L. D. (1980) Categorical trait sire evaluation by best linear unbiased prediction of future progeny category frequency. *Biometrics* **36**, 117–22.

Racine-Poon, A. (1985) A Bayesian approach to nonlinear random effects models. *Biometrics* **41**, 1015–23.

Racine-Poon, A., Grieve, A. P., Fluhler, H. and Smith, A. F. M. (1986) Bayesian methods in practice: experiences in the pharmaceutical industry (with discussion) *Appl. Statist.* **35**, 93–150.

Rao, C. R. (1965) The theory of least squares when the parameters are stochastic and its application to the analysis of growth curves. *Biometrika* **52**, 447–58.

Rao, C. R. (1966) Covariance adjustment and related problems in multivariate analysis. In *Multivariate Analysis*, pp. 87–103. Academic Press, New York.

Rao, C. R. (1967) Least squares theory using an estimated dispersion matrix and its application to measurement of signals. *Proceedings of the Fifth Berkeley Symposium on Mathematical Statistics and Probability*, Volume I, pp. 355–372. University of California Press, Berkeley.

Raz, J. (1989) Analysis of repeated measurements using nonparametric smoothers and randomization tests. *Biometrics* **45**, 851–871.

Rogers, G. S. and Young, D. L. (1977) Explicit maximum likelihood estimators for certain patterned covariance matrices. *Comm. Statist. Theory Methods A*, **6**, 121–33.

Rouanet, H. and Lepine, D. (1970) Comparison between treatments in a repeated-measures design: ANOVA and multivariate methods. *British J. Math. Statist. Psych.* **23**, 147–63.

Rowell, J. G. and Walters, D. E. (1976) Analysing data with repeated observations on each experimental unit. *J. Agric. Sci. Cambridge* **87**, 423–32.

Royall, R. M. (1986) Model robust inference using maximum likelihood estimators. *Internat. Statist. Rev.* **54**, 221–6.

Rudemo, M., Ruppert, D. and Streibirg, J. C. (1989) Random-effect models in nonlinear regression with applications to bioassay. *Biometrics* **45**, 349–62.

Sandland, R. L. and McGilchrist, C. A. (1979) Stochastic growth curve analysis. *Biometrics* **35**, 255–72.

Scheffé, H. (1959) *The Analysis of Variance.* Wiley, New York.

Searle, S. R. (1971) *Linear Models.* Wiley, New York.

Silvey, S. D. (1970) *Statistical Inference.* Penguin, London.

Srivastava, M. S. (1984) Estimation of interclass correlations in familiar data. *Biometrika*, **71**, 177–85.

Srivastava, M. S. and Keen, K. J. (1988) Estimation of the interclass correlation coefficient. *Biometrika* **75**, 731–9.

Stiratelli, R., Laird, N. and Ware, J. H. (1984) Random effects models for serial observations with binary response. *Biometrics* **40**, 961–71.

Stram, D. O., Wei, L. J. and Ware, J. H. (1988) Analysis of repeated ordered categorical outcomes with possibly missing observations and time-dependent observations. *J. Amer. Statist. Assoc.* **83**, 631–7.

Sigiura, N. (1972) Locally best invariant test for sphericity and the limiting distributions. *Ann. Math. Statist.* **43**, 1312–16.

Thall, P. F. and Vail, S. C. (1990) Some covariance models for longitudinal count data with overdispersion. *Biometrics* **46**, 657–71.

Vonesh, E. F. and Carter, R. L. (1992) Mixed-effects nonlinear regression for unbalanced repeated measures. *Biometrics* **48**, 1–17.

Walters, D. E. and Rowell, J. G. (1982) Comments on a paper by I. Olkin and M. Vaeth on two-way analysis of variance with correlated errors. *Biometrika* **69**, 664–6.

Wedderburn, R. W. M. (1974) Quasi-likelihood functions, generalized linear models and the Gauss–Newton method. *Biometrika* **61**, 439–47.

Wei, L. J. and Stram, D. O. (1988) Analyzing repeated measurements with possible missing observations by modelling marginal distributions. *Statist. in Medicine*, **7**, 139–48.

Weiss, R. E. and Lazaro, C. G. (1992) Residual plots for repeated measures. *Statist. in Medicine*, **11**, 115–24.

White, H. (1982) Maximum likelihood estimation of misspecified models. *Econometrika* **50**, 1–25.

Whittle, P. (1961) Gaussian estimation in stationary time series. *Bull. Internat. Statist. Inst.* **39**, 1–26.

Williams, D. A. (1984) Residuals in generalized linear models. In *Proc. 12th Internat. Biometric Conf.*, Tokyo, pp. 59–68.

Williams, E. J. (1959) *Regression Analysis.* Wiley, New York.

Wilson, S. R. (1991) *Workshop on Design of Longitudinal Studies and Analysis of Repeated Measures Data.* Proceedings of the Centre for Mathematics and its Applications, Vol. 28. Canberra: Australian National University.

Wishart, J. (1938) Growth rate determination in nutrition studies with the bacon pig and their analysis. *Biometrika*, **30**, 16–28.

Yates, F. (1982) Regression models for repeated measurements. *Biometrics* **38**, 850–3.

Zeger, S. L. (1988a) The analysis of discrete longitudinal data: commentary. *Statist. in Medicine*, **7**, 161–8.

Zeger, S. L. and Karim, M. R. (1991) Generalized linear models with random effects. *J. Amer. Statist. Assoc.* **86**, 79–86.

Zeger, S. L. and Liang, K.-Y. (1986) Longitudinal data analysis for discrete and continuous outcomes. *Biometrics* **42**, 121–30.

Zeger, S. L., Liang, K.-Y. and Albert, P. S. (1988) Models for longitudinal data: a generalized estimating equation approach. *Biometrics* **44**, 1049–60.

Zeger, S. L., Liang, K.-Y. and Self, S. G. (1985) The analysis of binary longitudinal data with time-independent covariates. *Biometrika* **72**, 31–8.

Zhao, L. P. and Prentice, R. L. (1990) Correlated binary regression using a quadratic exponential model. *Biometrika* **77**, 642–8.

Author index

Subject index